WILDLIFE IN ASIA

Animals assume a cultural importance throughout Asia. Tigers, monkeys, wild pigs and other animals feature in Asian proverbs, myths, religion, art and literature, while in many parts of Asia great cultural emphasis is placed on wild animals as a source of natural energy and health-giving vitality. But animals are also seen as a threat, both to human livelihoods and to human safety. *Wildlife in Asia* provides a rich and diverse collection of case studies of human representations of, and relationships with, wild animals in Asia.

Drawing on anthropological and historical data, this book examines human–wildlife relations in China, Tibet, Japan, Bhutan, Indonesia, the Philippines, Malaysia, India, Thailand and Vietnam. This volume initially focuses on the various ways in which wild animals are exploited as a resource, for food, medicine and crop-picking labour, before examining animals termed as pests or predators that are deemed to be harmful and dangerous.

Bringing together anthropologists and historians, this book analyses the range, variability and historical mutability of human sensibilities towards animals in Asia and will be of interest to Asianists and anthropologists alike.

John Knight is Lecturer at the School of Anthropological Studies, Queen's University Belfast and a former Research Fellow at the International Institute for Asian Studies, The Netherlands.

NORDIC INSTITUTE OF ASIAN STUDIES
Man and Nature in Asia
Series Editor: Arne Kalland
Professor of Social Anthropology, University of Oslo

The implication that environmental degradation has only occurred in Asia as a product of Westernization ignores Asia's long history of environmental degradation and disaster. The principle aim of this series, then, is to encourage critical research into the human–nature relationship in Asia. The series' multidisciplinary approach invites studies in a number of topics: how people make a living from nature; their knowledge and perception of their natural environment and how this is reflected in their praxis; indigenous systems of resource management; environmental problems, movements and campaigns; and many more. The series will be of particular interest to anthropologists, geographers, historians, political scientists and sociologists as well as to policy makers and those interested in development and environmental issues in Asia.

RECENT AND FORTHCOMING TITLES

JAPANESE IMAGES OF NATURE: CULTURAL PERSPECTIVES
Pamela J. Asquith and Arne Kalland (eds)

ENVIRONMENTAL CHALLENGES IN SOUTH-EAST ASIA
Victor T. King (ed.)

STATE, SOCIETY AND THE ENVIRONMENT IN SOUTH ASIA
Stig Toft Madsen (ed.)

ENVIRONMENTAL MOVEMENTS IN ASIA
Arne Kalland and Gerard Persoon (eds)

WILDLIFE IN ASIA
John Knight (ed.)

THE SOCIAL DYNAMICS OF DEFORESTATION IN THE PHILIPPINES
Gerhard van den Top

CO-MANAGEMENT OF NATURAL RESOURCES IN ASIA
Gerard Persoon, Diny van Est and Percy Sajise (eds)

FENGSHUI IN CHINA
Ole Bruun

NATURE AND NATION: FORESTS AND DEVELOPMENT IN PENINSULAR MALAYSIA
Jeya Kathirithamby-Wells

WILDLIFE IN ASIA

Cultural perspectives

Edited by John Knight

Routledge
Taylor & Francis Group

LONDON AND NEW YORK

First published 2004
by Routledge Curzon

This edition published 2012 by Routledge

4 Park Square, Milton Park, Abingdon, Oxon OX14 4RN
605 Third Avenue, New York, NY 10017

Routledge is an imprint of the Taylor & Francis Group

First issued in paperback 2013
Typeset in Baskerville by LaserScript Ltd, Mitcham, Surrey

British Library Cataloguing in Publication Data
A catalogue record for this book is available from the British Library

Library of Congress Cataloging in Publication Data
Wildlife in Asia : cultural perspectives / edited by John Knight.
p. cm. – (Man & nature in Asia ; no. 5)
Includes bibliographical references and index.
1. Human-animal relationships–Asia. 2. Wildlife utilization
–Asia. 3. Wildlife pests–Asia. I. Knight, John, 1960–.
II. Series: Man and nature in Asia ; no. 5.

QL85.W55 2004
304.2'7'095–dc22   2003015375

ISBN 978-0-700-71332-5 (hbk)
ISBN 978-0-415-86520-3 (pbk)

CONTENTS

List of illustrations   vii
List of contributors   ix
Preface   xii

Introduction   1
JOHN KNIGHT

PART I
Wildlife as resource   **13**

1 Attitudes towards wildlife and the hunt in pre-Buddhist
   China   15
   ROEL STERCKX

2 The chase and the Dharma: the legal protection of wild
   animals in premodern Tibet   36
   TONI HUBER

3 Representations of hunting in Japan   56
   JOHN KNIGHT

4 Japanese perceptions of whales and dolphins   73
   ARNE KALLAND

5 Cultural underpinnings of the wildlife trade in Southeast
   Asia   88
   DEANNA G. DONOVAN

6 Coconut-picking macaques in southern Thailand: economic, cultural and ecological aspects 112
LESLIE E. SPONSEL, PORANEE NATADECHA-SPONSEL AND NUKUL RUTTANADAKUL

PART II
Wildlife pests and predators **129**

7 Wildlife depredations in Jigme Dorji National Park, Bhutan 131
KLAUS SEELAND

8 Farming the forest edge: perceptions of wildlife among the Kerinci of Sumatra 147
JET BAKELS

9 Pigs across ethnic boundaries: examples from Indonesia and the Philippines 165
GERARD A. PERSOON AND HANS H. DE IONGH

10 'Primitive' tiger hunters in Indonesia and Malaysia, 1800–1950 185
PETER BOOMGAARD

11 The Raj and the natural world: the war against 'dangerous beasts' in Colonial India 207
MAHESH RANGARAJAN

12 Wolf reintroduction in Japan? 233
JOHN KNIGHT

Index 255

ILLUSTRATIONS

Figures

3.1   A wild boar-hunter holding the carcass of a young boar caught
      by his dogs   59
3.2   Two boar-hunters returning from a successful hunt   61
3.3   A wild boar carcass   62
3.4   A wild boar trophy in the foyer of the Mountain Village
      Development Centre in Hongū, Japan   64
4.1   Dolphins landed at a fishing port in Japan   75
4.2   A whale monument in Taiji, Japan   78
4.3   The post office in Ayukawa, Japan   81
4.4   A manhole cover in Taiji, Japan   82
5.1   A nineteenth-century representation of the butchering of a deer
      from a temple in Laos   90
5.2   Shops in China near the border with Vietnam selling a variety
      of wildlife products (mainly from Vietnam and Laos)   93
5.3   Natural products on sale at a National Park in Laos   94
5.4   A leopard skin for sale in a shop in Sapa, northern Vietnam   100
7.1   A farming settlement in Bhutan   133
7.2   Map of the Himalayan Kingdom of Bhutan   136
7.3   Size of households, agricultural land and livestock   137
8.1   A pig hunt using dogs and spears   152
8.2   A *ranjau* – for use against wild pigs or deer   154
9.1   From pig hunter to farmer   166
9.2   A section from the poster *Suids of Southeast Asia* produced by the
      Pigs and Peccaries Specialist Group of the World Conservation
      Union (IUCN) – Species Survival Commission   168
10.1 Map of the Malay world   187

ILLUSTRATIONS

10.2 Various traps of semi-sedentary Malaysian 'tribes'   195
12.1 A stamp commemorating the extinction of the Japanese wolf   235
12.2 A depiction of Mitsumine Shrine   236
12.3 An *ōkami kuyō* ('wolf memorial') ritual carried out in 1987 in the
      village of Takada on the Kii Peninsula   238
12.4 A pamphlet produced by a rural municipality on the Kii
      Peninsula, which claims to be the last place in which the wolf
      was sighted   240

Table

  6.1 Shifts in the cultural ecology of macaques   121

CONTRIBUTORS

Jet Bakels did fieldwork on the Baduy of West Java where research focused on the meaning of the tiger and the crocodile in Sumatra. She works as a freelance curator, and has organized a number of exhibitions in ethnographic museums. Her publications include: 'But his stripes remain: on the symbolism of the tiger in the oral traditions of Kerinci, Indonesia', in J. Oosten (ed.) *Text and Tales: Studies in Oral Tradition* (CNWS, 1994); and *Het Verbond met de Tijger: Visies op mensenetende dieren in Kerinci, Sumatra* (Leiden University, PhD, 2000).

Peter Boomgaard is a Senior Researcher at the Royal Institute for Linguistics and Anthropology (KITLV), Leiden, and Honorary Professor at the University of Amsterdam, The Netherlands. His main publications are *Children of the Colonial State: Population Growth and Economic Development in Java, 1795–1880* (Free University Press, 1989), and *Frontiers of Fear: Tigers and People in the Malay World, 1600–1950*, New Haven and London (Yale, 2001).

Deanna G. Donovan is a Fellow in the Environment Studies Group, East–West Center in Hawai'i. Her current research topics include the effects of globalization on forest use, the adaptation of natural resource valuation techniques for community use, and the driving factors in household exploitation of forest resources. Her current geographic focus is on Southeast Asia, though previous work has addressed forestry and conservation issues throughout tropical and subtropical Asia. She is the editor of *Policy Issues in Transboundary Trade in Forest Products in Northern Vietnam, Lao PDR and Yunnan PRC. Vol. II: The Country Reports* (East–West Center and World Resources Institute, 1998); the author of 'Strapped for cash', *Asia Pacific Issues no. 39* (1999); and the author (with J. J. Gutrich) of *Environmental Valuation and Decision-making* (US Forest Service Institute of Pacific Island Forestry, 2001).

Toni Huber is Senior Lecturer in Religious Studies at Victoria University (Wellington), a former Alexander von Humboldt Fellow (1997–1998), and is the author of both *The Cult of Pure Crystal Mountain* (Oxford, 1999) and *The Guide To India by Amdo Gendun Chöphel* (LTWA, 2000), as well as numerous scholarly articles on Tibetan culture and society.

Hans H. de Iongh is Head of the Africa Programme of the Centre of Environmental Science at Leiden University. He has been Chairman of The Netherlands Committee for IUCN during 1999–2000 and he is a member of the Species Survival Commission. Since 1998 he has been the acting deputy Director of the Tropenbos Research Programme, a collaborative research partnership with universities in Colombia, Guyana, Indonesia, Vietnam, Philippines, Cameroon and Ghana. He is the co-editor (with P. J. M. Hillegers) of *The Balance Between Biodiversity Conservation and Sustainable Use of Tropical Rain Forests* (Tropenbos Foundation, 2001), and the co-author (with P. E. Loth, Madi Ali and H. Bauer) of 'Management of fragile ecosystems in the North of Cameroon: the need for an adaptive approach', in *Proceedings of the International Conference held at CEDC* (Maroua, 2001).

Arne Kalland is Professor at the Department of Social Anthropology, University of Oslo, Norway. His main fields of interest are maritime resource management, people's perceptions of nature, and the environmental movement, with a particular focus on Japan. He is the co-author (with B. Moeran) of *Japanese Whaling: End of an Era?* (Curzon, 1992), the author of *Fishing Villages in Tokugawa Japan* (Curzon, 1995) and the co-author (with Frank Sejersen) of *Marine Mammals in Northern Cultures* (in press). He is also the co-editor (with Ole Bruun) of *Asian Perceptions of Nature: A Critical Approach* (Curzon, 1995), the co-editor (with Pamela Asquith) of *Japanese Images of Nature: Cultural Perspectives* (Curzon, 1997), and the co-editor (with Gerard Persoon) of *Environmental Movements in Asia* (Curzon, 1998).

John Knight is Lecturer at the School of Anthropological Studies, Queen's University Belfast, and a former Research Fellow at the International Institute for Asian Studies in Leiden, The Netherlands. He has carried out long term field research in mountain villages in western Japan, and has written on a variety of topics to do with rural Japan, including depopulation, forestry and wildlife. He is the editor of *Natural Enemies: People–Wildlife Conflicts in Anthropological Perspective* (Routledge, 2000), the co-editor (with J. W. Traphagan) of *Demographic Change and the Family in Japan's Ageing Society* (SUNY, 2002), and the author of *Waiting for Wolves in Japan: An Anthropological Study of People–Wildlife Relations* (Oxford, 2003).

Poranee Natadecha-Sponsel is Assistant Professor of Philosophy and Religion, Chaminade University in Hawai'i. Her interests include ecological anthropology, comparative religion, and Thailand. She is the author of (with L. E. Sponsel, N. Ruttandakul, and S. Juntadach) of 'Sacred and/or secular approaches to biodiversity conservation in Thailand', *Worldviews: Environment, Culture, Religion* 2(2) (1998); and (with L. E. Sponsel and N. Ruttanadakul) of 'Monkey business? The conservation implications of macaque ethnoprimatology in Southern Thailand', in A. Fuentes and L. Wolfe (eds) *Primates Face to Face: The Conservation Implications of Human–Nonhuman Primate Interconnections* (Cambridge, 2002).

Gerard A. Persoon obtained his PhD in anthropology at Leiden University, The Netherlands. At present he is head of the Programme for Environment and

Development of the Centre of Environmental Science at Leiden University. His main interests are in indigenous peoples and forest management in Southeast Asia. He is the co-editor (with D. M. E. van Est) of *The Study of the Future in Anthropology, Fokaal* no. 35 (2000); and the editor of *The Philippines: Historical and Social Studies*, a special issue of *Journal of the Humanities and Social Sciences of Southeast Asia and Oceania* 157(3) (2001).

Mahesh Rangarajan is an independent researcher based in Delhi, India. He is a Visiting Assistant Professor at Cornell University, Ithaca and a former Fellow of the Nehru Memorial Museum and Library, New Delhi. He is the author of *Fencing the Forest* (Oxford, 1996) and *India's Wildlife History* (Permanent Black, 2001). His most recent work, co-edited with V. K. Saberwal, is *Battles Over Nature: Science and the Politics of Conservation* (Permanent Black, 2002).

Nukul Ruttanadakul is Assistant Professor in the Biology Section of the School of Science and Technology, Prince of Songkla University in Thailand. His interests include biological ecology, human ecology, environmentalism, and southern Thailand. On publications, see joint publications in the Natadecha-Sponsel entry.

Klaus Seeland is a political scientist and sociologist who has been doing research in South Asia for more than twenty years in the fields of socio-cultural aspects of forests, comparative studies in resource management, perception and local knowledge. He is Senior Lecturer at the Chair of Forest Policy and Forest Economics at the Swiss Federal Institute of Technology (ETH) in Zurich, Switzerland since 1990 and Reader at the University of Konstanz, Germany. He is the editor of *Nature is Culture: Indigenous Knowledge and Socio-Cutural Aspects of Trees and Forests in Non-European Cultures* (I.T. Publications, 1997); the co-editor (with F. Schmithüsen) of *Man in the Forest: Local Knowledge and Sustainable Management of Forests and Natural Resources in Tribal Communities in India* (D.K. Printworld, 2000); and the co-author (with M. K. Jena, P. Pathi, J. Dash, K. K. Patnaik) of *Forest Tribes of Orissa, vol. 1: The Dongaria Kondh* (D.K. Printworld, 2002).

Leslie E. Sponsel is Professor of Anthropology, and Director of the Ecological Anthropology Program, University of Hawai'i. His interests include ecological anthropology, historical ecology, ethnoprimatology, Southeast Asia, and Thailand. He is the editor (with T. N. Headland, and R. C. Bailey) of *Tropical Deforestation: The Human Dimension* (Columbia, 1996); the author of 'The historical ecology of Thailand: increasing thresholds of human environmental impact from prehistory to the present', in W. Balee (ed.) *Advances in Historical Ecology* (Columbia, 1998); and the editor of *Endangered Peoples of Southeast and East Asia* (Greenwood, 2000); See also joint publications in the Natadecha-Sponsel entry.

Roel Sterckx is University Lecturer in Chinese Studies at the University of Cambridge and specializes in the cultural history of pre-imperial and early imperial China. He has taught at the Universities of Oxford and Arizona and is the author of *The Animal and the Daemon in Early China* (SUNY, 2002).

PREFACE

This book is based on a two-day conference entitled *Animals in Asia: Representations and Relationships* held at the University of Leiden in the Netherlands in September 1997. The conference was sponsored by the International Institute for Asian Studies (IIAS) in Leiden, and sincere thanks are due to the IIAS and its staff, including Wim Stokhof (IIAS Director), Sabine Kuypers (Assistant Director), Marianne Lengehenkel (who helped to organize the conference), Cathelijne Veenkamp and Kitty Yang. Much of the editorial work that went into this book also took place at the IIAS, and the final stages of manuscript preparation were carried out at the School of Anthropological Studies, Queen's University Belfast.

In the course of producing this volume, encouragement, help and advice were received from many people, including Cynthia Chou, Cen Huang, Alex McKay, Shoma Munshi, Gerard Persoon and Reed Wadley. The two reviewers appointed by the publisher made many useful suggestions and criticisms for which I am grateful. I am also grateful to Arne Kalland and Kay Milton for their comments on an earlier draft of the *Introduction* to this book. Finally, I would like to thank Liz Bramsen of the Nordic Institute for Asian Studies (NIAS) for her help in the preparation of the manuscript for publication.

John Knight

INTRODUCTION

John Knight

Wild animals assume an obvious cultural importance throughout Asia. Elephants, tigers, monkeys, birds and snakes are among the creatures that feature in Asian proverbs, myths, legends, augury, religion, art and literature. The twelve animals of the Chinese zodiac are prominent in the popular cultures of East and Southeast Asia, with people born in the year of the rat, the tiger, the monkey and so on attributed the character or personality of the animal in question. Martial art traditions such as Indonesian *silat*, Chinese *kung fu* and Japanese *ninjutsu* are influenced by, if not based on, the movements and postures of assorted wild animals. Images of the tiger appear on the Malaysian national crest, on banknotes in Bangladesh and on stamps in Laos; for the Olympics held in Seoul in 1988, South Korea chose a tiger as the symbol of the Games (Jackson 1999: 50). These are some of the more familiar expressions of animal symbols and emblems found in Asian cultures. The premise of this book is that wild animals in Asia also assume a less obvious cultural importance in the ostensibly material relations that exist between humans and wildlife.

Wildlife affects human livelihoods in various ways. Wild animals are an economic resource in the form of valuable products such as hides, meat and other body parts that are directly used and consumed or traded for cash income. Wildlife figures prominently in some Asian cuisines, while a large number and wide variety of wildlife products are to be found in Asian pharmacopoeia. Wild animals can also have a negative impact on human livelihoods. Asia is the site of a great many people–wildlife conflicts, including wild herbivores that threaten crops and wild carnivores that threaten livestock and human safety. In this book, this relationship between human livelihoods and wildlife is analysed from a cultural perspective, one that pays particular attention to the local contextual meanings that inform these material relations. Studies of animal symbolism often neglect the domain of practical relations with animals in favour of the more obviously symbolic domains of myth and ritual. But in the following chapters the concern is with the 'pragmatic' human relations with wildlife as sites of cultural meaning.

This book examines human representations of and interactions with wild animals in a number of Asian societies and cultures. Drawing on anthropological and historical data, the chapters look at human–wildlife relations in China,

Tibet, Japan, Bhutan, Indonesia, the Philippines, Malaysia, India, Thailand and Vietnam, involving animals such as wild pigs, tigers, wolves, monkeys and whales. Many of the wild animals mentioned are exploited as a source of food and medicines, but some are used in other ways, such as in agriculture (coconut-picking monkeys), in tourism (whales and dolphins in Japan) and in the pet trade. These case studies of people–wildlife relations focus largely on rural areas, especially the remote peoples living on the frontier with the forest, and only secondarily on the views of wildlife among urban Asians.

Wildlife in Asia appears to be in a state of crisis. Wildlife habitat is subject to considerable pressure from resource appropriation, development activity, and growing human populations. By the late twentieth century, much wildlife habitat in Asia had been destroyed through deforestation; between 1960 and 1990 one-third of the tropical forest cover in Asia was lost (Livernash and Rodenburg 1998: 22). Given that a majority of all land species are found in forests, and that tropical forests are particularly species-rich, the decline of Asian tropical forests represents an enormous loss of biodiversity. Population pressure on wildlife habitat in Asia is especially acute. In 1995 the population of Asia was estimated at 3.4 billion people, but this is projected to increase to 5.4 billion by 2050, a figure accounting for well over half of the world's projected 9 billion-plus people (Livernash and Rodenburg 1998: 11). Population growth in forest-edge regions tends to lead to the extension of the arable frontier at the expense of the forest and wildlife habitat therein.

The cultural perspectives of anthropology, and cognate disciplines such as history, can make a twofold contribution to the field of wildlife management and conservation. The research tools of social and cultural anthropology can provide knowledge of the human dimension of wildlife management through descriptive accounts of the local communities living at the wildlife interface. This input would be consistent with the new conservationist orthodoxy, variously known as 'community-based management' or 'participatory conservation', which emphasizes the involvement and participation of the local human population in conservationist initiatives in order to make conservation policy more equitable and more effective. Cultural perspectives can make a further contribution by taking wildlife management as an object of study in its own right, focusing critically on the cultural assumptions underlying management and conservation discourse. By means of ethnographically informed accounts of the cultural contexts in which it is applied, anthropology can contribute to the formation of a more locally sensitive wildlife management policy.

A book such as this must be wary of an overly rigid 'Asia' focus that neglects the local and global levels of connection and variation. First, geographical Asia is marked by an enormous cultural diversity that challenges simplistic pan-regional generalizations. Notwithstanding the tendency within and outside of Asia to focus on East–West difference, in many cases differences within Asia are at least as striking as contrasts with non-Asia. One of the aims of this book is to convey a sense of the diversity and variety of Asian views of wildlife rather than attempt to discover generic 'Asian' characteristics. Second, the wild animals of Asia are not confined to geographical Asia. Asian animals are well-known to non-Asian publics as the stars of wildlife documentaries on television, as popular attractions in zoos and as objects of conservationist concern. Many of the high profile 'celebrity' animal species of the international conservation movement such as the tiger, the giant panda and the orang-utan are Asian animals. On the other hand, non-Asian wild animals are well known to people in Asia. A familiar line-up of exotic wild animals can be found in Asian zoos and theme parks, including elephants, lions, zebras, giraffes and so on, many of which are African in origin.

Representations

A major critical challenge facing modern anthropology has been to come to terms with the larger constitution and production of human difference, a process that often takes place through the idiom of culture. The study of Asian societies and cultures has been subject to the legacy of Orientalism – the tradition of constructing the Orient as an alien 'Other' through scholarly and popular texts and discourses in the areas of literature, religion, politics and so on (Said 1985). Human–animal relations too may well be an important site of Orientalist representations of Asia, one that serves to reinforce the impression of essential Asian difference from the West. There is, of course, a larger tendency within human cultures to classify other peoples in terms of their relationships with animals and to use the animal relation as a symbolic marker. The precise human–animal relation that is foregrounded for symbolic deployment will vary, and includes cruelty to animals and the illicit consumption of animals. In the case of human–animal relations in Asia, we find both romanticizing and stigmatizing Orientalist representations.

Idealized or romanticized representations of Asia often arise in connection with wildlife. 'Eastern thought', 'Oriental wisdom' or 'Asian traditions' have long served as 'a conceptual resource for environmental philosophy' (Callicott and Ames 1989: xi). More specifically, Asian culture and religion are invoked as the basis of a traditional wildlife conservationism and compassion towards animals. Such claims are particularly pronounced in Hindu and Buddhist parts of Asia. Thai proponents of national parks claimed that the parks were culturally compatible because, as wildlife sanctuaries, they would be consistent with the Buddhist prohibition on taking life (Vandergeest 1996: 260). Some Japanese zoologists refer to Japan's 'traditional religious conservationism towards wildlife' (Yoneda 1991: 152) and develop approaches to wildlife management based on 'oriental philosophy' (Kuroda 1991). Some writers argue that the Asian emphasis on *ahimsa* or non-violence, as contained in the traditions of Hinduism, Jainism and Buddhism, can contribute to the solution of contemporary issues, including animal and environmental protection (Chapple 1993).

Other writers stress the elevated cultural status of particular wild animals in Asian countries. Sukumar suggests that '[t]he worship of the elephant god,

Ganesha, which originated in the third or fourth century, must have [henceforth] created a strong ethos against the killing of elephants' (Sukumar 1989: 4). The same author adds that in China 'elephants did not have any religious significance and were usually exterminated as vermin', but that '[e]lephants enjoyed a close cultural and religious association with Man in Sri Lanka, Burma, Thailand and Kampuchea' (ibid.: 7). R. K. Sinha, in an article claiming traditional Indian antecedents of biodiversity conservation, writes as follows:

> Many wild animals have been revered in Hindu mythology as vahanas (vehicles) of Gods and Goddesses. Mythology depicts the tiger as the Vahana of the Goddess Durga, and identifies the lion with the Goddess Kali, the peacock with Karthikeya, the swan with Saraswati, the owl and the elephant with the Goddess Lakshmi.... This [religious] identification underlines not only a sense of respect for these animals, but also a reason for protecting them.
>
> (Sinha 1995: 283)

Asia's status as a rhetorical counterpoint to a supposed Western human–animal dichotomy emerges clearly in connection with wild primates. The bonnet macaque, Hanuman langur and rhesus macaque are all said to have 'benefitted from India's tradition of veneration for monkeys; extermination campaigns are unthinkable in this predominantly Hindu country' (Malik and Johnson 1994: 234). Respect for monkeys (and even tolerance of their crop-raiding) on account of their association with the Hindu god Hanuman is reported for India (Pirta *et al.* 1997: 102) and for Bali (Wheatley and Harya Putra 1994: 246). In China it is 'partly due to Buddhistic culture' that, despite periodic famines, Tibetan macaques have 'survived in the vicinity of people' in the Mt. Emei area (Zhao 1994: 259). The practice of feeding wild primates as a means of merit-making is reported for Thailand (Eudey 1994: 273–274) and for China (Zhao 1994: 260), while in Japan Buddhist priests, through their feeding of wild monkeys (especially in the winter), are said to have contributed to the successful provisioning of wild monkey troops that formed the basis for modern Japanese primatology (Carpenter and Nishimura 1969: 17–18). Some writers go on to contrast the 'distancing devices', the 'strict opposition between human and animal' and the 'categorical boundary' of Western cultures, to 'the roles of apes and monkeys in non-Western cultures' such as Japan, China and India (Corbey 1993: 128).

On the other hand, human–animal relations in Asia can arouse disapproval and condemnation. Much of this criticism is directed at the wildlife trade and the wide range of animal species, terrestrial and aquatic, that it threatens. For its critics, the wildlife trade appears to combine superstition, cruelty and a conservation threat. As with whaling, concern with the wildlife trade straddles welfare and conservation considerations. Just as whaling is condemned on the grounds of cruelty to whales (because of the prolonged duration of the whale's death) and the (perceived) threat to whale populations, so the wildlife trade appears both cruel to individual animals (the bears on Chinese bear gall farms, say) and a danger to animal populations (overhunting of the species to supply the trade). In addition, the wildlife trade is invoked as evidence of Asian superstition or irrationality. 'A wholly irrational and inaccurate East Asian belief in the healing virtue of bears' gall is now causing the extermination of most of the bear species in the world' (Anderson 1996: 126).

Asian pragmatism or utilitarianism is a recurring theme in the discussions of wildlife in Asia. A frequent object of criticism is what might be called Asian *hyper*utilization of wildlife – in the sense that Asians utilize or consume animals that should not be consumed and that Asian utilization of wildlife tends to deplete wildlife populations. Here we can recall the Chinese expression, 'if it has four legs and it's not a table, we'll eat it' (see Chapter 5). But in recent decades Japanese utilitarianism has become notorious in some circles, generating a variety of negative images and stereotypes based on a perceived Japanese overexploitation of the natural world, especially marine wildlife. Japanese greed is a common theme in environmental debates, as in references to 'Japan's insatiable desire for whale meat' (Moulton and Sanderson 1997: 69). These stereotypes pose a challenge to those who study Asian societies and cultures. The local level case studies presented here take up this challenge by examining ostensibly utilitarian relations with wildlife in the specific cultural contexts in which they operate, to show how these relations often assume a multi-faceted complexity that defies simple stereotypes.

Wildlife as resource

The first three chapters of Part I focus on hunting in China, Tibet and Japan respectively. Hunting – the human predation on wild animals – and is an activity that has both practical and symbolic aspects. The motif of domination and control figures prominently in Western hunting traditions, where the hunter defines himself through his power over the animals he hunts (Cartmill 1993; Hell 1996). But aspects of this dominionistic disposition are also evident in some Asian hunting traditions. In Chapter 1, Roel Sterckx examines attitudes towards wildlife and the hunt in pre-Buddhist China by tracing references to animals in an assortment of early Chinese texts. Sterckx shows that there was no clear conceptual dichotomy between human and animal realms, but instead an emphasis on the changing character of the human–animal interface. He argues that in early China governmental control over the human realm was associated with control over the animal realm, and that the hunt, by allowing rulers to display mastery over the wilds, served to reinforce socio-political authority. In this case, the hunt represents both a *direct physical action* that impacts on the animals themselves and an *indirect expression or statement* to wider human society.

A key feature of the Asian traditions of Hinduism, Jainism and Buddhism is the emphasis on *ahimsa* – the Indian notion of non-violence or non-injury to other beings, which is linked to the ideas of continuity between life forms, karma and reincarnation (Chapple 1993: Ch.1). *Ahimsa* is the antithesis of *himsa*, the human killing of other forms of life for food or in sacrifice that formed the basis of Vedic religion (Jacobsen 1994: 287, 298). Hunting clashes with the principle of *ahimsa*, and the denunciation of hunting is a constant theme in the sacred texts of Hinduism, Jainism and Buddhism. In Chapter 2, Toni Huber examines hunting and the attempts to restrict it in the Buddhist society of premodern Tibet. Drawing on historical texts (and recent ethnographic fieldwork), Huber shows that hunting in Tibet continued despite the Buddhist injunction not to practice it, and that in response Tibetan rulers passed a range of anti-hunting laws. A main point in Huber's chapter is that this early legal protection of wildlife in Tibet should be understood in relation to the benefits, in terms of merit earned, that accrue to the human protectors as much as the concern with the animals themselves. The legal proscription of hunting in his kingdom conferred on the Tibetan king great merit and enhanced soteriological prospects.

In Chapter 3, I examine representations of hunting among hunters and non-hunters in western Japan. I describe three sets of representations: hunting as a contest with animals and among men, hunting as protection of the wider community from dangerous animals, and hunting as itself a danger both to the practitioners and to the wider public. In addition to the *physical* danger it poses to hunters and non-hunters, due to stray fire and other accidents, hunting in Japan can pose a *spiritual* danger to hunters and their families as a result of the Buddhist concern with violence and killing. Although hunting in Japan has become Westernized in many ways (weapons, dogs and so on) and is officially designated a sportive recreation (to the point where the English word 'hunter' has become widely diffused among hunters), the negative associations related to the *ahimsa* sensibility remain.

The morally problematic character of hunting in Asia is further expressed in popular beliefs about spirit retribution. People who kill animals become vulnerable to the revenge of the spirit of the slain animal. Hunting and killing, therefore, rather than ending the antagonism between hunter and prey, can be seen as generating a second phase of conflict in which the direction of predation is reversed, with human-on-animal physical violence giving way to reciprocal animal-on-human spirit violence. This theme of position reversal, of hunter becoming prey and prey becoming hunter, is evident in the various references to spirit revenge found in the chapters of this book. Tibetan hunters believe that nature spirits angered by the death of game animals exact revenge on them (Chapter 2), hunters in Japan fear the retribution of the spirits of the wild boars they kill (Chapter 3), Japanese whalers are vulnerable to the curse of the whale spirit (Chapter 4), pig-killers on Mentawai fear the anger of slaughtered pigs (Chapter 9) and Malay and Indian tiger-killers face the threat of revenge from the tiger spirit (Chapters 8 and 11). Human violence against animals (especially large, powerful animals) risks animal violence against humans *twice over* – both the direct physical violence of the living animal at the point of confrontation, and the posthumous indirect violence associated with the animal's restless spirit.

It is because of the *provocative* character of human hunting – which, while it may immobilize the animal, inadvertently mobilizes its spirit – that ritual specialists often intercede on behalf of human killers to help manage the consequences of the kill and to restore the order that the hunter disturbed. In short, violence against animals ceases to be the exclusive domain of the hunter, but comes to involve ritual specialists who mediate with the animal spirits. One way in which religious specialists can prevent such disaster is by stopping the kill beforehand, as in Chapter 2, where Tibetan monasteries close hunting grounds, rulers proscribe hunting and lamas even contrive to convert hunter and prey to Buddhism! But the more common scenario is for the religious specialists to be called in to deal with the consequences, rather than the causes, of killing. To this end, Buddhist or Shinto priests in Japan carry out annual memorial rites for the spirits of hunted animals on behalf of hunters or for the spirits of whales on behalf of whalers (Chapters 3 and 4), *dukun* in Sumatra appease the spirits of dead tigers lest they exact revenge on their human killers (Chapter 8) and *kerei* medicine men on Mentawai utter spells over the bodies of pigs about to be slaughtered in order to pacify their souls (Chapter 9).

Chapters 4–6 examine other aspects of the use of wildlife as a resource in Asia. In Chapter 4, Arne Kalland focuses on Japanese whaling and, in particular, Japanese views of whales and dolphins. The background here is the sustained Western criticism of Japanese whaling, in which the very legitimacy of whaling as an economic activity is challenged, and cetaceans represented as, in effect, animal *persons* that should be protected from immoral exploitation by human beings. Kalland argues that the Japanese perspective, on the other hand, stresses 'the interdependence of supernatural, human and animal worlds', such that whaling appears only partially predatory in character, as whalers emphasize that whales 'give themselves up' to them. Human consumption of whales does not preclude human 'compassion' towards whales because of the way people can become morally obligated to the animals they utilize.

In Chapter 5, Deanna G. Donovan examines the wildlife trade in Southeast Asia. Much attention is now focused on the wildlife trade and on the variety of terrestrial and aquatic animal species involved in it, including tigers, bears, rhinos, sharks, and turtles, which go to make up the range of wildlife products, from foods and medicines to furs, trophies and pets. This trade has a major impact on the wildlife populations of Asia and beyond, posing a serious threat to some wildlife species, while attracting much international criticism, not least for the consumption practices on which it is based. Donovan's chapter focuses on the cultural dimensions of demand and supply in this market. She shows how the demand for wildlife products, supported by rising incomes in China and elsewhere, substitution of imported pharmaceuticals by homegrown remedies and the revival of traditional medicine in Chinese-influenced cultures, is matched by an increasing supply of these products brought about by economic liberalization and infrastructure development. She warns that this trade, if it continues unchecked, threatens both 'biological and cultural loss' – that is, not just of the animal species themselves, but also the tribal groups in Vietnam, Laos and Thailand that depend on them.

Most of the examples above have involved the exploitation of dead animals, but in Chapter 6 we encounter an example of the exploitation of live animals. In parts of South and Southeast Asia, captive macaques are used by local communities to harvest commercial tree crops. Leslie E. Sponsel, Poranee Natadecha-Sponsel and Nukul Ruttanadakul discuss this phenomenon with reference to coconut-picking macaques in southern Thailand. Based on interviews with the owners and trainers of monkeys in Muslim and Buddhist villages, the three authors offer an outline of this practice and examine its ecological and cultural dimensions. They argue that the phenomenon of crop-harvesting monkeys represents a borderline instance of domestication that challenges the conventional dualist categories of wild and domestic or nature and culture.

Wildlife pests and predators

In Asia, as elsewhere, wildlife can be harmful as well as useful. The first three chapters of Part II focus on wildlife pestilence, drawing on examples from Bhutan, Indonesia and the Philippines. One form of people–wildlife conflict in Asia involves the protected wildlife of national parks that raids the crops of local farmers and preys on livestock. National parks and other wildlife reserves have often aroused strong, negative feelings among farmers and livestockers living nearby. In Chapter 7, Klaus Seeland describes such problems for Bhutan by means of a survey in a national park area of wildlife damage among villagers. Seeland shows how, in this people–park conflict, villagers are subject to the frequent and large-scale depredations of protected park animals, to the point where the park is seen as a threat to village livelihoods. This Bhutanese example, in which externally imposed conservation collides with local livelihoods, is clearly at odds with the recent trend in conservationist thinking which holds that conservation, to be sustainable in the long-term, should be consistent with the interests of local people. Seeland warns that, because of the suffering it causes in local communities, the problem of wildlife pestilence in Bhutan has the potential to cause political instability in the country's remote areas.

One of the contributions that anthropologists can make to the study of people–wildlife relations is to challenge axiomatic Western representations of them, and the assumptions on which they are based, by placing these relations in their local cultural context. In the case of wildlife pestilence, this takes the form of exposing the underlying utilitarian assumptions of pestilence discourse by showing how wildlife damage may be experienced in terms of a different set of assumptions. In Chapter 8, Jet Bakels shows that wildlife pestilence among forest-edge farmers in central Sumatra is not reducible to a simple conflict with animals, but is locally understood in terms of the larger relationship between village and forest. Among the Kerinci, forest wildlife is 'owned' by forest spirits, just as cattle are owned by

8

villagers, and it follows that wildlife actions affecting humans (such as crop-raiding) and human actions affecting wildlife (such as trapping) implicate the relationship between villagers and forest spirits, which should ideally be balanced and reciprocal. Hence, before they trap a destructive animal, Kerinci villagers obtain permission from the forest spirit; after trapping takes place they compensate this spirit so that he 'suffers no loss'. In this way, Kerinci cultivators strive to maintain a balance with the forest as they farm along its edge.

One of the most destructive wildlife pests in Asia is the wild pig, an animal that raids the crops of farmers across the continent with often devastating consequences. In Chapter 9, Gerard Persoon and Hans de Iongh focus on pigs in three different social contexts in Southeast Asia – central Sumatra, the island of Mentawai (off the coast of Sumatra) and the Philippines. In a survey that includes semi-domesticated pigs and wild pigs, the authors show how these animals mediate social relations between different groups of people, variously bringing them together or keeping them apart. They show how, in response to wild pig pestilence in central Sumatra, farmers and hunters of different ethnic groups cooperate to their mutual advantage, as the farmers have major crop pests removed and the hunters obtain bush meat. For the Muslim Sumatrans, pigs have a wholly negative relation to livelihood because they are 'unclean animals' as well as farm pests. Here 'culture' clearly plays a part in defining the (negative) utility of these animals by limiting the 'demand' for such meat. For non-Muslim farmers in the Philippines, by contrast, wild pigs are harmful to crops *and* valuable as bush meat – that is, they are a pest and a resource.

Here we can note an interesting contrast in the way that religious ideas inform the human relation to wildlife between these Muslim farmers in Southeast Asia and the Buddhist hunters of Tibet and Japan. If for Muslim farmers religion reinforces the negative economic specification of the wild pig, for Buddhist hunters religion would be at odds with the economic relation. In both cases, religion makes human predation on wild animals problematic: the Sumatran Muslim cannot hunt the wild pig because it is unclean and hunting implies contact with it, while the Tibetan Buddhist should not hunt wild animals because they are sentient beings similar to himself. Human predation on animals is therefore problematic in both cultural contexts, but for contrasting reasons. For the Muslim, the problem is contact rather than lifetaking; for the Buddhist, the problem is lifetaking rather than contact.

In many parts of Asia, farming is morally problematic because, like hunting, it is subject to the *ahimsa* sensibility. Formally, hunting would appear to be a very different activity from farming: hunting is the human predation on wild animals and agriculture is the human cultivation of domesticated plants. But in practice, farming, like hunting, implies violence towards other animals – first, because it is based on an original displacement of wild animals from what becomes the space of cultivation, and second, because this space must be constantly defended against animal intrusion. Human crops, in other words, must be cultivated *and* protected. The violent dimension of human farming is especially pronounced where cultivation is sited at the forest-edge. Sinhala swidden farmers in Sri Lanka relate the wildlife damage to their fields to the earlier killing of wild animals that occurs when scrub is burned in preparing the land for cultivation – this land-clearing amounts to an infringement of the Buddhist precept of non-violence (Sandell 1995: 156).

In addition to human conflicts with crop-raiding ungulates, human conflicts with carnivores are found in many parts of Asia. For wild predators, the domestic animals kept by human communities generally make for easier prey than wild herbivores. The threat of wild predators is not confined to livestock, but extends to human life. Although the mankilling tendencies of wild predators are notoriously prone to exaggeration, there are credible reports from across the predator ranges of Asia of human deaths caused by predator attack – by brown bears and Siberian tigers in the Russian Far East, by Bengal tigers in the Sundarbans, by crocodiles in tropical Asia, and so on. The known presence of large predators in an area tends to profoundly alter the human perception of space, instilling a routine vigilance among local people. Those who live in predator country seek to protect themselves from predators in a variety of ways, including by taking precautions when travelling in predator habitat (Japanese foresters carrying bells in bear country, Indian villagers wearing face masks on the back of their heads in tiger country, etc.) and the pre-emptive killing of predators (through poisoning, forest fires, destruction of cubs, and so on).

In Chapter 10, Peter Boomgaard surveys tiger-killing in Indonesia and Malaysia in the nineteenth and twentieth centuries, using colonial documents to trace the ways in which different ethnic groups living in tiger country interacted with the tiger. At a time when conservationist calls for human–predator 'coexistence' are ubiquitous, Boomgaard addresses the question, 'To what extent did people co-exist with tigers in the past?' His findings are that people–tiger relations varied significantly within the region, and that human intolerance of the tiger appears to be linked to sedentarization, such that sedentary cultivators in Java, Sumatra and Malaysia (in contrast to nomadic and semi-sedentary groups) 'feared and hated' tigers in a way that recalls negative European attitudes towards bears and wolves. It is in the light of these findings that Boomgaard rejects romanticized views of Asian 'co-existence' with tigers and, more generally, challenges any simple dichotomy between European and Asian dispositions to wild predators.

European colonialism has had a considerable impact on Asian wildlife, and in many cases continues to have a lasting legacy in terms of people–wildlife relations in Asia. In Chapter 11, Mahesh Rangarajan describes the 'war against vermin' conducted by the British Raj in India, with specific reference to wild predators such as the tiger and the wolf. Rangarajan draws a distinction between the 'animosity' towards and actual extermination of wild predators on the part of the Raj, and the Indian responses to predators, which were marked by 'religious and cultural objections' to large scale killing and 'a willingness to coexist' with these animals. While Indians too killed wild predators, this was not on the same scale, nor did it have the same systematic character as the colonial assault on the wild predators of the sub-continent. To this day, the Indian natural world is marked by the legacy of this earlier colonial specification of the wild predators as 'vermin' to be eradicated.

Finally, in Chapter 12, I examine the recent proposal to reintroduce wolves to Japan and local reactions to it in the mountainous areas where it would be carried out. With reference to the cultural meanings of the wolf in Japan prior to its extinction at the beginning of the twentieth century, the chapter describes how the wolf was seen as *both* a sacred animal that protected people and their crops *and* a dangerous animal that threatened human livelihoods and even, at times, human life. This earlier ambivalence towards the wolf informs the present-day debate on wolf reintroduction in Japan. But Western views of wildlife have an important bearing in this chapter too, in the form of the well-known American example of wolf reintroduction in Yellowstone National Park which serves as the model for the prospective return of the wolf to Japan. However, Japanese advocates of wolf reintroduction also invoke the tradition of wolf worship and reverence absent in the West to claim that wolf reintroduction in Japan would represent both a natural *and* a cultural restoration. Yet the proposal does not go uncontested at the local level.

References

Anderson, E. N. (1996) *Ecologies of the Heart: Emotion, Belief, and the Environment.* Oxford: Oxford University Press.

Callicott, J. Baird and Roger T. Ames (1989) 'Preface'. In J. Baird Callicott and Roger T. Ames (eds) *Nature in Asian Traditions of Thought: Essays in Environmental Philosophy.* New York: SUNY, pp. ix–xii.

Carpenter, C. R. and A. Nishimura (1969) 'The Takasakiyama colony of Japanese macaques (*Macaca fuscata*)'. In C. R. Carpenter (ed.) *Behavior: Proceedings of the Second International Congress of Primatology, Atlanta*, vol. 1. Basel and New York: S. Karger, pp. 16–30.

Cartmill, M. (1993) *A View to a Death in the Morning: Hunting and Nature through History.* Cambridge: Harvard University Press.

Chapple, C. K. (1993) *Nonviolence to Animals, Earth, and Self in Asian Traditions.* New York: SUNY.

Corbey, R. (1993) 'Ambiguous apes'. In P. Cavalieri and P. Singer (eds) *The Great Ape Project: Equality Beyond Humanity.* London: Fourth Estate, pp. 126–136.

Eudey, A. A. (1994) 'Temple and pet primates in Thailand'. *Revue d'Ecologie*, vol. 49, pp. 273–280.

Hell, B. (1996) 'Enraged hunters: the domain of the wild in north-western Europe'. In P. Descola and G. Pálsson (eds) *Nature and Society: Anthropological Perspectives.* London: Routledge, pp. 205–217.

Jackson, P. (1999) 'The tiger in human consciousness and its significance in crafting solutions for tiger conservation'. In J. Seidensticker, S. Christie and P. Jackson (eds) *Riding the Tiger: Tiger Conservation in Human-dominated Landscapes.* Cambridge: Cambridge University Press, pp. 50–54.

Jacobsen, K. (1994) 'The institutionalization of the ethics of "non-injury" toward all "beings" in ancient India'. *Environmental Ethics*, vol. 16, no. 3, pp. 287–301.

Kuroda, N. (1991) 'What is wildlife management? An application of oriental philosophy'. In N. Maruyama, B. Bobek, Y. Ono, W. Regelin, L. Bartos and P. R. Ratcliffe (eds) *Wildlife Conservation: Present Trends and Perspectives for the 21st Century.* Tsukuba: Japan Wildlife Research Center, pp. 1–2.

Livernash, R. and E. Rodenburg (1998) 'Population change, resources, and the environment'. *Population Bulletin*, vol. 53, no. 1, pp. 2–40.

Malik, I. and R. L. Johnson (1994) 'Commensal rhesus in India: the need and cost of translocation'. *Revue d'Ecologie*, vol. 49, pp. 233–243.

Moulton, M. P. and J. Sanderson (1997) *Wildlife Issues in a Changing World.* Delray Beach, FL: St. Lucie Press.

Pirta, R. S., M. Gadgil and A. V. Kharshikar (1997) 'Management of the rhesus monkey *Macaca mulatta* and hanuman langur *Presbytis entellus* in Himachal Pradesh, India'. *Biological Conservation*, vol. 79, pp. 97–106.

Said, E. W. (1985 [1978]) *Orientalism.* London: Penguin.

Sandell, K. (1995) 'Nature as the virgin forest: farmers' perspectives on nature and sustainability in low-resource agriculture in the dry zone of Sri Lanka'. In O. Bruun and A. Kalland (eds) *Asian Perceptions of Nature: A Critical Approach.* London: Curzon Press, pp. 148–172.

Sinha, R. K (1995) 'Biodiversity conservation through faith and tradition in India: some case studies'. *International Journal of Sustainable Development and World Ecology,* vol. 2, no. 4, pp. 278–284.

Sukumar, R. (1989) *The Asian Elephant: Ecology and Management.* Cambridge: Cambridge University Press.

Vandergeest, P. (1996) 'Property rights in protected areas: obstacles to community involvement as a solution in Thailand'. *Environmental Conservation*, vol. 23, no. 3, pp. 259–268.

Wheatley, B. P. and D. K. Harya Putra (1994) 'The effects of tourism on conservation at the monkey forest in Ubud, Bali'. *Revue d'Ecologie*, vol. 49, pp. 245–257.

Yoneda, M. (1991) 'The status of the Asian black bear in the western part of Japan'. In N. Maruyama, B. Bobek, Y. Ono, W. Regelin, L. Bartos and P. R. Ratcliffe (eds) *Wildlife Conservation: Present Trends and Perspectives for the 21st Century.* Tsukuba: Japan Wildlife Research Center, pp. 148–152.

Zhao, Q. (1994) 'A study on semi-commensalism of Tibetan macaques at Mt. Emei, China'. *Revue d'Ecologie*, vol. 49, pp. 259–271.

Part I

WILDLIFE AS RESOURCE

1

ATTITUDES TOWARDS WILDLIFE AND THE HUNT IN PRE-BUDDHIST CHINA

Roel Sterckx

Introduction

In contrast to the extensive literature on the perception of animals in the cultures of Mediterranean antiquity, animal culture and man's attitude towards the animal realm is a relatively understudied topic within the field of Chinese cultural history.[1] Evidence suggests, however, that animals figured prominently in early Chinese culture. In addition to their practical role in husbandry, the hunt, transport and human consumption, animals were used as victims in sacrificial religion, figured as agents and objects in ritual practice and served as symbols and metaphors in the creation of social models of authority. Moreover, the animal realm also provided a rich thesaurus for the expression of fundamental social, moral, religious and cosmological ideas. But whereas the student of early Greece and Rome may find recourse to a large body of primary texts that deal with animals in a more or less exclusive manner – ranging from the proto-zoological treatises of Aristotle to Xenophon's *Cynegeticus* ('Hunting Man') – the sinologist finds himself having to sift animal references from a large and disparate corpus of texts, including literary, historiographical and philosophical writings.[2]

A prominent feature of early Chinese texts is that they reflect an aporia on the animal world as a distinct realm of knowledge. In the extant sources from the pre-imperial and early imperial eras, there appears to have been no conscious effort to dissociate discourse on the animal realm from the literary contexts in which they appear (by, say, integrating them into separate canons). Part of the explanation for this has to be attributed to the broader paradigm in which the Chinese perceived the animal world and nature in general. The classic Chinese perception of the world did not insist on clear categorical or ontological boundaries between animals, human beings and other creatures such as ghosts and spirits. Consequently, the demarcation of the human and animal realm was not perceived to be permanent or constant, and the fixity of the species was not self-evident. The natural world was not understood as an 'objectified' reality that could be scientifically investigated or that functioned according to its own independent biological mechanics.

Instead, the animal realm was positioned as part of an organic whole in which the mutual relationships among the species were characterized as contingent, continuous and interdependent. Animals were rarely thought of as purely natural categories. As a consequence, a dichotomy between the 'moral' and 'natural' animal has rarely been articulated in early China, and it would prove difficult to unveil a development towards systematized theoretical discourse on the physical animal. The Chinese 'thought their animals' with morality; the aim was to elucidate issues of morality rather than the nature of animals. The animal world, in several ways, provided normative models and signs for the guidance of human society. More than any other category in the natural world, animals provided a lens through which the natural realm and the human social order converged.

In this chapter, I shall argue that the discourse on wild animals and the hunt in early China exemplifies a tendency to conflate moral and physical categories. Early Chinese writings emphasize that notions of humanity and bestiality, domesticity and wildness, and the hunter and the hunted were subject to ongoing change. Mastering these patterns of change in the interface between humans and animals was the prerogative of the sage or ruler–king. While asserting his socio-political authority through hunting wild animals and through the symbolic appropriation of animal paraphernalia, the ruler had to monitor a balance between the human and animal realms by correlating behavioural patterns in the surrounding wilds to the workings of human government. My analysis is based on a reading of received sources and archaeologically recovered manuscript material from the Warring States period and early imperial China (ca. fifth century BC–mid-second century AD).[3]

Defining animals

The domestication of the wilds and their fauna and flora is portrayed in many cultures as a development in which mankind distinguishes and elevates itself from the natural character of its surrounding through a process of gradual civilization. Traces of this idea are preserved in several early Chinese narratives. They relate how divine sages transformed a hunting and gathering society into an agricultural community and how humankind developed from a pristine state of either peaceful cohabitation or frenzied antagonism with the animal world into an ordered community in which the organized breeding, domestication, slaughter and consumption of animals redefined the relationship between human beings and the surrounding natural world.[4] Although early Chinese texts exposed different views on how to accommodate, assimilate or differentiate human beings from animals, a universally accepted theory that glorified the domination of animals by a superior human species did not exist. Nor did early Chinese thinkers develop natural philosophies which sought to explain the superiority of the human species according to physical or proto-biological premises: 'One of the most important consequences of this idea of the historical creation of humanity through separation from the animal world was that the fundamental distinctions between men and animals were not biological but technological and, above all, moral' (Lewis 1990: 171).

Unlike the Judeo–Christian tradition, where man's dominion over the animal world was canonized in a divine exhortation to subdue the animal species (*Gen.* 1: 28),[5] no textual source gives a detailed history of the domestication of animals in early China. The origins and development of a received term such as, for instance, 'the six domestic animals' (*liuchu*: horse, ox, sheep, pig, dog, chicken) cannot be traced with certainty. Animal husbandry played a minor role in traditional Chinese agriculture. Livestock were certainly kept by Chinese farmers, but in far smaller numbers than in Europe. (Among the domesticated animals, dogs and pigs had the longest history.) Meat itself played a relatively minor role in the traditional Chinese diet in which grain constituted the main food (Bray 1984: 3–9; Chang 1977: 25–83). In short, the absence of a record of animal domestication in the early Chinese textual corpus is significant. It clarifies the presence of a twofold theme in early Chinese writings, namely, a permanent tension between, on the one hand, the assertion that humans have a degree of control over the animal world and, on the other hand, the recurrent idea that humans should strive to harmonize the workings of human society with the rhythms and patterns of the animal realm. But before I discuss this tension in more detail, some of the distinctive features of the human–animal relationship as evinced in texts of the pre-Buddhist era deserve to be highlighted.

The frail distinction between the realms of the human and the bestial in early Chinese thought is exemplified, first, by an absence of attempts to develop ontological theories that set humans apart from beasts and, second, by the recurring idea that wild animals can be transformed into cultured subjects under the influence of moral government. Following this same logic, humans were thought to degenerate into a bestial existence when such proper conduct of human government was absent. A maxim in the book of Mencius (372–289 BC) reads: 'What differentiates man from birds and beasts is but a trifle. Ordinary people cast it aside, only the gentleman preserves it' (*Mengzi* 8A.10a). Mencius recognizes a distinction between the human and animal species, but also emphasizes the frailty of this distinction. In addition, he infers that human comprehension and sagacity determine whether the distinction between the human and the beast is upheld or obliterated. Similarly, in the *Analects*, Confucius (551?–479 BC) states: 'I cannot flock together with the birds and beasts. Am I not a member of this human species? With whom, then, can I associate myself?' (*Lunyu* 18.4a). Although Confucius dissociates himself from the animals, by indicating that he would only associate with morally superior specimens of the human species, he implies that the divide between the human and the bestial does not run parallel with the species distinction between animals and human beings.

One Han author, Han Ying (fl. 150 BC), goes even further and equates the physiology of morally inferior humans with that of animals by differentiating the petty man from the gentleman as someone 'whose structure of the limbs and body is joined like those of birds and beasts' (*Han shi waizhuan* 4.153). Such statements suggest that notions of bestiality and humanity were not perceived as fixed or permanent. They disclose an interaction between the two worlds in which the sage functioned as the agent able to draw boundaries or accommodate similarities between the animal realm and the human world. The philosopher Xunzi (300–237 BC) summarizes this idea by stating that 'what makes a man really human lies not merely in his being a hairless biped, but rather in his ability to draw boundaries' (*Xunzi* 5.65). Rather than defining the difference between humans and animals according to a set of innate physical or 'ontological' properties, human nature is presented as a faculty that is subject to permanent change and in need of constant moral cultivation to prevent it from lapsing into a bestial existence.

In various texts, the delineation of the species boundaries is tantamount to the establishment of a 'moral' taxonomy of the living species. This does not imply that animals were never defined in physical or proto-biological terms. The book of *Liezi*, for example, defines human beings as 'anything with a skeleton seven feet high, hands different from its feet, hair on its head and teeth inside the mouth, standing upright as it runs'. Birds and beasts are defined as 'anything with wings at its side or horns on its head, teeth apart and claws spread out, flying upwards or walking bent down'. However, the identification of an innate sense of morality distinguishing man from beasts constituted a much more prominent taxonomic difference, and human–animal distinctions in terms of morals or human virtues clearly outnumber biological or zoological differentiation. The same *Liezi* passage continues with a statement that there are no great differences in mind (*xin*) and intelligence (*zhi*) between the 'species endowed with blood and *qi* [vapour or energy]'. The latter comprises ghosts, spirits, human beings, as well as birds, beasts and insects (*Liezi* 2.21b; Graham 1991: 53–54).

'Blood and *qi*' constituted a distinctive criterion of animacy at least since the fourth century BC.[6] However, this notion of 'blood and *qi*' was more than a biological or physical property. In a number of texts it is said to underlie the faculty of emotions. This correlation is significant since it attributes a faculty similar to human emotions to non-human living beings. One Han text, for example, states that all species with 'blood and *qi*' possess the temperaments of joy and anger, and the inclination to advance towards benefit and to shun danger in the same way as humans (*Huainanzi* 19.645). According to a ritual text, the extravagant shedding of blood of sacrificial animals is seen as at odds with the respect that a gentleman should demonstrate towards the living blood species. The text stipulates that animals should not be killed 'without a reason' and prescribes that a gentleman (*junzi*) should keep a distance from the kitchen (that is, the abattoir).[7]

Another story from Mencius is an account of King Xuan of Qi, who cancels the slaughter of a bull for the blood consecration of a bell because he had seen the animal alive and could not bear the sight of a shivering animal on its way to its execution (*Mengzi* 1B.2b–3b). Although one might be tempted to detect an implicit notion of animal compassion in such stories, this does not seem to have been the intended message by the authors of the time. No school of thought in pre-Buddhist China explicitly advocated a theory of animal compassion or imposed universal religious or social taboos on the hunt, slaughter and consumption of animals. As will be illustrated below, the occasional references to such prohibitions show that a concern for moderation in the killing and consumption of animals was meant to reflect the moral cultivation of humans themselves. The scattered references to practices which approximate a notion of 'compassion' towards animals need to be read with the knowledge that such compassionate attitudes were thought to benefit human nature rather than the welfare of animals themselves. The needless killing of animals was seen as a symbolic slaughter of the officiant's moral integrity.

The association of a 'moral consciousness' or 'moral potential' with non-human living species gave way to a 'moralization' of the physical animal and its bio-behaviour. Possibly the most articulate theoretical distinction between humans and animals in moral terms occurs in Xunzi's famous 'Ladder of Souls' passage:

> Water and fire have *qi* but do not contain life. Herbs and trees contain life but have no knowledge, birds and beasts have knowledge but no righteousness. Man has *qi*, contains life, has knowledge and also has righteousness, therefore he is the most valuable being for the universe.
>
> (*Xunzi* 9.153; Needham 1985: 21–23)

The text continues by stating that a capacity for social organization enables humans to subdue and use animals that surpass them in physical strength. This ability to form social groups originates from man's ability to 'draw (social) distinctions', which in turn stems from his sense of righteousness. The *Huainanzi* (second century BC) alludes to the same theme: birds and beasts cannot form flocks together because their species are different, tigers and deer cannot gambol about together because their strength is unequal (*Huainanzi* 9.286). And the book of *Guanzi* states that 'although relations among a flock of crows may seem good, they are never really close ...' (*Guanzi* 1.7a–b).

The human–animal difference centres thus to a large extent around the idea that animals possess physical power (*li*), whereas humans have a sense of morality and ritual propriety (*li*). One text identifies physical power as a distinctive property of animals by stating that 'heaven serves with its seasons, earth with its material resources, man with his virtue, the spirits with their omens, and animals with their strength' (*Guanzi* 4.12a). Elsewhere, a gentleman who has lost his ritual propriety is equated with common folk, and common folk without ritual propriety with birds and beasts (*Yanzi chunqiu* 1.6, 2.170). Mencius argues that by dwelling in idleness without instruction one approximates the condition of wild birds and beasts. The same applies to humans who fail to reciprocate ritual propriety, benevolence and loyalty (*Mengzi* 5B.3b, 8B.5a–b). In several texts this lack of *li* or 'ritual propriety' is exemplified by the observation that animals as well as petty humans flock together or have instinctive physical intercourse, failing to make a distinction between young and old, or species and kin. A recurrent expression is that deer lack ritual propriety because stags and calves follow the same doe. The dynastic histories of the Western Han dynasty (206 BC–AD 8), for instance, interpret the appearance of tailed wild deer (recorded in an earlier chronicle) as an indication of illicit sexual behaviour that could undermine the affairs of state. Another Han text states that people who lack filial piety, food and clothes, rites and music are like deer that follow their own desires.[8] Certain gradations in moral consciousness were associated with different animals, and not all species occupied the same place on the moral ladder. One text for instance points out that the phoenix 'by birth' has a sense of benevolence and righteousness but that the tiger and wolf 'by birth' have a covetous and violent heart; 'both have a different reputation because of their mother'. In this case, a different susceptibility to morals is said to be innate (*Da Dai Liji* 3.7b).

A moral hermeneutic of the animal world was also evident in the practical management of animals. One such practice was animal physiognomy where an animal's character traits were inferred from external bodily signs such as plumage texture, bone structure or skin patterns. Moral virtues were literally projected onto the anatomy or physiognomy of animals. Physiognomy, which consists of finding a correlation between the physical organization of the animal body and its life course and/or its qualities for human use, is a process that draws the natural into the social. It applied to both mythical or sacred animals as well as animals known from daily life. The anatomy of the phoenix, a hybrid avian whose appearance was thought to herald the advent of a virtuous ruler, was associated with Confucian virtues: its head carries virtue, its cranium manifests righteousness, its back supports benevolence and its heart is entrusted to knowledge.[9] The rooster stands out among its animal peers for its timely crow that determines day from night and induces the bird species to 'resonate'. Its physiognomy was also associated with Confucian virtues: the virtues of culture (*wen*) symbolized by the rooster's crest, martialness (*wu*) embodied by its spurs, bravery (*yong*) because it fights its enemies, benevolence (*ren*) because it calls its companions upon finding food, and trustfulness (*xin*) because its crowing is perfectly timed (*Xin xu* 5.188–89; *Han shi waizhuan* 2.60–61).

Early Chinese texts describing the interaction between humans and animals were based on the assumption that animal behaviour was related to the human social world and that human ethics were related to the animal realm. These linkages between the human and animal worlds affected both worlds (if they could be separated at all). In the same way that animals could be domesticated for husbandry, it was thought that through various mechanisms they could also be 'cultured' or 'civilized' and drawn into the socio-religious spheres of society. This particular process of 'drawing the bestial into the social' and 'extrapolating the social into the bestial' reflected a certain mode of organic thought, one which focused on condensing or expanding the categories of the human and the animal in order to demonstrate the interdependence and mutual influence among all animate beings. This hermeneutic process of comparison crystallized around the idea that one could 'expand the categories' (*tui lei*). This process of projection, or method of induction, implied that aspects of action, behaviour and physical reality among animals were observed as having a spontaneous or induced impact on humans and vice versa. It also implied that behavioural changes in one realm were thought to lead to changes in the other.

To summarize, the idea of human–animal interdependency runs through most contexts of textual discourse on animals in early China. The portrayal of the animal world was therefore marked by a hermeneutic impulse to detect mutuality, congruence, and correlation between animals and humans. The same moral parameters which transformed human beings into cultured beings also exerted their influence over the animal realm. As a consequence, a direct correspondence was established between the ruling or governing of human society and wild animal behaviour.

Targeting the wilds

The importance of the hunt and the hunting park had to do with symbolism rather than economics. The seclusion of wild animals in a confined space such as a park or a court garden, aside from availing the ruler with a sufficient supply of sacrificial victims, was primarily an endeavour to sanctify the numinous powers of the ruler. Accounts of royal hunts and ritual killings of wild animals within such animal preserves describe how game animals symbolically represented all species within the ruler's realm. Sima Xiangru (ca. 180–117 BC) has captured the grandeur and exotic nature of these parks, as well as the scale of early imperial hunts in a rhyme-prose poem describing emperor Han Wudi's (141–87 BC) Shanglin park:

> The Son of Heaven stakes his palisades and holds his hunts,
> Mounted in a carriage of carved ivory
> Drawn by six spangled horses, sleek as dragons ...
> His attendants fan out on all sides
> As they move into the palisade.
> They sound the sombre drums
> And send the hunters to their posts ...
> Leopards and panthers they take alive;
> They strike down jackals and wolves.
> With their hands they seize the black and tawny bears,
> and with their feet they down the wild sheep ...
> And with short spears [they] stab the little bears,
> Snare the fabulous *yao-niao* horses

And shoot down the great boars.
No arrow strikes the prey
Without piercing a neck or shattering a skull;
No bow is discharged in vain,
But to the sound of each twang some beast must fall ...[10]

The ruler exerted his authority beyond his domestic realm by hunting wild animals. One chronicle recounts how a minister refuses to grant leave to an official for a hunting expedition in preparation for a sacrifice (the story can be dated to ca. 522 BC). He argues that only the ruler is permitted to use 'fresh animals' (that is, animals caught in the wilds) for the sacrifice, whereas ordinary officials are only allowed to provide domestic animals (*Zuozhuan* 40.12a). The 'Ritual Canon of the Zhou' contains a statement stipulating that among the common people those who do not raise animals are not permitted to use animal victims in sacrifice (*Zhouli* 7.7a). The hunting of game animals in the wild thus appears to have been the prerogative of the ruler and the nobility. By roaming through a park and contemplating or hunting exotic beasts in artificial landscapes, a ruler symbolically paced through his empire, in the same way that he engaged in inspection tours (*xunshou*) of his political realm. By subduing wild animals in parks, the ruler–hunter symbolically subdued all living species within his socio-political realm. For instance, one ritual codex states that the ritual display of tiger and leopard skins demonstrated a ruler's power to 'subdue' that which is wildly natured (*Liji* 25.12a). Such allusions to the wilds did not only refer to wild animals but also, and perhaps more importantly, to human barbarian vassals who lived on the periphery of the Chinese cultural epicentre. These nomadic outsider tribes were thought to have the inner disposition of animals, and comments about their physical appearance likewise equated them with wild animals. They are said to dress in animal hides, speak the language of birds and beasts, or 'squat on their heels and crouch down in a haughty manner, and not to differ at all from birds and beasts'.[11]

The decorative exhibition of exotic animal hides and their use as clothing asserted a ruler's power both over the animal specimens from which they were stripped and over the human populations which shared the habitat of these animals. This enactment of power through the symbolism of animal designs and animal hides can also be seen in descriptions of archery ceremonies. Archery targets and tally holders were decorated with animal designs:

As for targets, the Son of Heaven's target has a (picture) of a bear's (head) on a white background; the feudal lord has a tailed deer target on a red background; a great officer has a cloth target background, with (the heads) of a tiger and a leopard drawn on it; the ordinary officer a cloth target, with (the heads of) a deer and a wild boar drawn on it.

(*Yili* 13.10b; *Bohutong* 5.243)

The text seems to imply that the marksman was thought to be endowed with the controlling or exorcizing power of the animal pictured on the target he was aiming at. Elsewhere reference is made to animal hides and skins as the representation of the beast's essence on the targets they cover or decorate. According to one source, an officer with the title 'Manager of Furs' had to provide animal skins which served as targets during archery ceremonies. He supplied tiger skin targets, bear and leopard skin targets and deer skin targets according to the rank of nobility of the participants (royal: tiger, bear and leopard; feudal lords: bear and leopard; grandees: deer) (*Zhouli* 7.7a). One scholar has suggested that 'the skins of the bear, tiger and leopard [used to cover the target stands] . . . were meant, together with the humanoid configurations in which the targets were cut, to be manifestations of the erratic malcontents who dared to challenge the king's sovereignty' (Riegel 1982: 3).

The animal targets likewise served to identify the marksman with the beast's power. In other words, the marksman is put on a par with the power of the animal skin he targets. By exerting his skilful dominance over the target hide, he not only controls the forces of the represented animal but also adopts and enacts these forces. During royal archery contests the most ferocious animal specimens (tiger, bear and leopard) are targeted by the ruler who symbolically asserts his dominance over his subordinate officers and vassals (in the same way as the weaker species in the animal world are dominated by the fiercest predators). By aiming at the skin, the marksman aims at the whole animal and at the whole animal species it represents: '[From shooting at his target] each (marksman) takes on the power that enables him to subdue' (*Bohutong* 5.244). The association of power with shooting is further evident in several references in which the ruler is said to 'personally shoot' (*ipso manu*) a wild animal or sacrificial victim.[12] The targeting of animal hides was also tied in with the rhetoric that equated barbarians with animals because they were dressed in animal hides. By shooting at animal hides, the ruler transformed the animal targets (*hou*) into yielding vassals (*hou*).[13] This transformational power of the ruler over the wilds through the targeting of wild animals is further attested in an ode in the *Shijing* ('Book of Songs'), where the ruler-hunter is compared with a fabulous beast named *zouyu*. The poem states how the ruler-hunter kills five wild boars by discharging only one arrow (*Mao shi zhengyi* 1E.13b–15b; Legge 1991: 36–37). Although commentators disagree on whether the *zouyu* was a righteous beast appearing as a response to a virtuous government or whether it was the name of a hunting officer, both interpretations support the image of a ruler exerting his authority through 'targeting' the wilds. Both the *zouyu* ode and a lost piece entitled 'The Head of the Wildcat' were performed during archery ceremonies.[14]

While hunting wild animals provided the ruler with ritual and social authority, the ability to interpret and accurately respond to wild animal behaviour demonstrated a sage-ruler's comprehension and encompassing control over both the human and animal realms. A prominent theme which exemplifies this 'moral' biology of the wilds in early China are accounts of violent and predatory animal behaviour. First, instinctive animal behaviour was interpreted as being determined by moral principles in human society. Second, changes in human society were believed to spontaneously induce behavioural changes in the animal world. As a consequence, virtuous conduct in human society would generate a 'moral' disposition in animals which would transform their innate 'wild' instincts. A prose poem on the crane, 'He fu', attributed to Lu Qiaoru (second century BC?) illustrates this. The poem describes cranes wading at the edge of a pond, presumably in a park or courtyard of a ruler. The cranes – renowned as symbols par excellence of escapism and freedom – are described in all their natural movements. The poet suggests that despite their natural inclination and ability to fly off, they remain in the surroundings of the pond. He then concludes as follows:

> Therefore we know that
> These wild birds with their wild instincts
> Have not yet escaped their cage.
> Relying on the magnanimous love of our king,
> Even wild birds cherish his grace.
> While prancing they sing and dance,
> the red railings are their reason of joy.
> (Xi jing za ji 4.4a)

Although the cranes have wild instincts and are free to leave the king's park, the king's moral government makes them stay, their captivity ('red railings') being their reason for joy.

Alternatively, human–animal antagonism and anomalous animal behaviour was believed to be rooted in decadent human government, including the violation of the regular patterns of nature codified in calendrical texts. The unbridled physical chase of animals was in the first place a transgression of a moral code. The notion that human virtue constituted the ultimate principle of balance between man and beast occurs in narratives describing the legendary origins of the geo-political organization of China. The following piece is staged as an exhortation spoken by a forester at the time of King Wu of the Zhou (ca. 1121 BC):

> Far and wide Yu's footsteps reached.
> As they lined out the Nine Provinces,
> and traversed them to open up the Nine Paths.
> The people possessed chambers and temples,
> the animals possessed flourishing grasslands,
> each of them (man and beast) had a place to dwell,
> and because of virtue there was no distress.
> Then Archer Yi took over the emperor (Yu's) place,
> he coveted a desire for (the chase of) wild animals,
> neglected the care of his state,
> but (instead) thought of its does and stags . . . .
> (Zuozhuan 29.24b–25a)

Rather than being ascribed to a purposeful evil nature innate in animals, predatory killings and attacks on humans were explained as the result of a distorted balance within the human realm or between humans and animals. When sages ruled human society, the influence of good government spontaneously transformed the predatory disposition of wild animals, 'tigers and leopards could be pulled by the tail, vipers and snakes could be trod upon' (*Huainanzi* 8.253), or as the *Da Dai Liji* states: 'When the sages are ruling the state ... rapacious beasts forget to attack, and (predatory) birds forget their spurs. Wasps and scorpions don't sting young babies. Mosquitoes and gadflies don't bite young foals' (*Da Dai Liji* 9.12b–13a).

Ravaging raids by tigers and wolves on human settlements, usually referred to as 'tiger and wolf calamities' (*hu lang bao*), are the subject of many discourses. The constant threat of wild animal attacks also resulted in the establishment of legal measures stipulating the amount of predatory beasts that could be caught, their cash value, regulations on animal enclosures etc.[15] Against the background of idealized hunting regulations, devised by sage kings to balance the human need for game animals with the preservation of the rhythms in the animal realm, violent animal behaviour was explained as a consequence of a transgression of this balance. Such transgressions blurred the physical and moral distinctions between man and beast. This is attested in the theme that a 'bestial' government or ruler changes animals into predators, and that the lavish hunting of animals causes animals to chase humans. The latter, for instance, is articulated in a later poem, possibly written by Wang Su (AD 195–256), entitled 'Rhapsody on Remonstrating against Engaging with Tigers in Combat'. Echoing earlier rhapsodic prose poems, an interlocutor criticizes a ruler's overindulgence in hunting on the grounds that it distorts both the habitat of humans and animals ('You drive people into the forest glens and attack tigers in their realm ...') (*Kong Congzi* 7.1a–2a; Ariel 1996: 98–101). The moral failure of human rule reverses the disposition of the species: the (human) hunter becomes the (animal) victim; and just as humans trespass upon animal habitats, animals inflict similar bestial calamities in the human habitat. The Eastern Han critic Wang Chong (AD 27–100?) devotes a whole chapter entitled 'Encountering Tigers' to the ravaging tiger theme. Throughout this chapter Wang attempts to refute a correlation between tiger attacks and the depravity of officials in government but fails to posit an alternative explanation. While attributing tiger attacks to an arbitrary accidental encounter with humans, he likewise acknowledges that tigers enter human settlements as soon as human rule has degraded to the bestial level of the wilds, that is, the tiger's natural habitat (*Lunheng* 48.707–11).

Similar ideas survived in later times. For example, a famous piece entitled 'Proclamation to the Crocodile', written by Han Yu (AD 768–824), contains a similar reasoning. The emperor, having ascended the glorified throne of a unified Tang empire after China had suffered centuries of disunion, can no longer share his realm with the wild animals that had intruded the human realm following its moral disintegration. The reclamation of land and empire, that is, human civilization and imperial authority, is tantamount to the expulsion of wild animals. In a mandate addressed to these malevolent creatures, Han Yu declares war on the whole reptile tribe:

> The crocodiles and the governor cannot together share this ground. The governor has received the command of the Son of Heaven to protect this ground and take charge of its people; but you, crocodile, goggle-eyed, are not content with the deep waters of the creek, but seize your advantage to devour the people and their stock, the bears and boars, stags and deer, to fatten your body and multiply your sons and grandsons. You join issue with the governor and contend with him for the mastery. . . . To the south of the province of Ch'ao lies the great sea, and in it there is room for creatures as large as the whale or roc, as small as the shrimp or crab, all to find homes in which to live and feed. Crocodile, if you set out in the morning, by the evening you would be there. . . .[16]

Many stories illustrate how overindulgence in the hunt was seen as a sign of neglect in human government. One such story relates how Master Yan reprimands a duke for not having returned from the hunt for seventeen days. He argues that the duke's subjects considered their lord to hate his people and to prefer the beasts. Next he warns his lord of the pernicious consequences that this might have: 'Fish and turtles who reject the deep springs and come to the dry surface are consequently caught by hook or net. Birds and beasts who reject the dense mountains and come down to the cities and plains are therefore caught by hunters . . .' (*Han shi waizhuan* 10.358–59). The two mid-Eastern Han cases that follow are good examples of how bestial animal instinct is attributed to the course of human actions. One account relates how in the 110s AD, a command in what is now modern Hubei province was suffering from 'tiger and wolf plagues'. An edict from the hand of its governor argued that the intrusion into the natural habitat and the extravagant hunting of these animals was the cause behind such problems:

> In general, the residence of tigers and wolves in the mountains and forests is like the residence of human beings in cities and markets. In antiquity, in the age of complete transformation, wild animals did not cause any trouble. All this originated from the fact that grace and trust were wide-ranging and abundant, and benevolence reached the avian and running species. Although I, your governor, possess no virtue, how could I dare to neglect this righteous principle. (Therefore) when this note arrives, let cages and pit-traps be destroyed and do not recklessly go on the catch in mountains and forests.
>
> (*Hou Hanshu* 38.1278)

This passage refers to antiquity as an age of 'supreme Transformation', a time when the human virtue of benevolence had transformed the wild beasts and altered a situation of static species separation.

A second example features Song Jun (fl. AD mid-first century), a critic of shamanic practices, a sceptic of beliefs in demons and an ardent opponent of popular superstitions. When sent off to become governor of Jiujiang (modern Anhui province), a command plagued by tiger calamities, Song issued the following edict:

'Tigers and leopards live in the mountains, turtles and alligators in the water, each have their entrusted habitat. The presence of wild beasts in the Jiang-Huai region is similar to the northern territories having chickens and pigs. Now they harm the people, and the malign influence behind this lies with low-hearted officials who put all their efforts into catching them. This is not the basis of showing sympathy. If these officials were to devote their attention to removing detrimental poverty and think about promoting loyalty and goodness, then they could at once remove traps and pits, discharge taxes and restore order.' Later it was said that the tigers had moved (together with Song Jun) to the east and had crossed the Jiang (river).
(*Hou Hanshu* 41.1412–13; *Fengsu tongyi* 2.92–93)

Song starts his argument by establishing congruence between territorial habitat and its animal presence (cf. Sterckx 1996a). He then argues that wild beasts are endemic to the Jiang-Huai region and attributes their anomalous behaviour to the officials' indulgence in hunting for pleasure. However, Song's exemplary rule as governor transforms the tigers' bestial nature, and as a consequence they 'cross the Jiang' and leave their natural habitat. Moral authority has transformed their bestial instincts, and a susceptibility to human virtue has induced the tigers to transcend their biotopical boundaries and follow the human ruler.

A similar moral human–animal contingency underlies the depictions of wild animals in early Chinese calendrical texts. The calendar provided a schematized framework which monitored the actions of human society according to the changing patterns in nature, its fauna and its flora. As is clear from the data on animals in one such calendar, the 'Little Annuary of the Xia' – incorporated in a Han ritual handbook but in its title claiming origins back to the proto-historical Xia dynasty (2207–1766 BC) – the observation of animal behaviour was an important ingredient in the development of calendars (cf. Wilhelm 1930: 233–43). Prescriptions concerning the timing, methods and targeted species of the hunt served to maintain a moral balance between the human and animal realms over a temporal sequence of changing natural habitats (that is, the seasons). Thus the hunting season was opened when the game animals themselves displayed their 'innate' moral disposition. Fishermen are advised to start fishing only when 'the otters sacrifice fish' (that is, when there is an abundant catch). Hunts would start only 'when the wolves sacrifice prey'. And the nets were set out only 'when doves transformed into eagles' (that is, when seed-eaters assume a carnivorous condition).[17] So respect for the animal's 'innate' sense of sacrifice ultimately prevented these predatory animals from preying upon humans as their sacrificial meat. Respecting the prescriptive cycle of the calendar was a means of preventing human mores from being sanctioned by the natural world. Indeed, untimely appearances of animals were interpreted as omens for impending changes in human society. Early Chinese texts document a great number of creatures that cross their natural habitat to come and abide in unnatural proximity to humans. Such transgressions of the inner-outer or wild-domestic boundaries were interpreted as meaningful indicators of various social anomalies within the human realm. Numerous examples occur in the dynastic histories of the Western Han dynasty. They include a bear from the wilds entering the palace, wild birds entering the ancestral temples or palaces, pigs breaking out of the pen or stable and entering residential halls, snakes emerging from a palace or entering the capital, rats dancing at the palace gates and inside the court, rats nesting in trees, wild animals playing in the courtyards, wild birds perching on the trees of courtyards and many others (cf. Sterckx 1998; Sterckx 2002: 205–237).

A moral imperative also underpinned descriptions of hunting techniques and the conservation of game. These include statements stipulating that animals killed out of season should not be sold at the markets and that the Son of Heaven (emperor) does not eat or sacrifice pregnant animals. The moral bias behind such regulations occurs in numerous anecdotes. Confucius, for instance, states that the unseasonal killing of a single animal is 'contrary to filial piety'. One ritual codex states that the wrongful slaughter of the six domestic animals affects one's relatives. Elsewhere, an emperor is said to refrain from catching cranes since it was spring, the season during which the use of nets was not permitted. Another exemplary story tells of a duke who robs a sparrow's nest but puts the fledglings back when he sees that they are too weak. Confucius allegedly used a fishing line but not a net, and he used a stringed arrow but did not shoot at roosting birds. Literati are said not to use stringed arrows. One source states that those who mistreat herbs, trees, chickens, dogs, cattle and horses will, as a result, suffer and receive their just deserts. In another chronicle, a ritual imperative underlies the hunt. It states that in ancient times the dukes would not shoot birds and beasts unless the meat was used in sacrificial stands, and their skins, hides, teeth, bones, horns, hair and feathers were used to decorate ritual implements.[18] The close relationship between conservation and hunting is also exemplified in the homophony of the Chinese character for the 'winter hunt' (*shou*) and the verb 'to preserve' (*shou*) (*Bohutong* 12.590; *Chunqiu fanlu* 10.262). Another standard expression that occurs throughout the corpus of Warring States and Han texts warns that the depletion of natural resources will result in the absence of numinous and auspicious animals. It states that if one cuts wombs and kills foetuses, the unicorn will not come to the outskirts of the city. If one exhausts the marshes and their fish, the dragon will not reside in its springs. If one tramples nests and smashes eggs, the phoenix will not fly over the region.[19]

As these extracts show, the ambiguous nature of the hunt as both an act of killing and a moral gesture of animal preservation – a theme which figures prominently in Western discourses on the hunt [20] – was a topic of concern in a number of writings. However, as stated above, these fragmentary assertions were primarily concerned with the moral well-being of human society. Ultimately, the discourse on the wilds was a discourse on the self-perception of human nature. It provided a window on how humans were thought to relate to nature rather than a focus on the animal world itself. Ideally, rulers would only connect with wild animals in an indirect way. While the ordinary hunter chases and subdues animals through the use of physical violence, the sage–ruler connects with animals in a moral way. An anecdote from Mencius recounts the story of a tiger hunter named Feng Fu. He was renowned for being an expert in seizing tigers. But when, having become an 'eminent gentleman', he once bared his arms to assist a crowd to catch a tiger, 'the crowds (*zhong*) were delighted but those who were gentlemen (*shi*) laughed at him' (*Mengzi* 14A.11b). The moral of the story is clear: a gentleman does not bare his arms to counter a wild beast. For the sage, the transformation of the wild beasts is effective through his harmonization with or balancing of the animal's natural instincts rather than through brute force, technology or the implementation of the yoke.

Thus the skilful animal tamer Liang Yang is successful because he neither gives the tigers their way nor thwarts them:

> Now in my heart I neither comply with them nor oppose them, thus the birds and beasts regard me as their equal. Therefore when they roam in my garden, they do not think about their lofty forests and desert marshes, and when they sleep in my courtyard, they never desire to be deep in the mountains or hidden away in the valleys. This principle is only natural.
>
> (*Liezi* 2.9b–10b)

Similarly, the book of Zhuangzi argues that the incompatibility between different species can be encompassed when the animal breeder 'accords' himself with the disposition of the animals: 'Tigers are a different species from man but when they fawn on the person who rears them, it is because of (this person's) accordance (with their dispositions). Therefore, when they kill him, it is because he opposes (their dispositions)' (*Zhuangzi* 4.167). Since the distinction between human beings and animals was to a large extent based on moral rather than biological premises, the domestication of the animal realm was primarily expressed in terms of a moral transformation.

Craft analogies involving animals (such as charioteering, horse-bridling, archery, and fishing) emphasized that such techniques were only successful through harmony between man and animal rather than through physical domination. The *Huainanzi* exemplifies this in several cases. One passage maintains that there exists an incompatibility between humans and wild horses, but admits that horses are docile and susceptible instruction if one 'connects with their *qi* [energy] and intent':

As for the fact that fish leap about and magpies are piebald, this is similar to the reason why man and horse are man and horse. Muscles, bones, shape and body are what they receive from Heaven. They cannot be transformed. From this point of view man and horse do not resemble each other [that is, they belong to different species]. But when a horse is a grazing foal, it jumps and leaps with its hooves up in the air, it hoists its tail up in the air and runs about. Man cannot control it ... [until] the stable officer trains it to obedience and the skilful cavalier instructs it.... Therefore, although its outward appearance makes it a horse, and this 'horse-ness' cannot be changed, the fact that it can be driven is due to instruction. A horse is but a daft animal. But if one connects with its energy and intent, it will likewise await instruction and accomplish itself. How much more is this the case for humans!

(*Huainanzi* 19.638)

Another passage refers to a renowned archer named Bo Juzi and the famous angler Zhan He who 'could connect with birds flying a thousand feet above', and 'cause fish [to dart towards the hook] from the depths of the great springs', because they had obtained an all-pervading harmony with the Dao (*Huainanzi* 6.194). The sage accords with rather than manipulates the movement of the animal. Elsewhere the 'shepherding of the people' (that is, the ruling of a state) is compared to the raising of domestic animals. By locking off an animal enclosure and hindering the animals in their natural movements, one only induces them to be 'wild at heart' (*Huainanzi* 7.241; Sterckx 2002: 151). Other craft analogies involving animals likewise attribute the successful human–animal cooperation to a mutual moral understanding rather than a relationship of physical domination or subordination. In the *Analects*, Confucius epitomizes this by stating that 'a swift horse is not praised for its [physical strength] but for its virtue' (*Lunyu* 14.13b). Confucius does not respect the animal in its 'bare animal state'. When his stables catch fire he inquires whether any human being was hurt, but 'did not ask about the horses' (*Lunyu* 10.10a). Another passage relates how Confucius chooses to respect animals in their ritualized state. He refuses to bow after having been presented a carriage and horse because it was not (as worthy) as sacrificial meat. Confucius advocates that the use of an animal is ultimately subject to a moral imperative.[21] When a disciple suggests that the sacrifice of sheep at the announcement of the new moon be abolished, Confucius reprimands him for loving the sheep rather than being concerned with the disappearance of the rite (*Lunyu* 3.10a). In the end, the sage is able to connect with animals in a moral way. Or, as one source puts it, the skilful charioteer or cavalier aligns himself with the natural movement of his horse(s), 'the horse knows that behind there is a chariot but considers it to be light. It knows that there is a man in it but loves him. The horse adores (the charioteer's) righteousness and likes being used by him' (*Han shi waizhuan* 2.43). An innate respect for human virtue thus transforms a wild animal into an obedient and domesticated companion.

Conclusion

This chapter has addressed certain aspects of attitudes towards the wilds and the hunt in early China. The analysis has primarily focused on the way in which hunting practices and attitudes towards wildlife are reflected in early textual discourse. Material artefacts, pictorial sources such as tomb murals and physical animal remains need to be studied to complete the picture. Furthermore, the following observation by Stanley Tambiah holds equal truth for the historian: the meaning of animals in a particular society cannot be unravelled without outlining the contours of the society's 'entire system of collective representations, and of how animals enter into the social practices and cultural concerns of a people in their lives, in both everyday and extraordinary events' (Tambiah 1985: 8–9). Textual sources do not allow us to reconstruct with precision the ways in which ordinary peasants and village dwellers in ancient China dealt with wild animals in daily life. However, detailed examination of the topic of hunting and the wilds as it appears in canonical texts, as well as later references to hunting in antiquity, suggests that it was more than a marginal subject in literati circles. Descriptions of imperial hunts, animal parks and exotic wild animals would remain important *topoi* in Chinese literature. The attention paid to discourse on hunting and the wilds in early texts is significant for our understanding of Chinese attitudes towards animals, since inquiries into the animal kingdom primarily reflected the way in which humans perceived their place in relation to other living species. The picture that emerges from this corpus of texts is marked by a peculiar assessment of the human–animal relationship. It presents an animal realm that functions as a catalytic medium for the conception of human morality, and portrays animal behaviour in the surrounding wilds as an indicator of the workings of human society. While hunting and the ideal of a balanced harmony with the wilds constituted an important aspect of socio-political authority for the sage–ruler, it also provided a means of sanctioning human conduct.

Notes

1 For a bibliographical introduction, see Sorabji (1993: 221–232). Other studies I have found particularly informative include Toynbee (1996) and Anderson (1985). For a detailed study and bibliography of secondary scholarship on animals in ancient China, see Sterckx (2002).

2 Compare, for example, the brief survey on China in Petit and Théodoridès (1962: 15–22), with its data on Greece and other areas.

3 This roughly corresponds with textual evidence from the pre-Buddhist period. Buddhism did not infiltrate into China until the 1st century AD, when it travelled from India along the Silk Road. For a thorough account of the socio-political history of the period referred to in this text as 'early China', see Loewe and Shaughnessy (1999); Twitchett and Loewe (1986). This text was originally presented as an outline for a non-sinological readership. I have reduced the number of technical Chinese references as well as notes that are of strictly sinological relevance.

4 For versions of such civilization histories see *Zhouyi* (8.4b–8a); *Xinyu* (1.1–21); *Huainanzi* (13.421–24, 19.629–31); *Bohutong* (2.49–52); *Hanshu* (91.3679–80); *Fengsu tongyi* (1.1–15); *Mengzi* (5B.2b–3b, 6B.3a–4a, 13A.10a). Mozi (479–438 BC) describes how the sages emancipated humankind from a primitive, bestial existence and invented weapons to counter wild beasts. See *Mozi* (2.109–110, 3.116, 6.255, 8.382).

5 One of the most popular depictions of animals in early Christian art is the portrayal of Adam naming the beasts. God gave Adam dominion over the animals, and by naming them Adam showed that he understood their nature and was able to control and use them. See Salisbury (1994: 6–7); George and Yapp (1991: 37–41).

6 For the concept of *qi* in Chinese thought, see Graham (1989: 101–104 *et partim*); Schwartz (1985: 179–184). On the role of blood see Sterckx (2002), 73–78.

7 *Liji* (29.8b). For similar ideas, see *Da Dai Liji* (3.4a); *Hanshu* (48.2249); *Xin shu* (5.4b, 6.3a).

8 *Yanzi chunqiu* (7.430); *Liji* (1.11a); *Xin xu* (6.205); *Lü Shi chunqiu* (16.2a, 20.2a); *Hanshu* (27.1396) (referring to *Zuozhuan* 9.14a); *Chunqiu fanlu* (6.156).

9 *Shuoyuan* (18.455); cf. *Han shi waizhuan* (8.277). Discussions of the phoenix include Diény (1989–90), Hachisuka (1924) and Hargett (1989).

10 Translated in Watson (1961: 314–316). The symbolism of the hunt as evinced in Sima Xiangru's rhapsodies is treated in detail in Hervouet (1964: 215–244, 256–258, 271–286, *et partim*). For the Shanglin park and a discussion of animal parks in later times, see Wu (1995: 165–176); and Schafer (1962) and (1968).

11 *Hou Hanshu* (25.876). For identifications of barbarians with wild animals, see, for example, *Zuozhuan* (11.1b, 21.21a–23a, 29.22a); *Zhouli* (36.16b); *Hanshu* (49.2285, 52.2398, 94A.3743). See further Sterckx (2002), 158–161.

12 See, for example, *Hanshu* (25A.1225); *Shiji* (28.1392); *Hou Hanshu* 'zhi' (5.3123, 8.3182). See further Sterckx (2002), 186–194.

13 For the association of 'target' and 'vassal' see (Riegel 1982: 5–10). For another comparison of the subjugation of barbarians with skilfulness in fishing and shooting birds, see *Lü Shi chunqiu* (2.10a–b).

14 *Zhouli* (23.3b); *Yili* (18.14a–b); *Liji* (39.13b–14a, 62.2a); *Hanshu* (57A.2573).

15 Traces of such measures occur in Qin legal documents. See, for example, Hulsewé (1985: A64, C16). Laws regarding animals keep recurring throughout legal codes in later Chinese history (see Ikeda 1984). For the theme of the tiger plague in the post-Han period, see Eichhorn (1954). See further Sterckx (2002), 137–158.

16 *Changli xiansheng ji* (36.5a–6a); Translated in Sommer (1995: 174–75). As Sommer notes, the admonitions to the crocodiles could be directed at corrupt officials. See also Hartman (1986: 91–93); and Schafer (1967: 217).

17 *Liji* (12.5b, 14.14a, 16.13a, 17.1b); *Lü Shi chunqiu* (1.1b, 9.1a); *Yi Zhoushu* (6.2a, 6.3a–3b); *Huainanzi* (5.160, 5.173, 5.177); *Da Dai Liji* (2.4b, 2.8a, 2.9b). See also *Shuoyuan* (19.490); *Wenzi* (2.24a–b); *Xin shu* (6.3a). The opening lines of the treatises on sacrifice in the dynastic histories of the Western and Eastern Han evoke the innate sense of sacrifice in these animals. See *Hanshu* (25A.1189); *Hou Hanshu* 'zhi' (7.3157). The predator referred to in the calendar as *chai* has been identified with the dhole, a small Asian wild dog (see Schafer 1991).

18 For the above described statements on animal conservation and hunting regulations, see *Guoyu* (4.178); *Huainanzi* (9.308–9); *Yi Zhoushu* (4.9a); *Zhouli* (16.14a–b); *Liji* (4.15a, 12.5a–6a, 13.10a, 25.1a, 48.6a); *Xunzi* (9.154); *Lü Shi chunqiu* (1.3a, 4.8a); *Wenzi* (2.24a–b); *Da Dai Liji* (4.11b, 5.2a); *Hanshu* (6.211); *Yanzi chunqiu* (5.312–13); *Shuoyuan* (5.101); *Lunyu* (7.8b); *Han Feizi* (12.692); and *Zuozhuan* (3.23a–b). For a short survey of the technical aspects of the hunt and hunting terminology, see Böttger (1960).

19 See, for example, *Huainanzi* (8.245–46); *Da Dai Liji* (13.9a); *Shizi* (1.5b); *Wenzi* (2.38b); *Shuoyuan* (5.104, 13.313); *Hanshu* (51.2371); *Shiji* (47.1926).

20 '... [t]hroughout European history, hunters have tended to see themselves as enemies of the individual animals but friends of the animal *kinds* – and by extension as friends of the wild, nonhuman realm that the animals inhabit' (Cartmill 1993: 31).

21 *Lunyu* (10.11a). Elsewhere (Sterckx 1996b) I have suggested that this disinterest towards the physical animal may explain the paucity of textual references to animal cults and zoolatry in early China.

References

Primary sources

Bohutong shu zheng. Compiled by Ban Gu (AD 32–92), annotated by Wu Zeyu. Beijing: Zhonghua, 1994.

[*Zhu Wengong jiao*] *Changli xiansheng ji.* Attributed to Han Yu (AD 768–824). *Sibu congkan* edition.

Chunqiu fanlu jinzhu jinyi. Attributed to Dong Zhongshu (179–104 BC), annotated by Lai Yanyuan. Taipei: Shangwu, 1992.

Da Dai Liji. Attributed to Dai De (fl. 72 BC). *Han Wei congshu* edition.

Fengsu tongyi jiaoshi. Compiled by Ying Shao (d. ca. AD 206), annotated by Wu Shuping. Tianjin: Renmin, 1980.

Guanzi. Sibu beiyao edition.

Guoyu. Shanghai: Guji, 1978.

Han Feizi jishi. Attributed to Han Fei (d. 233 BC), annotated by Chen Qiyou. Gaoxiong: Fuwen, 1991.

Han shi waizhuan jishi. Attributed to Han Ying (fl. 150 BC), annotated by Xu Weiyu. Beijing: Zhonghua, 1980.

Hanshu. Compiled by Ban Gu (AD 32–92). Beijing: Zhonghua, 1962.

Hou Hanshu. Compiled by Fan Ye (AD 398–445). Beijing: Zhonghua, 1965.

Huainanzi honglie jijie. Compiled under the auspices of Liu An (d. 122 BC), annotated by Liu Wendian. Taipei: Wenshizhe, 1992.

Kong Congzi. Sibu beiyao edition.

Liezi. Sibu beiyao edition.

Liji. Shisanjing zhushu edition (Collated by Ruan Yuan [1764–1849]; reprint Taizhong: Landeng, n.d.).

Lü Shi chunqiu. Compiled under the auspices of Lü Buwei (290–235 BC). *Sibu beiyao* edition.

Lunheng jiaoshi. Attributed to Wang Chong (ca. AD 27–100). Beijing: Zhonghua, 1990.

Lunyu zhushu. Attributed to Confucius (551–479 BC). *Shisanjing zhushu* edition.

Mao shi zhengyi. Shisanjing zhushu edition.

Mengzi zhushu. Attributed to Mencius (ca. 372–ca. 289 BC). *Shisanjing zhushu* edition.

Mozi jiaozhu. Attributed to Mo Di (ca. 480–ca. 390 BC), annotated by Wu Yujiang. Beijing: Zhonghua, 1993.

Shizi. Sibu beiyao edition.

Shuoyuan jiaozheng. Attributed to Liu Xiang (79–8 BC), annotated by Xiang Zonglu. Beijing: Zhonghua, 1987.

Wenzi. Sibu beiyao edition.

Xi jing za ji. Attributed to Ge Hong (ca. AD 283–343). *Gujin yishi* edition.

Xin shu. Attributed to Jia Yi (201–168 BC). *Sibu beiyao* edition.

Xin xu jinzhu jinyi. Attributed to Liu Xiang, annotated by Lu Yuanjun. Taipei: Shangwu, 1975.

Xinyu jiaozhu. Attributed to Lu Jia (Third–Second centuries BC), annotated by Wang Liqi. Beijing: Zhonghua, 1986.

Xunzi xin zhu. Attributed to Xun Kuang (313–238) and his disciples. Taipei: Liren, 1984.

Yanzi chunqiu jishi. Annotated by Wu Zeyu. Beijing: Zhonghua, 1962.

Yi Zhoushu. *Sibu beiyao* edition.

Yili zhushu. *Shisanjing zhushu* edition.

Zhouli zhushu. *Shisanjing zhushu* edition.

Zhouyi zhengyi. *Shisanjing zhushu* edition.

Zhuangzi jishi. Annotated by Guo Qingfan (AD 1844–1997). Taipei: Guanya, 1991.

Zuozhuan zhengyi. *Shisanjing zhushu* edition.

Secondary sources

Anderson, J. (1985) *Hunting in the Ancient World*. Berkeley: University of California Press.

Ariel, Y. (1996) *K'ung-Ts'ung-Tzu: A Study and Translation of Chapters 15–23 with a Reconstruction of the Hsiao Erh-ya Dictionary*. Leiden: E.J. Brill.

Böttger, W. (1960) *Die Ursprünglichen Jagdmethoden der Chinesen*. Berlin: Akademie Verlag.

Bray, F. (1984) *Science and Civilisation in China*, vol. 6, part II 'Agriculture'. Cambridge: Cambridge University Press.

Cartmill, M. (1993) *A View to Death in the Morning: Hunting and Nature through History*. Cambridge: Harvard University Press.

Chang, K. C. (ed.) (1977) *Food in Chinese Culture: Anthropological and Historical Perspectives*. New Haven and London: Yale University Press.

Diény, J. P. (1989–90) 'Le fenghuang et le phénix'. *Cahiers d'Extrême-Asie*, vol. 5, pp. 1–15.

Eichhorn, W. (1954) 'Das kapitel *tiger* im T'ai-P'ing Kuang-Chi'. *Zeitschrift der Deutschen Morgenländischen Gesellschaft*, vol. 104, no. 9, pp. 140–162.

George, W. and B. Yapp (1991) *The Naming of the Beasts: Natural History in the Medieval Bestiary*. London: Duckworth.

Graham, A. (1989) *Disputers of the Tao*. La Salle: Open Court.

—— (1991) *The Book of Lieh-Tzu*. London: Mandala.

Hachisuka, M. U. (1924) 'The identification of the Chinese phoenix'. *Journal of the Royal Asiatic Society of Great Britain and Ireland*, pp. 585–589.

Hargett, J. M. (1989) 'Playing the second fiddle: the Luan-bird in early and medieval Chinese literature'. *T'oung Pao*, vol. 75, pp. 235–262.

Hartman, C. (1986) *Han Yü and the T'ang Search for Unity*. Princeton: Princeton University Press.

Hervouet, Y. (1964) *Un Poète du Cour sous les Han: Sseu-ma Siang-ju*. Paris: Presses Universitaires de France.

Hulsewé, A. F. P. (1985) *Remnants of Ch'in Law*. Leiden: E.J. Brill.

Ikeda, O. (1984) 'Chūgoku kodai no mōjū taisaku hōki'. In *Ritsuryōsei no shomondai*. Festschrift for Takigawa Masajirō (Tokyo: Kyūko shoin), pp. 611–637.

Legge, J. (1871) *The She King*. Taipei: SMC Publishing (1991 reprint).

Lewis, M. E. (1990) *Sanctioned Violence in Early China*. Albany: State University of New York Press.

Loewe, M. and E. Shaughnessy (eds) (1999) *The Cambridge History of Ancient China*. New York: Cambridge University Press.

Needham, J. (1985) *Science and Civilisation in China*, vol. 2. Cambridge: Cambridge University Press.

Petit, G. and L. Théodoridès (1962) *Histoire de la Zoologie*. Paris: Hermann.

Riegel, J. K. (1982) 'Early Chinese target magic'. *Journal of Chinese Religions*, vol. 10, pp. 1–18.

Salisbury, J. E. (1994) *The Beast Within: Animals in the Middle Ages*. London: Routledge.

Schafer, E. H. (1962) 'The conservation of nature under the T'ang Dynasty'. *Journal of the Economic and Social History of the Orient*, vol. 5, pp. 279–308.

—— (1967) *The Vermilion Bird: T'ang Images of the South*. Berkeley: University of California Press (1985 reprint).

—— (1968) 'Hunting parks and animal enclosures in ancient China'. *Journal of the Economic and Social History of the Orient*, vol. 11, pp. 318–43.

—— (1991) 'The Chinese Dhole'. *Asia Major* (New Series), vol. 4, no. 1, pp. 1–6.

Schwartz, B. I. (1985) *The World of Thought in Ancient China*. Cambridge: Harvard University Press.

Sommer, D. (1995) *Chinese Religion: An Anthology of Sources*. Oxford: Oxford University Press.

Sorabji, R. (1993) *Animal Minds and Human Morals: The Origins of the Western Debate*. London: Gerald Duckworth and Co.

Sterckx, R. (1996a) 'Transcending habitats: authority, territory and the animal realm in warring states and early imperial China'. *Bulletin of the British Association for Chinese Studies*, pp. 9–19.

—— (1996b) 'An ancient Chinese horse ritual'. *Early China*, vol. 21, pp. 47–79.

—— (1998) 'Debating the strange: records of animal anomalies in early China'. *Working Papers in Chinese Studies* (Center for Chinese Studies, National University of Singapore) 1.

—— (2002) *The Animal and the Daemon in Early China*. Albany: State University of New York Press.

Tambiah, S. J. (1985) *Culture, Thought, and Social Action*. Cambridge: Harvard University Press.

Toynbee, J. M. C. (1973) *Animals in Roman Art and Life*. Baltimore: Johns Hopkins University Press (1996 reprint).

Twitchett, D. and M. Loewe (eds) (1986) *The Cambridge History of China. Volume I: The Ch'in and Han Empires (221 BC–AD 220)*. Cambridge: Cambridge University Press.

Watson, B. (1961) *Records of the Grand Historian of China, Vol. 2*. New York: Columbia University Press.

Wilhelm, R. (1930) *Li Gi: Das Buch der Sitte des Älteren und Jüngeren Dai*. Jena, n.p.

Wu, H. (1995) *Monumentality in Early Chinese Art and Architecture*. Stanford: Stanford University Press.

2

THE CHASE AND THE DHARMA

The legal protection of wild animals in premodern Tibet

Toni Huber

Introduction

In premodern Tibet, as in many similar societies, people categorized and related to wild animals in many different ways. By 'premodern' here I mean pre-1959, a date which not only marks the full Chinese colonial occupation of Tibet but also the beginning of a period of intensive structural changes, such as social and land reforms, technical modernization and later infrastructure developments. By 'Tibet' here I mean 'ethnographic Tibet' as described by Goldstein (1994: 76–77) and Samuel (1993: Chapters 1–8). It is an area comprising the Tibetan plateau, its eastern marches and various high-altitude Himalayan valley systems, and inhabited by peoples with a manifestly high degree of linguistic similarity who share cultural and social patterns and historical experience. But it is not coterminous with any historical or modern political boundaries.

The characteristic large Tibetan herbivores (wild yak, wild ass, antelope, deer, wild sheep, and so on) of the high mountains and plateau grasslands have long been associated with ideals of strength, purity or intelligence. For example, the massive wild yak bull (*'brong*) is legendary for its immense power, and the human ability to capture or kill one has always been the measure of a hero. The elusive deer are believed to have a sensitive intelligence, and often feature in Tibetan ritual and symbolism. The collective designation for all such wild animals that subsist on plants and water is *ri-dwags* (also spelled *ri-dag/dags/dwag*), which also signifies 'game animals' in general. One possible etymology for this term is 'mountain' (*ri*) + 'purity' (*dag-pa/dwangs*),[1] which agrees with Tibetan thinking since ancient times about the essential, pristine quality of high and remote places in the natural environment, where deities dwell and which are undefiled by human activity. In certain areas, wild animals were considered to be the embodiments of deities, or to be under the ownership of local spirit powers. Their sudden appearance or particular behaviour patterns have often been interpreted by Tibetans as portents or divinatory signs.

Although there is no shortage of positive or powerful associations made with wild animals in the traditional Tibetan worldview, these appear never to have stopped Tibetans from killing them for survival, profit or pleasure. Indeed, all records (archaeological, historical and ethnographic) for premodern Tibet suggest that hunting has long taken place and was widespread, and that wild animal products were important. It is also true that the essential ethical systems of historical Tibetan religions (both Buddhism and Bon) are completely opposed to any intentional taking of life, such as hunting. From the religious point of view, practices like hunting wild animals are considered in the most negative moral light, and seen as soteriologically detrimental to those involved. Thus, unlike many other societies which practice hunting, premodern Tibetans hunted (and still hunt) in an atmosphere of stark contradiction between ideals and practices. A few preliminary observations have to be made about this tension before introducing the main topic of the chapter: hunting laws and the protection of game animals in premodern Tibet.

Buddhist values are not originally Tibetan. They were introduced in formal and well articulated forms from India over a thousand years ago. The extent to which they have influenced the Tibetan world appears, from all the evidence, to be extremely variable. This variability is itself socially recognized in Tibetan discourse. Anthropologists have pointed to a fundamental value dichotomy in Tibetan life and religious culture which is expressed as the difference between that which is considered 'wild' (*rgod-pa*) and that which has been 'tamed' (from the verb '*dul-ba*: 'to tame', 'subdue', 'civilize', 'convert', etc.). Within this framework, figures such as the human hunter or the local mountain cult deities (who are martial in nature) are clearly considered 'wild', while the ordained, vow-holding Buddhist monk is 'tame'. Those persons who attempt to heed the moralizing sermons of elite Buddhist practitioners such as the lama, or the local deities who are subjected to the lama's superior ritual powers, are considered as 'never being tamed'.[2] It is the lama's business to go about taming that which is not in conformity with the Tibetan form of Buddhism. Thus, on the ethical front, we find abundant historical evidence of influential lamas actively attempting to 'tame' or 'convert' professional life-takers, such as pastoralists, hunters or butchers, as part of their role in society. They preach on the evils and dire consequences of killing, they undertake exemplary good deeds by purchasing and freeing (*tshe-thar*) or ransoming (*srog-blu[s]*) the lives of animals destined to be slaughtered and they also encourage lay people to take vows to abstain from killing (*srog mi-gcod-pa'i sdom-pa*).[3]

While there is ample evidence of active Tibetan Buddhist opposition to practices like hunting, within the framework of institutionalized Tibetan religion the results of this opposition were never a foregone conclusion. For one thing, values and concomitant prohibitions are never uniformly or consistently subscribed to or internalized in relation to competing discourses, survival needs and other interests (such as economic gain). Also, Tibetan religion offers its followers a multitude of ritual means for dealing with the spiritual burden of ethical breaches. Practitioners can either gradually cleanse themselves of the cognitive and physical defilements ([s]grib) thought to result from immoral actions, or obviate them altogether. Thus the hunter, as a Buddhist, could continue to kill game with the hope of, at some point (usually in later life), combatting or avoiding the soteriological consequences of his (only males hunt) negative actions with ritual. A hunter's personal feelings of guilt, and any social reprobation, had to be dealt with in other ways – for example, by choosing to moderate the number of animals taken and hunting discreetly or in secret.

It has long been clear to the ethically concerned Buddhist elite in Tibetan societies that no amount of preaching or good examples could ever restrain or eliminate hunting, especially when practised for economic gain – such as the lucrative trade in valuable animal products (e.g., musk, bear gall, stag horn, leopard skins, antelope belly hair, etc.) of the plateau and Himalayan regions. To keep hunting in check, Buddhist clerics and lay rulers resorted to alternative methods. To augment the internalized ethical controls which they attempted to socialize into the lay population, a complement of externalized prohibitory measures against hunting were developed. Such measures, largely legal in nature, are the subject of this chapter.

The aim of this brief study is twofold. First, I shall give a historically based outline of the type and development of traditional Tibetan concepts and legal institutions used to prohibit hunting and to protect game. Second, using more recent documentary evidence and ethnohistorical data gained from interviews with Tibetans, I shall briefly review how these were applied in Tibet in the twentieth century, prior to and during the 1950s. For reasons of space, the closely related topics of private non-legal initiatives to ban hunting, prohibitions of hunting associated with non-Buddhist local cults and the prohibition of fishing in Tibet will not be treated. The data on which this chapter is based are drawn from a close examination of historical texts on Tibet and from ethnographic fieldwork carried out in Central Tibet and Amdo between 1986 and 1996.

Historical development of game and hunting laws

In Tibetan historical documents, we find that legal attempts to stop hunting were formally addressed at various levels of society: the monastery, the agricultural estate, the cult territory, the local principality and the centralized state itself. I shall use the expression 'anti-hunting laws' here as a general term to discuss a number of different premodern legal institutions occurring at all of these levels. Yet the phrasing of many of these laws reveals that they were not only concerned with the hunters but also referred to the wild game animals. The historical trend has been towards emphasizing the protection of the game itself, rather than specifically designating the hunter's actions (although this is always implicit). The reason for this, as we shall see, is that Tibetan anti-hunting laws became increasingly Buddhist in the rhetoric of their presentation and justification.

The earliest hunting laws that we know of in Tibet occurred at the state level. The first centralized state during the Yar-klungs dynasty era (ca. AD 622–842) differed considerably from the later Buddhist influenced period, and its culture and legal systems reflect this. The contemporary records of the Yar-klungs (Yarlung) state period show that the early kings, or *btsan-po* ('mighty ones'), the political elite and the general populace all hunted, and there is no evidence of opposition to hunting. Indeed, the grand battue-style hunts, known as *lings*, were an important part of court and social life. The most sophisticated and lengthy Tibetan hunting laws are also known from this period. However, they only concern compensation resulting from hunting accidents and other regulations observed during the *lings* hunts.[4] There is no relationship between these early hunting laws and any known from later Tibetan history during the Buddhist period.

Following the demise of the Yar-klungs dynasty, no centralized state authority existed on the Tibetan plateau until Mongol suzerainty was imposed during the thirteenth and fourteenth centuries. At this time, the first anti-hunting laws were introduced and enforced by Yuan administrators. Imperial edicts, in both 'Phags-pa script and Tibetan issued in 1277 (or 1289) and 1324, prohibited hunting, but in non-religious terms. There is no mention of ethics or any hint of Buddhist motives in these laws. Directives not to hunt were included in general rules of behaviour for the population in certain districts and were associated with the rights of official control over a range of environmental resources (grazing, timber, water, etc.) which had been sanctioned by imperial decree for certain large monasteries.[5] At this time, codes of monastic rights and regulations began to be formulated by the Yuan administration throughout China as well, and general Yuan legal codes of the period also contained regulations to control hunting.[6] So the earliest Tibetan laws against the hunt do not represent any uniquely Tibetan development during this period.

'Sealing the hills and sealing the valleys'

It is not until the fifteenth century that we find purely Tibetan anti-hunting laws which are phrased in relation to Buddhism. By this time, under a strong Buddhist cultural influence, a new and very different set of models of Tibetan kingship and statehood became well developed. The model ruler, whether lay or religious, was now based upon the bodhisattva ideal to varying degrees, and was supposed to embody Buddhist principles, such as ethics, compassion, sagacity, etc.[7] The first Buddhist anti-hunting laws were issued by local Buddhist lay rulers of kingdoms in Central and Western Tibet, such as the so-called 'religious-kings' (*chos-rgyal*) of rGyal-rtse and Gung-thang.[8] They are phrased in terms of Buddhist ethical and soteriological concerns and certain Mahayana doctrines, but are still related to control over, and access to, natural resources. Most of the features of the fifteenth-century anti-hunting laws, including their method of promulgation, became routinized and recur in all the later Tibetan legal documents and practices concerning prohibition of the hunt and protection of game.

An important feature of these first Buddhist-styled laws is their use of the concept of 'sealing the hills and sealing the valleys' (*ri-rgya klung-rgya sdom-pa*, or in short, *ri-klung rgya sdom*, hereafter 'sealing the hills'). This concept continued to be employed in anti-hunting laws right up until the 1950s. It is also known from many pre-fifteenth-century Tibetan historical sources, but in nearly all these instances it refers to an institution that had nothing directly to do with prohibition of hunting and protection of game. Early forms of application of 'sealing the hills' actually express a system of territorial control used by politically influential lamas, particularly in times of social conflict. In the later codes of monastic regulations (*bca'-yig*), enforcing a 'sealing the hills' law clearly comes to function as a fundamental legal claim of rights over territory, as much as it does to stop hunting and protect game.[9] Both the idea and terminology of 'seals' and 'sealing' are found in some Buddhist sutra texts and in the tantras translated into Tibetan. Early Tibetan lamas applied various types of external sealing, such as the 'command seal' (*bka'-rgya[-ma]*), which refers to the restriction of access a lama puts on certain teachings by commanding that they remain closed, and not be revealed or 'opened' and read until specified.

'Sealing the hills and sealing the valleys' is very closely related to, and often found in combination with, the idea of 'sealing the roads' (*lam-rgya*). The earliest use I have found of all such types of territorial sealings is in relation to Bla-ma Zhang's (1123–93) worldly activities, including fighting against bandits and going to war.[10] A clear example of the 'sealing the hills' law functioning as a tool for political control of territory in times of war or social breakdown is found in the actions of the Third Zhwa-dmar-pa Chos-dpal Ye-shes (1406–52). He applied a 'sealing the hills' edict to the area of Kong-lung Ral-gsum in southeast Tibet during 1434 to control the local Tibetans who were fighting amongst themselves and with neighbouring Klo-pa and Mon-pa tribes.[11] Unfortunately, the sources give no detailed indication of how these territorial closures were applied and enforced. We assume that they depended greatly upon respect for the lamas' moral authority and even more upon fear of the lamas' magico-ritual powers – the practice of sealing in later times invokes powerful wrathful deities who are 'entrusted with the function' (*'phrin-bcol*) of punishing those who violate seals.

It was a logical step for politically powerful Buddhist leaders who wanted to close an area to hunters and thus protect the game there, to apply a 'sealing the hills' decree as an act of law. When this is described in the anti-hunting laws, it is done so explicitly in relation to various Buddhist ideas.

'The gift of fearlessness'

Another new feature of the fifteenth-century Tibetan anti-hunting laws is the explicit linkage made between the practice of 'sealing the hills and sealing the valleys' and the idea of 'the gift of fearlessness' (*mi-'jigs-pa shyin-pa*) being offered to wild animals. The Tibetan expression *mi-'jigs-pa shyin-pa* here is a translation of the Sanskrit *abhayadana*, a familiar concept from Mahayana Buddhist literature. *Dana* is the first of the six perfections (*paramita*) and consists of three types: (i) the gift of material goods, (ii) the gift of fearlessness and (iii) the gift of the Buddha's teachings. It is an old idea in Tibet, and one finds *abhayadana* fully treated in some of the earliest indigenous Tibetan systematic expositions of Buddhism, including in relation to wild animals. For instance, the twelfth chapter of sGam-po-pa's (1079–1153) *Thar pa rin po che'i rgyan* states: '[*Abhayadana*] means to be a refuge for those who are frightened by robbers, wild animals, diseases and floods … The gift of fearlessness is to be known as being a refuge for those who are frightened by lions, tigers, crocodiles, kings, robbers, floods and other disasters.'[12]

Such Tibetan expressions of 'the gift of fearlessness' depict wild animals as the threat from which one requires protection. Yet elsewhere in the Buddhist literature and in the Tibetan anti-hunting laws, the *abhayadana* concept is fully extended to cover the wild animals themselves, including carnivores, which are frightened and on the run from the human threat of hunting, and which are given refuge by way of 'sealing the hills and the valleys'. The anti-hunting laws contain many expressions of the position of the hunted game animals in relation to the sealings, which are said '[to] offer the protection of fearlessness to defenceless living creatures',[13] or that 'the living creatures of the hills and valleys should breathe freely without fear for their lives'.[14] Similarly, territory sealed from hunters' access to wild animals is described as being 'an island of freedom (*thar-pa-gling*) [for game animals], offering the gift of fearlessness'.[15] Such expressions were stock phrases in the Tibetan anti-hunting laws used up until the 1950s. The moral imperative of offering the gift of fearlessness was also cited by lamas as a reason for other related practices, such as purchasing and freeing the lives (*tshe-thar*) of animals destined for slaughter.[16]

Benefits and motivations

Merely to offer the gift of fearlessness by way of a sealed territory for protecting game was in itself a highly positive act in Buddhist terms, since it partly entailed cultivating the Mahayana perfections. Yet other Buddhist justifications were also introduced in relation to the early anti-hunting laws, and these specifically concern the benefits of issuing such laws. They clearly indicate that, whatever a ruler may have thought about protecting defenceless game animals because of their intrinsic value, the ruler himself also stood to gain directly from the issuance. For example, by proclaiming and enforcing such laws a king is said to have 'accumulated merit by way of the ten virtuous actions' (*dge-ba'i bcu*) and to have 'cleansed the effects of negative actions',[17] both of which are fundamental prerequisites for attaining better rebirths in Tibetan Buddhism. There was therefore a personal soteriological imperative for rulers to impose such laws upon their subjects. It is also likely that Buddhist interpretations of Indian Arthasastra theories of statecraft and kingship, which circulated in Tibet, influenced rulers to formulate anti-hunting laws. We know of only one, rather late reference to this in the legal documents. In the twenty-one article administrative code of the regent sDe-srid Sangs-rgyas rGya-mtsho (1653–1705), the 'sealing the hills' section is immediately preceded by a quote which sets out the general duties of a king as an environmental steward, and which is taken from the *Canakyarajanitisastra* translated in the Tibetan Buddhist canon.[18]

Influential lamas and clerics, who often went about privately setting up their own sealed territories to protect wild animals from hunters, had a range of other motivations and benefits in mind as well. The west Tibetan lama bSod-nams Blo-gros (1456–1521) read verses in several tantric texts about the benefits – such as being spared evil rebirths – of directly saving the lives of hundreds of beings. On the basis of these teachings, he states that he established preserves to protect game animals from hunters around two monasteries which he founded.[19] Another lama, 'Jigs-med Gling-pa (1730–98), who often expressed his great love for animals, purchased and sealed the territory of a whole mountain as an act of compassion when he learned of the destructive hunting practices of the local community there.[20]

Monastic and state applications of laws

A further precedent set already by the fifteenth century was the development of two parallel traditions of 'sealing the hills' laws. This can be seen in the anti-hunting laws issued by the king of rGyal-rtse, Rab-brtan Kun-bzang 'Phags-pa (1389–1442), between the years 1415 and 1440.[21] One is in the form of general edicts issued by the king, and proclaimed in public. The other is in a monastic code of rights and regulations applied to the territory controlled by the local monastery. This parallel application links the 'sealing the hills' laws with specific units of space and time, a parallel found in all later laws up to the 1950s. In the monastic regulations (*bca'-yig*), the law applied to a specifically defined unit of territory over which the monastery had rights and control. Here time was the constant, while the precise space covered by the regulations varied. Different units of territory were sealed to hunters, but always in perpetuity. Monastic territories which were sealed against hunters were defined and delimited in different ways. Some had precise boundaries fixed in relation to natural landmarks or cultural features, while others existed within 'a distance as far as the sound of a conch trumpet can be heard' or in 'an area as far as the sound of a horn travels surrounding the monastery', and 'in the area within sight of the precious sangha in the environs of the monastery'.[22]

In the public edicts (*rtsa-tshig*, *bca'-tshig*, *bka'-shog*, etc.), by contrast, space is the constant since the laws theoretically applied to the whole area of the kingdom or state. But they are also related to abstention from taking life according to the annual Buddhist ritual calendar: hunting is prohibited everywhere but only during certain important calendar months or ritual periods. The following examples show how different time frames for sealing were related to different purposes. In the fifteenth-century rGyal-rtse kingdom, the hills and valleys were sealed specifically at the time of the four great festivals (*dus-chen bzhi*) celebrating events in the Buddha's life: his miracles at Sravasti on the fifteenth day of the first Tibetan month; his awakening, death and final passing beyond suffering on the fifteenth day of the fourth month; his first teaching on the fourth day of the sixth month; and descent from the realm of the gods on the twenty-second day of the ninth month.[23] The main reason for this timing is that any positive deeds (in Buddhist terms) performed on these holy days, such as abstaining from taking life or even sealing areas from hunters, are believed to generate thousands of times more merit for the practitioner than at other times. The reverse is believed to apply with the consequences of negative acts, which are multiplied greatly.

We find a much more extensive sealing of hill and valley areas against hunters applied in the laws of the seventeenth-century gTsang kingdom: 'from the time of the festival of the Sravasti miracles [i.e. the *sMon-lam Chen-mo*, on the fourth to fifteenth days of the first month] up until the tenth month'.[24] This sealing period offered blanket cover for all major rituals in the year. In addition, it also coincided with the breeding, carrying and birthing season for game animals, and more recent evidence confirms that animal biology was a consideration in the prohibition of hunting. At least one of my informants, a nomadic hunter from mNga'-ris in western Tibet, stated that in the 1950s antelope and other large game, especially females, were not traditionally hunted during breeding and pregnancy periods. A second informant, from Amdo in the eastern plateau and also speaking of the 1950s, stated that local monasteries performed a popular dance known as the 'Stag and Hunting Dog Ritual Dance' (*Sha-ba dang sha-khyi 'cham*) in the early autumn.[25] The performance narrative relates the story of a famous Tibetan lama (Mi-la Ras-pa, 1040–1123) who saves a frightened deer from a hunter and his dog and then 'tames' them all by conversion to Buddhism. It was said to have been staged at this time because the following period was the optimal time for local hunting of musk deer (musk pod quality and size were optimal), and the dance was intended to dissuade hunters from hunting in this crucial season.

A final example of a sealing period is found in several of the main state law codes for Central Tibet. In one of these codes, used from the seventeenth to the twentieth centuries, is stated the following: 'During the religious festival of the fifth month, the hills and the valleys are to be sealed'.[26] The festival of the fifth month is the popular 'Universal Incense Offering' (*'Dzam-gling spyi-bsangs*), a major occasion for the worship of the many local protective deities said to have been tamed by an early tantric lama in Tibet and then converted to become protectors of Buddhism. These deities are mainly nature spirits redefined within the Tibetan Buddhist pantheon and cosmology. They often dwell on mountains, in rivers and springs and in the subsoil. Some of them are invoked by lamas to guard sealed territories, others rule over pilgrimage mountains or monastic precincts. Others, dwelling in remote places, are popularly believed to be owners of the game animals, or to use these animals as their mounts; hunters believe that the deities can sometimes become angered when 'their' game animals are caught and killed. These deities are also highly offended by certain types of pollution of their abodes, such as the blood a hunter might accidentally spill in a spring, lake or stream when processing a game carcass, or the smoke from flesh roasted over a campfire. All Tibetans consider these deities capricious and dangerous. They are deemed (mostly retrospectively) a major cause of illness, natural disaster and bad fortune, and people therefore devote careful ritual attention to them. It is due to these concerns, on the part of the religious state itself, that the hills and valleys must be sealed against hunters to protect against upsetting the deities during their special time of annual worship. In this syncretistic aspect of Tibetan religion, which provided frequent grounds for sealing areas against hunters (as we shall see below), the protection of game had little or nothing to do with Buddhist ethics or soteriology.

Anti-hunting laws based on the combination of 'sealing the hills' and *abhayadana* became incorporated into the formal legal codes (*khrims-yig*) of successive Buddhist states in Tibet until the 1950s. They are found in the related thirteen-, fifteen- and sixteen-article *zhal-lce* codes, which elaborate few new features beyond those already established in the fifteenth century. One notable exception is a tendency towards enumerating the classes and species of wild animals to be protected, as well as those exempt. This enumeration began in the Zhal lce bcu gsum of the early dGa'-ldan Pho-brang state in Central Tibet (that is, the Dalai Lama era) and developed until a more or less blanket cover is given for the two main Tibetan categories of large wild animals hunted as game, the 'herbivores' (*ri-dwags*) and 'carnivores' (*gcan-gzan*). In addition, it extended to lesser creatures of harvest and control, like birds, fish, otters and even eggs and bees. The stock phrase for inclusive protection is 'all creatures great and small dwelling on dry land and in water'. The only exceptions are wolves, which may always be hunted due to the threat they pose to livestock.[27]

Enforcement and punishment

How were hill and valley seals against hunting enforced and what happened to hunters caught violating them? Although the area of enforcement and punishment by the state is never elaborated in the main state law codes and proclamations, we do know of them from recent oral testimonies and I shall discuss them when reviewing the twentieth-century data below. In the monastic regulations, however, specific details are often listed. Some monastic laws state that the responsibility for superintending the sealed areas lies with the resident lamas and monks of the monastery or hermitage. This was, in most cases, the only option since many religious communities were intentionally located in areas remote from the lay populace, a fact that also made it easier to hunt discreetly or secretly in such areas. Some monastic codes explicitly warn administrators that hunters should be regarded as persons not to be trusted, and one should be wary of them 'telling lies of their ignorance that [hunting] is not allowed'.[28] The hunters are also liable for punishment without even being caught in the act of making a kill; merely to see a hunter in the monastery's sealed area or to hear his horn, was enough.[29] This seems to be exactly the spirit of the 'sealing the hills' as a form of territorial closure: unauthorized or suspicious entry was an automatic violation in itself, whether one hunted or not.

A range of possible punishments could be applied to apprehended hunters, depending upon the traditions at individual monasteries. Fines in kind were common, but they were always described as 'offerings'. A guilty hunter might have to offer one or more communal tea servings (mang-ja) for the assembly of monks in the monastery or offer ceremonial butter lamps and scarves (in specified amounts).[30] Fines in cash are reported from very large monasteries,[31] and confiscation of hunting equipment was also practised.[32] More interesting is the use of the local protective deities, who usually have a special shrine (mgon-khang) dedicated to them within the main temple complex. One monastic code for a dGe-lugs-pa establishment states the following: 'When itinerant game hunters appear, they should be punished by gathering their weapons in the protector's temple and in addition exhorted once again to observe lawfulness'.[33] The logic behind this is that one major type of offering to protector deities are weapons of all kinds – their shrines are usually full of assorted antique arms and armour that reflect their martial and wrathful natures. Judging from the deference and care that present-day rural Tibetan men display when entering such shrines (women are usually barred from entry), one can easily imagine how any punishment connected to these deities might serve as an edifying experience for the guilty hunter. Some codes mention hunters having to recite religious texts in the protector's temple.[34]

One assumes that Tibetan hunters were always laymen, but that was not so. It is known that very occasionally Tibetan clerics and lamas who were extremely lax about observing their own vows, or about Buddhist ethics in general, sometimes hunted. The punishments mentioned for monks who break the monastic anti-hunting laws were generally much harsher than those for laymen. In a monastic code written in 1846 by the Fourteenth Karma-pa lama, Theg-mchog rDo-rje (1799–1869), such a punishment, probably based upon the precedent of an actual incident, is recorded:

> In the case of a monk with bad mental intentions of killing wild animals of the hills and valleys, the offender is to be apprehended and immediately receive a warning by way of heavy physical punishment. He has to offer to the Buddha, Dharma and Sangha a Hundred Offerings (usually one hundred butter lamps), four communal tea servings [for the assembly of monks] and make a confession of sins, together with such things as one hundred prostrations; and in no case may these punishments and obligations be reduced or avoided, not even upon the intervention of a third party. After this, the offender must make a vow not to repeat the offence.
>
> (Schuh and Dagyab 1978: 246, l.6–7)

Hunting and protection of game in the twentieth century

The survey above does not, for the most part, include twentieth-century Tibetan historical sources. We have little corroborating evidence that what is recorded in the documents was put into practice. We do know that nearly all of the concepts and institutions outlined were still in evidence and put into practice in various populated areas in Tibet up until the colonial occupation of the plateau by China in the late 1950s. Between 1986 and 1996, I gathered oral information about hunting from elderly Tibetans who were adults in the premodern society that existed prior to Chinese occupation. Drawing upon some of this data, and also written eyewitness accounts, I shall briefly survey what is known about Tibetan efforts to prohibit hunting and protect game animals in the final period before premodern institutions were suddenly abolished (or transformed) under Chinese colonialism.

Aside from occasional claims in the biographies of earlier lamas, we are left to speculate about the actual operation and success of sealing areas off from hunters before the twentieth century. What was often written in the earlier law codes represented an ideal that was perhaps seldom, if ever, realized. In certain cases, the laws could also have been set down as a pious effort to earn religious merits and maintain a morally upright appearance, but with little real interest in the ultimate fate of the game animals. There is also the problem of how such laws were enforced on the vast, sparsely populated Tibetan plateau, in which premodern administrative systems were minimal in most areas. Many questions remain unclear, though more recent eyewitness accounts reveal that some individuals and institutions did indeed take the protection of game from hunters very seriously. We also know some of their motives and why their attempts were sometimes successful, sometimes not.

The later Tibetan Buddhist state, the *dGa'-ldan Pho-brang*, was headed by the Dalai Lamas and their regents and functioned in various forms from 1642 up to 1959. The twentieth century is remarkable in the history of this state as it witnessed a continuity of strong religious heads. These men, committed to upholding and cultivating Buddhist traditions and values, took active roles in leadership: the Thirteenth Dalai Lama (r. 1895–1933), the sTag-brag regent (in office 1941–1950) and the Fourteenth Dalai Lama (r. 1950–59). Frequently in the past, this had not been the case. For example, between the time of the strong Fifth Dalai Lama (1617–82) and the Thirteenth Dalai Lama, there is virtually no evidence of state edicts and other legal measures to ban hunting. The complex reasons for this have partly to do with variability in the political culture. Ensuring that Buddhist values prevailed throughout the polity took a much lower priority under rulers with greater interests in power politics, commerce and so on. In addition, it must be considered that the behaviour (such as passing legislation) of the ruler, both at the level of the centralized state and in more local political units, were not entirely dominated by Buddhist models and agendas. Indeed, since the period of the early kings – the *btsan-po* or 'mighty ones' – the more martial 'big man' (*mi che-ba*) type of leader has remained strongly evident in Tibetan cultural history. This model lived on as an ideal in folk stories and epic literature, and in the martial nature of the local mountain gods and territorial deities, but also in the style of leadership of historical lay heads of state, numerous tribal or clan chiefs and even village headmen, particularly in eastern Tibetan areas. The more 'wild' (*rgod-pa*) model of lay ruler, one associated with martial ability, valued prowess in and enjoyment of the hunt.[35] Thus in the mid-eighteenth-century, when a layman headed the Central Tibetan state, the ruler himself went on huge annual hunting expeditions with government troops.[36] This was something unthinkable during the twentieth century, with strong Buddhist rulers intent on 'taming' (*'dul-ba*) the populace.

The twentieth century was thus a period when the Buddhist heads of state consistently sought to prohibit hunting within their sphere of influence. But it was also a period during which hunting became increasingly common, and more effectively practised, due to various factors. Since the nineteenth century, Tibetan luxury animal products, like musk, stag antler, bear gall and fox or leopard pelts, fetched high prices in colonial and regional Asian markets, which demanded an ever increasing supply. Hence Tibetan harvests and trade in these valuable items boomed. At the same time, a steady supply of modern rifles began to be smuggled and imported into Tibet, often by the Lhasa government or minor local states and kingdoms for defence. The traditional Tibetan muzzle-loading matchlock guns used by hunters were fairly inaccurate, effective only at close range and good for only one shot before the alerted game was dispersed. With modern weapons, the possible kill rates became much higher. Already before mid-century, localized extinctions, mass killings of game animals and environmental problems due to predator species depletions were being reported as a result of over-hunting in Tibetan areas.[37] Although most such problems were more evident in eastern Tibet, where the Lhasa-based state exercised a much lower (if any) degree of administrative influence, many such areas also had local rulers, lamas and monasteries committed to sealing hills and valleys and punishing hunters.

To keep hunting in check, the Dalai Lamas and regents periodically issued a form of public edict known as the 'Hill and Valley decree' (*Ri-klung rtsa-tshig*), as well as other more general edicts. These often composite documents could cover topics such as law and order, ethical issues, ritual and public behaviour. They contained sections addressing the prohibition of hunting and the sealing of hills and valleys.[38] Although issued in the capital Lhasa, multiple copies of the edicts where circulated and posted throughout the network of government administrative centres (*rDzong*). From there, the public proclamation and enforcement of the edicts were the duty of the district administrative officer (*rDzong-dpon*) and his agents, who further made their contents known to local estates and settlements. My informants from former administrative districts of southeastern Central Tibet remember that the annual proclamation of these decrees or edicts by local officials or headmen was often the only regular contact they had with the traditional legal system.

When edicts against hunting were circulated, hunters who did not want to get caught simply desisted for some time until the proclamation had been forgotten or overshadowed locally by other events, or they hunted in secret in remote areas. Secret hunting alliances existed between men as a form of horizontal relationship in some villages.[39] According to most informants, if one really wanted to hunt in spite of regulations and seals, one just had to be very careful and practise restraint by not taking large numbers of animals. Many areas used by hunters were sparsely populated, and it was easy to conceal one's activities. Apart from pure bad luck and coincidence, or being informed on by one's enemies, many hunters who got caught only did so because they pursued game too close to monasteries with hill and valley seals in place. The reason for this was that many game animals gathered in these preserves since they knew from experience they would be safer around the monasteries. In some monastic sanctuaries, game animals became undisturbed by human presence and even semi-tame when monks regularly put food out for them. This of course was a great temptation to hunters.

However, when the anti-hunting laws were circulated in local communities it was much more difficult to get around the regulations. These were the so-called 'obstacle years' (lo-'gag) – astrologically inauspicious or negative years occurring every thirteenth year in the life of the Dalai Lama or other high lamas. The government was scrupulous about announcing and enforcing anti-hunting laws to everyone during these critical years, as any offence caused to the powerful protective deities dwelling at certain special locations could bring bad fortune to the Dalai Lama and thus to the state as a whole. Good examples of such locations are sacred pilgrimage mountains (gnas-ri), like Tsari or Ti-se (i.e. Mount Kailas), where it was commonly believed that the resident deities could become embodied as wild game animals. Lamas therefore warned against harming the animals and the government sealed the areas against hunters.[40] Such cases of sensitivity towards hunting were not directly related to Buddhist ethics, but rather to specific Tibetan views of nature and notions of causality. Normally, a hunter caught operating in his home area got off with a warning the first few times, especially if he was friendly with the village leaders. But during inauspicious lo-'gag years, headmen were under great pressure from government officials to be strict about the laws and seals, and to hand over culprits to the local authorities for formal punishment.

Punishment in the form of physical violence for breaking the anti-hunting laws is not mentioned in any of the state law codes. However, it is reported by all informants as having been normal. Hunters who came before the authorities received a sound beating with a whip or cane administered by the local officials. This was also the normal method of punishing hunters caught violating monastic preserves during the twentieth century.[41] In cases where a stronger public example was to be made, confiscation of weapons, imposition of fines, short-term imprisonment and severe beatings were used. Even harsher punishments could be imposed. Mutilation punishments are listed for serious offences (murder, treason, violence against the clergy, etc.) in the main Tibetan law codes, but were rarely imposed in the twentieth-century Lhasa state. However, in small states ruled by local kings and lamas in eastern Tibet, we have at least two records of mutilation punishments for breaking anti-hunting laws, both involving the removal of a hunter's hands at the wrist.[42]

Edicts issued by the Tibetan Buddhist head of state were directed as much at administrative officials as they were towards the general populace; this was also the case with anti-hunting laws. A common feature of premodern state administration in Tibet was the prevalence of corruption, especially in outlying areas. Official corruption was a principal reason why anti-hunting laws were often ineffective or not even brought to the public's attention.

Administration of *rDzong* districts was undertaken by both lay and clerical officials, mainly drawn from a small body of aristocratic families. Officials were normally given the position of district administrator or *rDzong-dpon* as a three-year contract, with the primary task of collecting revenues on behalf of the government. They received no direct salary, but instead, outside of sending the expected annual taxes to Lhasa, they could derive various personal profits from the position.

One way to increase these profits illegally was for a *rDzong-dpon* to disregard the anti-hunting laws issued by the state when the edicts were forwarded to him. Instead, he would demand that part of the revenues due from the local populace be paid in valuable animal products, which could then be traded at great profit. For example, at the beginning of the 1950s, the monk official who was *rDzong-dpon* of sKyid-grong district in western Tibet required local residents to hunt leopards for their skins, which he collected and sold for a good price.[43] This sort of practice was tacitly accepted until the Thirteenth Dalai Lama's reign, when he began to issue strong warnings against it as a part of the anti-hunting laws in his edicts:

> Generally, although specific edicts [against hunting] are periodically issued, officials at remote *rDzong* and estates must not be negligent in stringently applying them because they put extortion and obtaining personal advantages first. Also, those in authority and the general populace have not consistently observed these regulations... The restrictions on not harming, even for a moment, the lives of all living creatures, large or small, must be publicized and earnestly exhorted.
>
> (Dalai Lama XIII 1901: 1.42–46)

Another related practice mentioned by all my informants was the purchase of large quantities of valuable game animal products from peasants by influential Tibetan lama corporations (*bla-brang*) and monastic business managers (*phyag-mdzod*) for trading to China and India. Thus, the harvest of game animals was always being directly or indirectly encouraged by the same institutions that were publicly opposed to it.

A final example which reveals some of the dynamics surrounding twentieth-century Tibetan state attempts to ban the hunt concerns a special prolonged 'sealing the hills' edict imposed by the Thirteenth Dalai Lama from the mid-1920s. It is reported that hunting was prohibited in all districts for a seven year period 'of special prayer by the Dalai Lama on behalf of his country'.[44] This came immediately following the critical political struggle between the pro-British, pro-modernizing government officers (favoured by the Dalai Lama) and the conservative monastic body, in which the former lost out completely. The events, a major blow to the Dalai Lama and his policies, destabilized the situation and nearly triggered a coup. It was normal that the pious Dalai Lama should encourage the scrupulous observation of ritual and Buddhist ethics in order to increase the spiritual and political welfare of his country at this time. Since the political turn of events was essentially against foreign influences, the activities of foreign visitors and residents in Tibet came under critical scrutiny. It was common for foreigners to engage in shikar, or shooting, while in Tibet. The conservative Tibetan government reacted by putting strong pressure on them not to hunt. Travel documents issued to visitors at this time stated that shooting was strictly prohibited in Tibet, and officials reminded them of this as they entered through the border regions.[45] Yet reports of such encounters reveal how ineffective and even contradictory the prohibitions were in practice. Not only did foreigners continue to hunt freely in the vast tracts of country where no officials could check on them, but some Tibetan government officials who served as their escorts even 'blasted away' at wild animals themselves.[46] It was noted by foreign observers in some areas that, despite the special anti-hunting prohibitions announced locally by the highest ranking government officers from Lhasa, local people continued their hunting in secret, and that game numbers were decreasing.[47]

Conclusion

The materials briefly summarized in this chapter demonstrate that the Buddhist-inspired anti-hunting laws and institutions which existed in twentieth-century pre-colonial Tibet have a long history (of at least six or seven centuries). They are clear evidence of a continual willingness on the part of certain Buddhist religious and secular leaders to take measures to protect the lives of a wide range of wild animals from human harvest. All such measures were motivated by a combination of formal Buddhist doctrines and ethics and other Tibetan beliefs about causality and the local deities associated with the natural environment. Within this framework of ideas, the benefits of protecting wildlife from hunters were believed to accrue to both the wild animals and their human protectors. But there was no sense that wild game had any intrinsic value – that is, as uniquely evolved species or as quantifiable units whose ecological viability depends upon maintaining certain population levels.

The variability in Tibetan political culture and value orientation has been consistently underestimated by scholars because of an undue emphasis upon Buddhism and its major institutions. In spite of mainly Buddhist inspired measures against hunting, Tibetan societies always maintained a positive interest in hunting, whether this was in connection with ideals of leadership or simply had to do with survival needs or economic gain. The promotion and effectiveness of measures to prohibit hunting in a specific period and area depended upon the value orientation of the ruling elite and their administrative agents, and the range of other interests and needs of the population at large. At least during the twentieth century, anti-

and pro-hunting forces were both strong and in competition. In addition, Tibet's large land area, sparse population and low level of premodern administrative infrastructure militated against the effectiveness of anti-hunting measures. It would appear that localized prohibitions, such as the Tibetan Buddhist innovation of 'sealing the hills and sealing the valleys' around individual monastic communities or discrete natural landscapes (such as holy mountains), were far more effective against hunting than the blanket prohibitions issued by the state.

Acknowledgements

This article is summarized from parts of my forthcoming book, *Tibetan Hunting Culture*, under preparation with generous support by the Alexander von Humboldt-Stiftung, Bonn and the Zentralasien-Seminar, Humboldt Universität, Berlin. I am also grateful both to the Orientabteilung, Staatsbibliothek zu Berlin and the Library of Tibetan Works and Archives, Dharamsala. The alphabetic transliteration used for Tibetan words in this chapter is the standard Wylie system.

Notes

1 It is also possible that *d[w]ags* here is related to *gdag[s]* ('light' or 'daylight'), since this has long been a Tibetan designation for the sunny (i.e. south-facing) slopes of the mountains (as opposed to *srib*, 'shade/dark'), which are the preferred habitat of both wild and domestic animals. *Ri-dwags* is normally opposed to *g.yung-dwags* ('domesticated animals'; g.yung means 'tame').
2 Samuel (1993: 217–222).
3 See, for example, Ricard (1994: 447, 542).
4 See Pelliot Tibétain in Spanien and Imaeda (1979: 1071, 1072; Richardson 1990).
5 Bod kyi lo rgyus, Document 2, cf. also Documents 1, 3, 5, 13, 14; Tucci (1949: 750).
6 See Ratchnevsky (1985: 371–374).
7 See especially Ruegg (1995: 17–92).
8 For example, see 'Jigs-med Grags-pa (1987: 198, 222, 262, 271–272) for rGyal-rtse, and Schuh (1981: 350, l.16–17; 364, l.16–17) for Gung-thang.
9 Similar types of institutions, often linked to hunting and harvesting, were common in many premodern societies. For example, on rahui preserves among the Maori of New Zealand, see Best (1904) and Firth (1929: 258–262).
10 In the Shog dril chen mo of Zhang g.Yu-brag-pa brTson-'grus Grags-pa (1972: 140); dPa'-bo II gTsug-lag 'Phreng-ba (1989, vol. 1: 808).
11 Si-tu Pa-chen Chos-kyi 'Byung-gnas and 'Be-lo Tshe-dbang Kun-khyab (1972, vol. 1: 494); dPa'-bo II gTsug-lag 'Phreng-ba (1989, vol. 2: 1047).
12 Guenther (1986: 157).
13 'Jigs-med Grags-pa (1987: 272, l.13–14).
14 sDe-srid Sangs-rgyas rGya-mtsho (1987: 210).
15 Schuh and Phukhang (1976: 111).
16 See 'Jigs-med Gling-pa (1971: 405).
17 'Jigs-med Grags-pa (1987: 271, l.1–10; 272, l. 13–14).
18 sDe-srid Sangs-rgyas rGya-mtsho (1987: 210); Tsa-na-ka: 25–27.
19 Snellgrove (1992: 119).
20 'Jigs-med Gling-pa (1971: 393); I am grateful to Janet Gyatso for this reference.

21 'Jigs-med Grags-pa (1987: 198, 222, 262).
22 See, respectively: 'Jigs-med Grags-pa (1987: 198); Dalai Lama VII (1976: 461, l.5–6); Schuh and Dagyab (1978: 247, l.27–28); Schuh and Phukhang (1976: 111).
23 'Jigs-med Grags-pa (1987: 222).
24 Zhal lce bcu drug: 90.
25 This dance is well documented by Draghi (1980; 1982) and Montmollin (1988; 1990).
26 Zhal lce bcu gsum: 157; Zhal lce bcu drug: 108.
27 See Zhal lce bcu gsum: 156; Dalai Lama XIII (1901; 1981: 334, l.5–7); French (1995: 235).
28 Kong-sprul Blo-gros mTha'-yas (1975: 306, l.1–3).
29 Dalai Lama XIII (1981: 184, l.1–3).
30 For example, see 'Jigs-med Grags-pa (1987: 198), Dalai Lama VII (1976: 377, l.2–3), Schuh and Dagyab (1978: 247, l.27–8), and Dalai Lama XIII (1981: 184, l.1–3).
31 Das (1969: 80).
32 Dalai Lama XIII (1981: 82, l.1–3; 184, l.1–3).
33 Dalai Lama XIII (1981: 82, l.1–3).
34 Kong-sprul Blo-gros mTha'-yas (1975: 306, l.1–3).
35 For examples of lay rulers hunting in Tibetan societies, see mGon-po rNam-rgyal (d.1865) of Nag-rong (Tsering 1985: 205), Chagur Thubten (c.1940) of sDe-dge (Carnahan 1995: 95), 'Jigs-med rNam-rgyal (1825–1881) of Bhutan (Aris 1994: 66–67), and Nyi-ma rNam-rgyal (r. 1694–1729) of Ladakh (Bray and Butters in press); cf. n.38 below.
36 On 'Gyur-med rNam-rgyal's (r. 1747–1750) hunts, see mDo-mkhar-ba Tshe-ring dbang-rgyal (1981: 61–62).
37 On localized extinctions of deer species by Tibetan government troops with modern weapons, see Allen (1938: 282; 292–93); on mass Tibetan kills of wild yak and wild sheep with repeating rifles, see Clark (1955: 154, 253); on environmental problems due to over-hunting of predator species by nomads with modern weapons, see Duncan (1952: 192, 209; 1964: 246).
38 For examples mentioning anti-hunting laws, see Dalai Lama XIII (1901; 1981: 332, l.1–337, l.2), sTag-brag sPrul-sku (1944), Rwa-sgreng sPrul-sku (1939), French (1995: 233–235).
39 See also Dargyay (1982: 49, 78).
40 See, for example, Huber (1998: 181, 198–200).
41 Denma Locho Rinpochey (1994: 86); cf. also Stubel (1958: 59).
42 Dolan (1938: 168), Duncan (1964: 245).
43 Tashi Khedrup (1986: 65); cf. also Sherring (1906: 27).
44 Hayden and Cosson (1927: 128); cf. Roerich (1931: 385) for the same area and period, and McKay (1994: 382–82) for British responses to the policy.
45 Roerich (1931: 309, 409).
46 Hayden and Cosson (1927: 126), describing mKhyen-rab Kun-bzang sMon-grong and his staff. On private tolerance of foreign hunting by government officials in spite of the prohibitions, see Macdonald (1932: 148–149).
47 Roerich (1929: 367), describing Nag-chu-kha and Ka-shod-pa Chos-rgyal Nyi-ma's (b. 1903) visit there as Hor sPyi-khyab.

References

Tibetan Sources (in Tibetan alphabetical order)

Kong-sprul Blo-gros mTha'-yas (1975) 'Dpal spungs yang khrod kun bzang bde chen 'od gsal gling gi sgrub pa rnams kyi kun spyod bca' khrims blang dor rab gsal phan bde'i 'byung gnas'. In *Rgya-chen bKa'-mdzod. A Collection of Writings of 'Jam-mgon Ko-sprul Blo-gros-mtha'-yas*, vol. 11. Paro, Ngodrup, ff.257–319.

'Gos Lo-tsÕ-ba gZhon-nu-dpal (1984) *Deb ther sngon po*, 2 vols. Chengdu, Si-khron Mi-rigs dPe-skrun-khang.

'Jigs-med Grags-pa (1987) *Rgyal rtse chos rgyal gyi rnam par thar pa dad pa'i lo thog dngos grub kyi char 'bebs*. Lhasa, Bod-ljongs Mi-dmangs dPe-skrun-khang.

'Jigs-med Gling-pa (1971) 'Yul lho rgyud du byung ba'i rdzogs chen pa rang byung rdo rje mkhyen brtse'i 'od zer gyi rnam thar pa legs byas yongs 'du'i snye ma'. In *The Collected Works of Kun-mkhyen 'Jigs-med-gling-pa*, vol. 9. Gangtok, Sonam T. Kazi, pp. 1–502.

sTag-brag sPrul-sku, Ngag-dbang gSung-rab Grub-thob bsTan-pa'i rGyal-mtshan (1944) 'Mon rta dbang gi rtsa tshig'. Handwritten document no. 161, Manuscript Department of the Library of Tibetan Works and Archives, Dharamsala.

Dalai Lama VII, bsKal-bzang rGya-mtsho (1976) *The Collected Works (gSun 'bum) of the Seventh Dalai Lama Blo-bzan-bskal-bzan-rgya-mtsho*, vol. 3 [bca'-yig and rtsa-tshig]. Gangtok, Dodrup Sangye.

Dalai Lama XIII, Thub-bstan rGya-mtsho (1901) 'Lcags mo glang lo'i rtsa tshig'. Handwritten document no. 170, Manuscript Department of the Library of Tibetan Works and Archives, Dharamsala.

Dalai Lama XIII (1981) *Collected Works of Dalai Lama XIII*, vol. 4 [*bca'-yig* and *rtsa-tshig*]. New Delhi, Jayyed Press (Sata-Pitaka Series, 286).

mDo-mkhar-ba Tshe-ring dbang-rgyal (1981) *Bka' blon rtogs brjod*. Chengdu, Si-khron Mi-rigs dPe-skrun-khang.

sDe-srid Sangs-rgyas rGya-mtsho (1987) Blang dor gsal bar ston pa'i drang thig dwangs shel me long nyer gcig, in: Bod kyi dus rabs rim byung gi khrims yig phyogs bsdus dwangs byed ke ta ka. Lhasa, Bod-ljongs Mi-dmangs dPe-skrun-khang, pp. 203–279.

dPa'-bo II gTsug-lag 'Phreng-ba (1989) *Dam pa'i chos kyi 'khor lo bsgyur ba rnams kyi byung ba gsal bar byed pa mkhas pa'i dga' ston*, 2 vols. Beijing, Mi-rigs dPe-skrun-khang.

Bod kyi lo rgyus (1995) Bod kyi lo rgyus yig tshags gces bsdus. sGrol-dkar (ed.) Lhasa, Rig-dngos dPe-skrun-khang.

Tsa-na-ka (1958) 'Canakyarajanitisastram [Tsa na ka'i lugs kyi bstan bcos]'. In *Visva-Bharati Annals* vol. 8, pp. 1–78.

Zhang g.Yu-brag-pa brTson-'grus Grags-pa (1972) *Writings (bKa' thor bu) of Zhang g.Yu-brag-pa brTson-'grus-grags-pa*. Tashijong, Sungrab Nyamso Gyunpel Parkhang.

Zhal lce bcu gsum (1989) 'Ta la'i bla ma sku phreng lnga pa'i dus su gtan la phab pa'i khrims yig zhal lce bcu gsum'. In Chab-spel Tshe-brtan Phun-tshogs (ed.) *Bod kyi snga rabs khrims srol yig cha bdams bsgrigs*. Lhasa, Bod-ljongs sPyi-tshogs Tshan-rig-khang-gi Bod-yig dPe-rnying dPe-skrun-khang, pp. 146–184.

Zhal lce bcu drug (1989) 'Gtsang pa sde srid karma bstan skyong dbang po'i dus su gtan la pheb pa'i khrims yig zhal lce bcu drug'. In Chab-spel Tshe-brtan Phun-tshogs (ed.) *Bod kyi snga rabs khrims srol yig cha bdams bsgrigs*. Lhasa, Bod-ljongs sPyi-tshogs Tshan-rig-khang-gi Bod-yig dPe-rnying dPe-skrun-khang, pp. 82–145.

Rwa-sgreng sPrul-sku, Thub-bstan 'Jam-dpal Ye-shes rGyal-mtshan (1939) 'Sa mo yos lo'i rtsa tshig'. Handwritten document no. 151, Manuscript Department of the Library of Tibetan Works and Archives, Dharamsala.

Si-tu Pan-chen Chos-kyi 'Byung-gnas and 'Be-lo Tshe-dbang Kun-khyab (1972) Bsgrub rgyud karma kam tshang brgyud pa rin po che'i rnam par thar pa rab 'byams nor bu zla ba chu shel gyi phreng ba [History of the Karma bKa'-brgyud-pa Sect], 2 vols. New Delhi, D. Gyaltsen and Kesang Legshay.

Other sources

Allen, G. M. (1938) 'Zoological results of the second Dolan expedition to western China and eastern Tibet, 1934–1936: part III, mammals'. *Proceedings of the Academy of Natural Sciences of Philadelphia*, vol. 90, pp. 261–294.

Aris, M. (1994) *The Raven Crown: The Origins of Buddhist Monarchy in Bhutan*. London: Serindia Publications.

Best, E. (1904) 'Notes on the custom of Rahui'. *Journal of the Polynesian Society*, vol. 13. no. 50, pp. 83–88.

Bray, J. and C. Butters (In press) 'An eighteenth-century Bhutanese Lama's journey to Ladakh'. In M. van Beek, K. Brix Bertelsen and P. Pedersen (eds) *Recent Research on Ladakh*, 8. Aarhus: Aarhus University Press.

Carnahan, S. (1995) *In the Presence of my Enemies: Memoirs of Tibetan Nobleman Tsipon Shuguba*. Santa Fe: Clear Light Publishers.

Clark, L. (1955) *The Marching Wind*. London: Hutchinson.

Dargyay, E. K. (1982) *Tibetan Village Communities: Structure and Change*. Warminster, Aris and Phillips.

Das, S. C. (1969) *Autobiography: Narratives of Incidents of My Early Life*. Calcutta: Indian Studies Past and Present.

Denma Locho Rinpochey (1994) 'My life in the land of snows'. *Cho-yang*, vol. 6, pp. 84–104.

Dolan, B. (1938) 'Zoological results of the second Dolan expedition to western China and eastern Tibet, 1934–1936: part I, Introduction'. Proceedings of the Academy of Natural Sciences of Philadelphia, 90, pp. 159–184.

Draghi, P. A. (1980) 'A Comparative Study of the Theme of the Conversion of a Hunter in Tibetan, Bhutanese and Medieval Sources'. Unpublished PhD dissertation. Bloomington: Indiana University.

—— (1982) 'The stag and the hunting dog: a Bhutanese religious dance and its Tibetan source'. *The Journal of Popular Culture*, vol. 16, no. 1, pp. 169–175.

Duncan, M. H. (1952) The Yangtze and the Yak: Adventurous Trails in and out of Tibet. Alexandria, Va.

—— (1964) *Customs and Superstitions of the Tibetans*. London: The Mitre Press.

Firth, R. (1929) *Primitive Economics of the New Zealand Maori*. London: Routledge.

French, R. R. (1995) *The Golden Yoke: The Legal Cosmology of Buddhist Tibet*. Ithaca: Cornell University Press.

Goldstein, M. C. (1994) 'Change, conflict and continuity among a community of nomadic pastoralists: a case study from western Tibet, 1950–1990'. In R. Barnett and S. Akiner (eds) *Resistance and Reform in Tibet*. London: Hurst and Co., pp. 76–111.

Guenther, H. V. (1986) *The Jewel Ornament of Liberation*. Boston, Shambhala.

Hayden, H. and C. Cosson (1927) *Sport and Travel in the Highlands of Tibet*. London: Richard Cobden-Sanderson.

Huber, T. (1998) *The Cult of Pure Crystal Mountain: Popular Pilgrimage and Visionary Landscape in Southeast Tibet*. New York: Oxford University Press.

Macdonald, D. (1932) *Twenty Years in Tibet: Intimate and Personal Experiences of the Closed Land among all Classes of its People from the Highest to the Lowest*. London: Seeley, Service and Co.

McKay, A. (1994) 'The other "great game": politics and sport in Tibet, 1904–1947'. *The International Journal of the History of Sport*, vol. 11, no. 3, pp. 372–386.

Montmollin, M. de (1988) 'Some more on the *sha ba sha khyi 'cham*: a Bhutanese *'cham* on the conversion of the hunter mGon po rDo rje by Mi la ras pa'. In H. Uebach and J. Panglung (eds) *Tibetan Studies*. Munich: Kommission für zentralasiatische Studien, Bayerische Akademie der Wissenschaften, pp. 293–300.

—— (1990) 'Sha ba sha khyi 'cham: A Bhutanese 'cham on the conversion of the hunter mGon po rdo rje by Mi la ras pa'. *Acta Orientalia Hungaricae*, vol. XLIV, nos. 1–2, pp. 89–96.

Ratchnevsky, P. (1985) *Un Code des Yuan*, Tome 4. Paris: College de France.

Ricard, M. (1994) *The Life of Shabkar: The Autobiography of a Tibetan Yogin*. Albany: State University of New York Press.

Richardson, H. E. (1990) 'Hunting accidents in early Tibet'. *The Tibet Journal*, vol. 15, no. 4, pp. 5–27.

Roerich, G. N. (1931) *Trails of Inmost Asia*. New Haven: Yale University Press.

Roerich, N. (1929) *Altai Himalaya: A Travel Diary*. New York: Frederick A. Stokes.

Ruegg, D. S. (1995) *Ordre spirituel et ordre temporel dans la pensée bouddhique de l'Inde et du Tibet*. Paris: Collège de France.

Samuel, G. (1993) *Civilized Shamans: Buddhism in Tibetan Societies*. Washington, D.C.: Smithsonian Institution.

Schuh, D. (1981) *Grundlagen tibetischer Siegelkunde: Eine Untersuchung über tibetische Siegelaufschriften in 'Phags-pa-Schrift*. Sankt Augustin: VGH Wissenschaftsverlag.

Schuh, D. and L. S. Dagyab (1978) *Urkunden, Erlasse and Sendschreiben aus dem Besitz sikkimesischer Adelshäuser und des Klosters Phodang*. St Augustin: VGH Wissenschaftsverlag.

Schuh, D. and J. K. Phukhang (1976) *Urkunden und Sendschreiben aus Zentraltibet, Ladakh und Zanskar, Teil 2*: Edition der Texte. St Augustin, VGH Wissenschaftsverlag.

Sherring, C. A. (1906) *Western Tibet and the British Borderlands*. London: Edward Arnold.

Snellgrove, D. L. (1992) *Four Lamas of Dolpo: Tibetan Biographies*. Kathmandu: Himalayan Book Seller (2nd edition).

Spanien, A. and Y. Imaeda (1979) *Choix de Documents Tibétains Conservés à la Bibliothèque Nationale*, Tome II. Paris: Bibliothèque Nationale.

Stubel, H. (1958) *The Mewu Fantzu. A Tibetan Tribe of Kansu*. New Haven: HRAF Press.

Tashi Khedrup (1986) *Adventures of a Tibetan Fighting Monk*. Compiled by H. Richardson, Bangkok: Tamarind Press.

Tsering, T. (1985) 'Nag-ro mGon-po rNam-rgyal: a nineteenth century Khams-pa Warrior'. In B. N. Aziz and M. Kapstein (eds) *Soundings in Tibetan Civilization*. Delhi: Manohar, pp. 196–214.

Tucci, G. (1949) *Tibetan Painted Scrolls*, vol. 2. Rome: La Libreria Dello Stato.

3

REPRESENTATIONS OF HUNTING IN JAPAN

John Knight

Introduction

Hunting is readily defined in terms of the primary relationship between the human hunter(s) and the hunted animal. Human hunting centres on an elemental confrontation between hunters and unrestrained wild animals that results in the violent killing of these animals (Cartmill 1993: 29–30). But there is also a secondary set of hunting relations in the form of the social context in which the activity of hunting takes place. This wider set of relations is especially significant in the case of recreational hunting in urban–industrial societies. As an activity that combines violence and sport, recreational hunting is often subject to disapproval and moral critique in the wider human society. As a result of the intensity and ubiquity of such criticism, hunting ceases to be simply a physical activity and tends to develop a capacity for rhetorical self-defence. Recreational hunters do not just hunt, but must also justify or rationalize hunting to the wider society in which they live. Hunters are often obliged to represent their hunting as consistent with the larger public interest. This is the background to the familiar utilitarian justification of hunting as a form of pest control found among hunters and shooters in many societies, including English fox-hunters (Marvin 2000) and snake and pigeon shooters in rural America (Weir 1992; Song 2000).

This chapter examines hunting in modern Japan, with specific reference to representations of the hunt and its dangers among hunters and non-hunters in rural areas. Three sets of representations of hunting in Japan are discussed: hunting as a contest with animals and among men, hunting as protection of the wider rural community from dangerous animals and hunting as itself a danger both to the practitioners and to the wider rural public. In Japan, hunting (especially wild boar-hunting) is marked by a symbolism of contest and conquest, while also represented by Japanese hunters in instrumentalist terms as a means of protecting the wider rural society from wildlife pests. On top of these symbolic and utilitarian representations of the hunt, however, hunting in Japan is viewed as a source of danger. This includes both the physical danger posed by the hunters' guns and hounds to hunters and bystanders and the spiritual danger (including spirit revenge) that arises from the violence intrinsic to the hunt. The ethnographic focus for the following discussion is the Kii Peninsula in western Japan, and specifically the municipality of Hongū-chō, where research has been carried out since 1987. This chapter also draws on secondary, Japanese-language sources, including books and articles by folklorists and other scholars and newspaper databases.

Hunting in Japan

Two-thirds of the Japanese land area is made up of mountain forests known as *yama*. The Japanese mountains have long been the site of hunting, and Japan has its celebrated regional traditions of hunting. In remote mountainous parts of northern Honshu are found the famous *matagi* hunting villages, the customs of which have been extensively documented by Japanese folklorists. However, for the great majority of the registered hunters in present-day Japan, including those on the Kii Peninsula, hunting is an expensive recreational pastime rather than an income-yielding occupation. Hunting in Japan today has become a Westernized activity in which hunters use shotguns, rifles, two-way radios, radio collars for their hounds and heat sensors to detect wounded game, with most of this hunting technology imported from Western countries. Japanese hunters make use of Western dog breeds, wear camouflage fatigues and peaked caps and read hunting magazines which regularly feature hunting in foreign countries and advertize foreign-made hunting accessories.

Hunting in Japan is usually divided into two categories: 'large-scale hunting' (*ōgataryō*), which consists of the hunting of wild boar and deer, and 'small-scale hunting' (*kogataryō*), which consists of the hunting of birds and small game animals. Japanese hunters are allowed to hunt twenty-nine species of bird, including pheasants, crows, sparrows and ducks, and seventeen species of animals, including wild boar, deer (stags), bears, hares, foxes, raccoon dogs, weasels and squirrels (Environment Agency 1997: 438). In 1992 around three million birds were captured by hunting (over one quarter of which were sparrows), while 330,000 animals were hunted (over half of which were hares) (ibid.). Most Japanese hunting takes the form of chase- or pursuit-hunting involving armed men with dogs, but trapping also takes place. In 1995 there were over 200,000 registered shotgun- and rifle-users, or 85 per cent of all registered hunters in Japan, while the number of trappers stood at around 20,000, less than one-tenth of the total number of hunters (the remaining 6 per cent were airgun users) (Environment Agency 2000).

In Japan, hunters must join a local branch of the hunting association known as *ryōyūkai* (literally 'hunting friends society'). In 1994 the membership of the Hongū *ryōyūkai* stood at 76 registered hunters, of which 63 were shotgun or rifle holders and 13 trappers. The local *ryōyūkai* disseminates information on hunting, holds regular meetings and elects officials who link with prefectural and national levels of the *ryōyūkai* and who liaise with local municipal and police authorities. Hunting in Japan is tightly regulated. In order to obtain a hunting licence, hunters must pass the state examination in which they are tested on safety and their ability to instantly identify game species. Once they have their licence, hunters must pay registration and hunting fees to the prefectural government. There is a winter hunting season and there are areas in the forest where hunting is prohibited. Rules restrict the methods and devices used in hunting and the number of game that can be caught per day and per season.

Hunting as contest

This chapter focuses largely on boar-hunting. Two basic methods of boar-hunting on the Kii Peninsula can be distinguished (Otake 1987: 84). One is *neyadome*, or 'lair-stopping', in which the wild boar is stopped at or near its lair by the hound. *Neyadome* is often synonymous with *tandokuryō*, or 'solo hunting', involving a lone hunter and his hound. The hound is released to find and engage the boar in order to prevent its escape, while the hunter, alerted by the hound's barking, runs to the scene to finish off the boar with his gun. But solo hunting has become rare. The second, more common method of boar-hunting is *machibauchi*, the ambushing of the fleeing wild boar by a group of hunters. This kind of hunt involves a division of labour between two principal roles: the *seko* and his hound who flushes out the wild boar from the mountains and chases it in the direction of the shooters; and the *machi*, the shooters, who wait along the likely escape routes of the wild boar (valley stream or mountain ridge). Typically, there will be one or two *seko* with two or three hounds and from one to ten *machi* strategically placed around the mountain. In principle, the larger the hunting group, the greater the chances of success. The *seko* is the key player, the one on whom all the others depend to track down the wild boar and drive it in their direction. The *machi* is expected to consummate the work of the *seko* by shooting accurately when he gets the chance.

The boar-hunt or deer-hunt typically involves a group of men who act as a team to flush out and shoot the animal. There are between ten and fifteen such large game hunting groups in Hongū. These groups are of different sizes and have different degrees of organization and cohesion. With most groups, membership is fixed over time. Groups often consist of members of the same family and other relatives (for example, fathers and sons, brothers, uncles and nephews, cousins and so on) and contain members from the same village or an adjoining village. Sometimes a group includes guests participating in the hunt, such as a relative visiting from the city. It is often said that hunting groups today have lost their cohesion because of the greater demands of family life. Some groups go out every weekend in the winter hunting season (weather permitting), but others convene only occasionally. Regularity in hunting is deemed important, because it allows a group to learn about game availability in the mountains and because it ensures that other groups do not have the chance to move in and take over a group's hunting ground.

Figure 3.1 A wild boar-hunter holding the carcass of a young boar caught by his dogs

In the past, village hunting contributed much more to local livelihoods. It provided meat for the family diet, clothing and medicine. Meat, hides and organs could also be sold, representing an important source of cash income. But for most hunters today the purpose of hunting is not to provide economic benefits. Rather, hunting is an intrinsically enjoyable activity which affords excitement, relaxation, camaraderie and, when successful, a sense of collective and individual achievement. It is variously characterized by hunters as a 'sport' (*supōtsu*), as 'play' (*asobi*), as 'exercise' (*undō*) and as a 'hobby' or 'interest' (*shumi*). For many upland men, hunting is an all-consuming passion, one which occupies them even outside the winter hunting season through the year-round care of hounds, through the reading of hunting magazines and books and watching videos (of hound trials etc.) in their spare time, and through routine discussion and conversation about hunting matters with fellow hunters. Hunting allows men to come together in a coordinated way and pit themselves in a contest with 'nature'. For the hunter, as one veteran put it, 'nature is the opponent' (*aite ga shizen*).

Hongū hunters point out that the boar-hunt is a special experience, and refer to its 'sweet taste' (*daigomi*), its 'pull' (*tegotae*) and its 'thrill' (*suriru*). They hold that those who have never faced a charging wild boar – including the foreign anthropologist among them – cannot really understand it. As one boar-hunter from the Hongū village of Ōtsuga put it, to catch one wild boar is better than catching ten deer. The successful boar-hunt creates a sense of 'self-satisfaction' (*jiko manzoku*) in the hunter that other forms of hunting cannot match. The challenge of the boar-hunt has to do with the particular attributes of the wild boar. The fleeing wild boar rarely allows itself to be cornered and can escape in the most perilous of situations. One of the wild boar's tactics is to move to the edge of its territory and then discreetly reverse back to bypass its pursuers (who may well forlornly continue the chase into the remote mountainous interior). Another boar trick, according to Hongū hunters, is to move in a figure of eight, to the great confusion and ultimate frustration of hunters and hounds. Boar-hunters also pit themselves against the fighting spirit of the wild boar. The wild boar's spirited engagement with the hounds is one of the most thrilling moments of the hunt. The wild boar stands out from other forest animals because of the danger it poses both to the hunter and to his dogs. It is an ill-willed (*ijiwarui*) animal with an aggression that the deer completely lacks.

Another aspect of the appeal of the boar-hunt is its war-like character. As an occasion for men to display courage, strength and teamwork, the boar-hunt is routinely likened to warfare, and there exists an extensive lexicon of militaristic terminology employed by hunters. Boar-hunters refer to the hunt as a 'battle' (*tatakai*), the hunting ground as a 'battlefield' (*senjō*), hunting plans as 'battle plans' (*senpō*), a pause or stand-off in the hunt as an 'armistice' (*kyūsen*), hunting success as 'war gains' (*senka*), past hunting experience as a 'war record' (*senreki*) and so on. Boar-hunters even apply military nomenclature to themselves; the lead hunter in one Hongū hunting group I observed was known as the 'captain' (*kyaputen*), while elsewhere overbearing hunters, who always think that they know best, are 'mountain generals' (*yama no taishō*) (Yamamoto 1995: 64). The war idiom readily extends to the wild boar itself. Hunters refer to the boar as the 'forest war veteran' (*yama no mosa*), to the boar's tusks as 'weapons' (*buki*) and to trophies from the boar-hunt as 'war trophies' (*rekisen no torofii*). The hunter's boar war is in part a patriotic war. Boar-hunters sometimes celebrate a successful kill with shouts of *banzai* or 'Long Live the Emperor' (literally 'ten thousand years'), as though celebrating a great military victory (AS 19/1/1996)![1]

The 'war' with the boar, or *shishi*, has its human casualties. Through their encounters with charging boars, hunters suffer serious injuries, incur long-term

Figure 3.2 Two boar-hunters returning from a successful hunt (Courtesy of Mori Eizō)

disabilities (in some cases becoming crippled), and even lose their lives. Many a boar-hunter has scars to show from past battles, usually on one of his limbs, especially the legs; some even have permanent metal plates in their limbs to support bones fractured in past hunts. The male *shishi*'s sharp, lacerating tusks are likened to 'swords' (*katana*) and 'razor blades' (*kamisori*) wielded at the level of a man's thighs and threaten to castrate him. But female boars are dangerous too because of their sheer physical power when charging and their ability to bite. Although direct boar attacks on hunters seldom occur, every boar-hunter knows that they are a possibility. Nakano Jōji from the Hongū village of Ōstuga expressed the risk run by the boar-hunter as follows:

> For a hunter to be done by a wild boar hardly ever happens. For a wild boar and a man to have a go at each other like this [i.e. 'be done by the boar'] occurs, well, once in a hundred times. Ninety-nine per cent of the time, it will be victory for Man.... But then you are left with that 1 per cent, or maybe even 5 per cent ...

The death of a hunter can occasion a 'war of revenge' (*tomurai gassen* or *fukushūsen*) against the wild boar. In January 1997 a hunter from the village of Takada (on the Kii Peninsula) was killed by a wild boar in the course of a hunt. As Takada is nearby, this incident was widely recounted and discussed by hunters in Hongū, and rumour had it that the other hunters in the Takada group had vowed to

Figure 3.3 A wild boar carcass (Courtesy of Matsumoto Tadami)

avenge their comrade's death and indeed subsequently pursued the wild boar responsible (though, apparently, without success).

The qualities of courage, determination and defiance associated with the wild boar make it an object of admiration and even identification among Japanese hunters. One contributor to the *Shuryōkai* hunters' magazine states that hunters view the wild boar as 'a manly fearless animal' (*danseiteki de daitan na dōbutsu*) (Satō Y. 1979: 111). The 'fearless' boar in turn provides a test of a man's courage. The veteran hunter Inui Mitsunori stressed this point to me in the course of an interview. He suggested that encounters with the wild boar reveal the weakness of many of his fellow hunters. This is especially the case when the boar is charging towards the position of a hunter.

If you make a mistake at that moment, it could be the end. Because it could be devastating, people run away. They run away as soon as they see it [the boar] – those kinds of people. The worst ones even leave their guns behind – they just chuck them away in the mountains [and flee] ... They do not have the strength to hunt. Those sort of people really exist. From out of their mouths they speak of the great things [they have done]. But they are no good for anything, they never catch a thing ... Some people will never be able to hunt wild boar ... Remember one thing. *When you shoot a wild boar you shoot it with your courage*.... You shoot it with the gun, but you [really] shoot it with courage. If you don't have that, you won't catch anything ...

Another aspect of the hunter's identification with the wild boar is his physical appropriation of the animal. Through the consumption of its flesh and body parts, hunters can partake of the power of the wild boar. Boar meat, in addition to being praised for its taste, is attributed a special powerful quality. Another part of the wild boar consumed by people is the gall bladder; hunters on the Kii Peninsula take the gall of the wild boar (in the form of a boiled extract) as a regular tonic to revive themselves when they are feeling unwell. Human beings can also appropriate the sexual power of the wild boar; older hunters on the southern part of the Kii Peninsula consume the testicles of the wild boar to enhance their diminishing *seiryoku*, or 'vitality' (WKRNS 1979: 61). The wild boar's association with masculinity is further evident in the fact that, for boar-hunters, the archetypical *shishi* – the supreme 'enemy' – is the aggressive male boar in mating season. Hunting season coincides with the mating season of the wild boar (in the middle of winter), the time of year when it is at its most aggressive. At this time, male boars fight each other for females, fights that can lead to the death of the defeated animal. These truculent males, known as *gari* or *garippo*, pose a special danger to hunters and hounds; some hunters even characterize them as potential 'dog-killers' (*inugoroshi*) because of the exceptional ferocity they show towards the hounds.

There are limits to the hunter's identification with the wild boar. Men who become 'hot-blooded' (*kekki*) during the hunt are said to resemble the headstrong wild boar, but this is in a negative sense. There are tales of boar-hunters who fight wild boars with their bare-hands (for example, Mainichi Shinbunsha 1965: 26)! But such reckless hunters are the cause of hunting accidents and confrontations. For most hunters the 'hot-blooded' hunter is a fool who has forgotten that men can only succeed in hunting if they use their wits. Being boar-like in temperament, such a man cannot outwit the animal. In Japanese hunting circles, the language of *bushidō*, or 'The Way of the Warrior', is sometimes invoked to express this difference. In contrast to the impulsiveness of the wild boar, the hunter should remain 'calm' (*reisei*) and display a 'samurai spirit' (*samurai no kimochi*) in the course of the hunt (Furuya 1998: 71). 'The Way of the Hunter' (*shuryōdō* or *ryōdō*), like 'The Way of the Warrior', should be based on this retention of equanimity (ibid.). To be a successful hunter, a man must have the

Figure 3.4 A wild boar trophy in the foyer of the Mountain Village Development Centre in Hongū, Japan

courage of the wild boar *and* the presence of mind to deploy this courage at the right time. Although there is considerable overlap between wild boar and the men who hunt them, ultimately men should have a capacity for reflection that wild boar lack.

Hunting as protection

Despite the 'recreational' character of the hunt, hunters readily attribute to it a serious purpose and give the impression that, while it may not strictly speaking be a job, it is a labour of sorts. Hunters sometimes characterize hunting as a form of public service, using the term *hōshi*, or 'service', consisting of the characters for 'obey' and 'serve'. This idea of hunting as a kind of public service refers first of all to the hunter's role of culler, when the town hall calls on local hunters to catch and kill problem animals. In Japan large numbers of birds and animals are killed each year. In 1992, for example, 1.28 million birds and 100,000 animals were captured in the course of such pest control operations (Environment Agency 1997: 438). But for many hunters, there is little practical difference between culling and hunting in this respect because winter season hunting, while different in terms of primary motivation, likewise reduces the number of wildlife and therefore benefits the wider community. From this perspective, all wild animals are potential pests and hunting represents, in effect, a kind of precautionary cull. Hunting may not nowadays be a significant livelihood activity in its own right, but it may be viewed as a vital or necessary support to other village livelihoods – that is, those based on growing commercially valuable plants (for food or timber) that are vulnerable to wildlife depredations. In other words, the rationale of hunting in upland Japan is protective rather than productive. As a number of Japanese anthropologists put it, hunting in Japan has largely taken the form of 'defensive hunting' (*bōgyoteki shuryō*) in which farmers hunt to protect their crops from wild animals, rather than 'offensive hunting' (*kōgekiteki shuryō*) in which they hunt to obtain animal products (Taguchi 1992: 229; Hagihara 1996: 196). There has long been a concern with keeping down the numbers of wild animals in upland areas of Japan.

The war idiom extends to this representation of hunting. The background here is longstanding characterization of the crop-raiding wild boar as a war enemy in upland areas of Japan. 'Since long ago there has been a desperate war between wild boar and peasants' (Mainichi Shinbunsha 1965: 27); 'the wild boar was [for the village farmer] the same as an enemy soldier [*tekihei*]' (Hayakawa 1982: 239); and the wild boar 'invades' (*shinnyū suru*) village farmland (Sutō 1991: 152). The farmer's conflict with the wild boar is a 'battle' or 'war' (*tatakai*), the animals themselves form a 'wild boar army' (*inoshishi gundan*) and the measures and tools deployed against them are so many 'weapons' (*heiki*) in the war. The wild boar 'war' appears to have worsened in the era of large scale rural depopulation, as depopulated villages prove to be that much easier for wild boar to raid. The decline in human numbers, especially of the young, means that human resistance to boar encroachment diminishes. Sutō Isao cites a 1969 newspaper report from Nagano Prefecture which describes the effect of depopulation as having reduced the number of village 'soldiers resisting the wild boar' (*inoshishi ni taikō suru senshi*) (Sutō 1991: 163). The new boldness of the wild boar is reflected in the headline of the newspaper article that reads 'Arrogant Wild Boars – Beasts Rule' (*ibaridasu inoshishi – kemono no jōi*) (ibid.). The role of the hunter as a defender of the village against harmful wild animals from the forest becomes even more important in the present-day era of rural depopulation.

The link between hunting and pest control is enshrined in Japanese hunting law. According to the 1918 Wildlife Protection and Hunting Law, hunting is allowed for 'special purposes' which 'contribute to "... the improvement of the human living environment", and "... the promotion of agriculture, forestry, and fisheries"' (Mano and Moll 1999: 128). In contrast to Western law, 'Japanese law shows little concern for wildlife as a renewable natural resource' (ibid.: 129). 'It can be considered that pest control ... has been the dominant concept guiding wildlife administration systems this century' (Hazumi 1999: 210). In Japanese law, therefore, wildlife appears not as a positive resource but as a danger which potentially threatens human livelihoods. In the absence of a strong hunting tradition that accorded wildlife a positive value as game animals, wildlife has been viewed negatively. Although in the modern period wild animals were actively hunted for their economic value, this de facto status as a resource was at odds with their *de jure* status as a pest. This association of hunting with pest control is further evident among hunters themselves in the way they represent and justify hunting. Hunters often characterize their hunting by using the term *mabiki* or 'thinning'. *Mabiki* originally referred to the removal of inferior rice stalks from the paddyfield, and was used for infanticide in Japan. The use of the term *mabiki* in relation to wildlife would appear to be a clear expression of the way this agrarian logic spills over into the space beyond the farm – how the forest must necessarily become a site of some degree of human control and management.

Hunting as danger

While hunters themselves may claim to be defenders of the village from the danger posed by wildlife, in wider society they themselves may be seen as a danger. This perception of hunting as dangerous centres on the guns used by hunters. Japanese citizens are not legally entitled to hold guns, except for the purpose of hunting. In the largely gun-free society of Japan there is a fear that hunting weapons could be obtained and used by criminals. Hunting shotguns are sometimes used in armed bank robberies, and occasionally hunters themselves have been involved in such crimes (for example, Satō M. 1979: 111). When they occur, such incidents only reinforce the negative image of hunters widely held in Japanese society. But it is not just the fear of shotguns falling into the wrong (criminal) hands that causes anxiety; shotgun possession on the part of the hunter may in itself be a cause for concern. For the most part, the physical danger posed by hunting is inadvertent rather than malicious in character. During the winter, newspapers regularly report hunting accidents involving human injury or death. Often the victim is one of the members of the hunting group who, in the darkness of the forest, gets mistaken for a game animal and shot. Guns sometimes go off mistakenly, for example when a man with a loaded gun slips on the mountainside and pulls the trigger, hitting the man behind him, or indeed killing himself (*Asahi Shinbun* 18/11/1994). Hunters' shot or bullets have hit passing cars or have been fired near housing estates and other populated areas (ibid. 20/11/1991). Other victims include forestry labourers and village housewives out collecting forest plants (ibid. 6/12/1993). The ageing of rural hunters is seen as contributing to the spate of hunting accidents in recent years and has only reinforced this perception of hunting as dangerous.

The danger of hunting is not confined to the forest but extends to the village in the form of the hounds. Sometimes the hounds, in pursuit of a wild animal on the edge of the forest, get the scent of livestock animals in the village or attack village pets. But hounds can also pose a threat to people and have attacked forestry labourers, herb-gatherers, hikers, and even villagers in their fields. In one much publicized incident in Shimane Prefecture, a 94-year-old woman tending her vegetable patch was attacked and killed by a hound (*Mainichi Shinbun* 18/12/ 1995). Boar-hounds have also attacked children in a number of highly publicized incidents, sometimes with fatal consequences.[2] Although rare, when such attacks do occur they attract enormous publicity. One recent incident in the news occurred in February 1998 when two hounds attacked a nine-year-old boy in Kagoshima Prefecture, causing serious injuries for which he had to have 130 stitches (MNSC 10/2/1998). This led the headmaster of the boy's school to publicly complain about the presence of hunters in the area: 'Why do we have to live in fear of gunshots and hounds?' (ibid.).

Another problem with hounds arises when hunters abandon them in the mountains. Hounds are discarded either because they fail to reach the required hunting standard or because they are past their prime. Japanese hunters are generally reluctant to kill their hounds themselves or to have them put down.[3] Consequently, there are large numbers of dogs in the mountains that can pose a danger to forestry labourers, forest gatherers and hikers. As a threat to human safety, feral and stray dog populations are the object of culling initiatives. But it is also the case that hunters can unwittingly make wild animals more dangerous. Wild boar chased by hunters and hounds have been known to attack forest hikers (for example, *Asahi Shinbun* 18/11/1991) and a wounded wild boar is especially dangerous and prone to attack passers-by (ibid. 16/2/1994).

In addition to the *physical* danger it may pose, hunting in Japan is associated with what might be called *spiritual* danger. This is indicated by one of the Japanese words for hunting – *sesshō*. *Sesshō* is written with two Chinese characters, 殺 for 'kill' and 生 for 'life', which together gives the meaning of 'killing' or 'life-taking'. Similarly, the word for hunter in Japanese is *sesshōnin* – a noun used more by earlier generations of hunters than today – which gives the meaning of 'killer' or 'life-taker'. *Sesshō* has strong negative religious connotations, as Buddhism expressly forbids the taking of life. *Fusesshōkai*, the precept of not taking life, is a central precept and refers to hunting as well as other kinds of killing such as the slaughter of domesticated animals and the taking of human life. Many Mahayana Buddhist texts condemn meat-eating and hunting, warning of the terrible consequences of such actions. Applied to hunting, the term *sesshō* foregrounds the act of killing. *Sesshō* is a relationship between two beings in which the life of one is taken by the other. The force of the term is to stress that hunting (or any other killing) can never be a discrete activity; life-taking is not a final outcome or an end-state because it, in turn, provokes a reaction on the part of the being (or its spirit) whose life has been taken. As an act, *sesshō* is also the start of further acts, the next phase of the relationship. This vengeful reaction of the spirit, known as *tatari*, or spirit retribution, to the hunter's violence takes a variety of forms, including illness, death and loss of wealth on the part of the hunter or other family member.

In a folk legend (from Saga Prefecture), a boar-hunter approaching his hundredth kill is repeatedly told by his mother to stop his *sesshō* for fear of the consequences, but he ignores her. He duly goes on to catch his hundredth victim, but when he goes over to look at the carcass of the felled animal he discovers that it is his own mother wearing a wild boar skin (Chiba 1995: 141–2). The mother's warning about the danger of hunting had become tragically and poignantly true! In 1995 I spoke to a hunter's wife in Hongū who had long opposed her husband's '*sesshō*' and took ritual precautions to protect the family. Whenever her husband was successful in the mountains, she would soon afterwards go to the temple to perform a *kuyō* or 'requiem' for the spirit of the animal killed. But in early 1997 I learnt that the wife had finally managed to persuade her husband to give up hunting – 'I have given up *sesshō*' (*sesshō o yameta*), was how he expressed his decision, in the manner of one who was repenting for past wrongdoing. A recent cinematic representation of the wild boar's curse appears in the 1997 hit animation film *Mononoke Hime* (*The Princess Mononoke*) by Miyazaki Hayao. The film begins with a scene in which a wild boar attacks a village and is slain by the hero – but who in the process incurs the animal's *tatari*. The rest of the film consists of the hero's quest to lift the wild boar's life-threatening curse.[4]

Japanese hunters do not necessarily accept the characterization of their hunting as *sesshō*. The folklorist Chiba Tokuji gives the example of a seventeenth-century hunter who found himself being condemned by a Buddhist priest for killing many wild boar and warned by the priest that he risked descending to hell in his afterlife. The hunter responded by claiming that the wild boars he catches are the reincarnated form assumed by 'people who do not attain enlightenment', who, forced to live as lowly wild boars, lead lives of great misery and suffering. By hunting these wretched boar-people, the hunter claims that he is putting an end to their lives of suffering. The hunter, in other words, sends the (human) soul of his animal prey to the Buddhist paradise and is therefore, despite the violence, engaged in an act of compassion (Chiba 1995: 130–1)! Present-day hunters too may well reject the moral condemnation of hunting as *sesshō*. In the hunting magazine *Zenryō*, for example, a boar-hunter from Yamanashi Prefecture argues that hunting makes hunters appreciate the value of life all the more.

> Although it is called *sesshō*, it [hunting] involves learning, through action, about the value and transience of the life of an animal that cannot speak. In my case, I consider hunting to be something that does not simply end with the taking of the animal's life.
>
> (Suzuki 1994: 127)

Japanese hunters can neutralize the threat of *sesshō* through ritual actions. *Ryōyūkai* hold annual requiems for the spirits of the game animals they have killed over the hunting season. Although this is usually a Buddhist observance known as a *kuyō*, Shinto priests carry out a similar rite known as an *ireisai*. At a special memorial stone, offerings of food (fruit, vegetables and ricecakes) and *sake* are made, and hunters 'console' (*nagusameru*) and express 'gratitude' (*kansha*) to the spirits of the slain animals. Hunters state that the short, simple ceremony is a way of saying *mōshiwakenai* ('there is no excuse', 'sorry') to their prey. The formal purpose of the rite is to expedite the posthumous wellbeing of the spirits and their attainment of Buddha status (*jōbutsu*). The rite can be seen as a pacification measure that transforms the spirit from a state of restless suffering, highly dangerous to those people (the hunters) responsible for causing it, to a state of repose. In fact, only a minority of Hongū hunters attends this annual rite. In 1991, for example, just one in five members of the Hongū *ryōyūkai* (eighteen people out of the then membership of around eighty) attended the memorial. Some hunters state openly that it is a waste of money (because a priest must be paid to carry it out). Indeed, a common reaction among hunters to the mention of *kuyō* is to dismiss any concerns about *tatari* as indications of a faint-heartedness that should not exist in the true hunter.

The definition of hunting as a 'sport' also has the effect of discursively negating the life-taking character of hunting. Present-day hunters seldom, if ever, use the term *sesshō*, opting instead to characterize hunting as a 'sport' (*supōtsu*). The English word 'hunter' or *hantā* is today commonly used for hunter and this word tends to suggest that hunting is a leisure pursuit rather than a livelihood activity. Far removed from the unpleasant, disturbing connotations of *sesshōnin*, the word *hantā* feels modern and Western and provides a new linguistic wrapping for an old activity. Arguably, the sportive *hantā* is less directly subject to Buddhistic censure than is the *sesshōnin*. But the respective moral statuses of the modern *hantā* and the traditional *sesshōnin* are not as clear-cut as this might suggest and are potentially subject to a radical moral re-evaluation. This is because the *sesshōnin*, though nominally a 'killer', at least contributed to his family's livelihood, whereas the dissociation of the *hantā*'s hunting from livelihood can have the reverse effect of making recreational hunting more problematic than *sesshō* hunting. The *hantā* who kills for enjoyment is potentially in a more morally parlous state than the earlier *sesshōnin* who killed for necessity.

This is, to some extent, supported by the fact that even younger hunters complain that when injuries or other misfortunes occur during the hunt the likely public reaction to their misfortunes is gossip about *tatari*. In Hongū critics of hunting characterize it with the word *asobi*, literally meaning 'play'; I have often heard non-hunters describe hunters as 'idlers' (*himajin*) who opt to waste their time in the mountains rather than spend their time productively tending family forest landholdings, finding extra waged work, or taking up an improving pastime. Moreover, for some critics the fact that hunters actually derive enjoyment from chasing and killing animals makes it a specially disturbing form of *asobi*. Given the existence of this kind of criticism of hunting among mountain villagers, it is easy to see why, when hunters do get injured or worse in hunting accidents, public sympathy for them can be somewhat muted, and why some of their neighbours may well infer that, to a large extent, they brought it on themselves.

Conclusion

This chapter has outlined three representations of hunting in Japan: hunting as contest, hunting as protection and hunting as danger. There is considerable tension between these different representations. First, there is a tension between hunting as contest and hunting as protection. The language of 'war' is applied both to the hunter's engagement with the wild boar and to the wild boar's threat to village farmland, and is therefore shared to some extent by hunter and farmer. However, this apparent unity of farmer and hunter is ultimately misleading, as the idiom of war obscures very different understandings of its significance and purpose. Although hunters claim to be catching harmful animals and therefore reducing their numbers, they are also concerned to sustain game animal populations and may even make efforts to increase animal numbers. Hunters on the Kii Peninsula reportedly release boar–pig hybrids (known as *inobuta*) into the forest in an effort to increase game numbers for the coming hunting season. At this point, the interests of the hunter and the farmer cease to coincide. If the Japanese hunter is at 'war' with the wild boar, this is a 'war' that should recur in each hunting season. By contrast, the farmer's 'war' should be a transient conflict that results in the removal of the 'enemy' and therefore lead to its own termination.

Second, there is a clear tension between the claim that hunting is a source of protection of the village and the perception, especially among non-hunters in upland areas, that hunting is a threat to human safety. In other words, hunting raises serious concerns about the danger of misdirected violence, violence not against the wild boar 'enemy', but against other parts of rural society! Given the motif of hunter identification with his prey (particularly marked in the case of the wild boar), this foregrounding of the hunter as a violent danger to the wider society is not without irony. The perception of the hunter as a dangerous, violent figure overlaps, to some extent, with the hunter's own self-image *vis-à-vis* the wild boar. But the violence of the hunt can also pose a more indirect threat to the Japanese hunter by rendering him morally vulnerable to a kind of retribution. In Japan, the violence of hunting is extended into a secondary phase of reverse predation whereby the spirit of the slain animal can exact revenge on the hunter and those close to him.

Notes

1 This recalls a claim reported by the folklorist Hayakawa Kōtarō that the Japanese national flag, the rising sun (*hinomaru*, literally 'round sun'), derives its imagery from boar-hunting. This is because the hunters' offering of the wild boar's heart (or other organ) to the mountain spirit, in recognition of the spirit's benevolence in granting them the animal, was originally made by placing the bloody heart on a white material, creating the flag image of a red circle against a white background, which hunters then display as a banner as they make a 'pledge' (*chikai*) to the mountain spirit (Hayakawa 1982: 242–3).
2 See, for example, Kurita (1999: 87), *Mainichi Shinbun* (18/11/1997), and *Asahi Shinbun* (13/12/1985).

3 In Japan, there are many tales of dog-curses incurred by dog-killers (for example, Iwasaka and Toelken 1994: 91; Nishiura 1989: 179). In the Hongū area too, there are stories of dog vengeance (WKMK 1981: 54, 100).

4 It turns out that the cause of the *tatari* is not really Ashitaka's defensive actions, but a wound earlier incurred by the wild boar which made it violent and vengeful. This earlier wound was caused when a more technologically advanced society in the west of Japan shot the wild boar with a primitive kind of gun. The climax of the film involves a war between forest animals against human society which leads to devastation. The ecological message of the film is that human society, through its violent actions towards the forest, is the cause of nature's *tatari* which threatens to destroy the world.

References

Cartmill, M. (1993) *A View to a Death in the Morning: Hunting and Nature through History.* Cambridge: Harvard University Press.

Chiba, T. (1995) *Ōkami wa naze kieta ka* [Why did the wolf disappear?]. Tokyo: Shinjinbutsu Ōraisha.

Environment Agency (1997) *Quality of the Environment in Japan – 1995.* Tokyo: Environment Agency, Government of Japan.

—— (2000) 'Shubetsu shuryō menjō kōfu jōkyō' [The different categories of hunting licences]. Available at: http://www.asahi-net.or.jp/~zb4h-kskr/licence.html [Accessed on 28 September, 2000].

Hagihara, S. (1996) 'Kurashi to dōbutsu' (Living with animals). In Akada M., Katsuki Y., Komatsu K., Nomoto K. and Fukuda A. (eds) *Kankyō no minzoku* [Folklore of the environment]. Tokyo: Yūzankaku, pp. 195–220.

Hayakawa, K. (1982) *Hayakawa Kōtarō zenshū 4* [Collected Works of Hayakawa Kōtarō, Volume 4]. Edited by T. Miyamoto and N. Miyata. Tokyo: Miraisha.

Hazumi, T. (1999) 'Status and management of the Asiatic black bear in Japan'. In C. Servheen, S. Herrero and B. Peyton (eds) *Bears.* Gland, Switzerland: IUCN, pp. 207–211.

Iwasaka, M. and B. Toelken (1994) *Ghosts and the Japanese: Cultural Experience in Japanese Death Legends.* Logan: Utah State University Press.

Kurita, Y. (1999) 'Inoshishi no wanaryō – 6' [Wild boar trapping – 6]. *Shuryōkai*, vol. 43, no. 2, pp. 86–90.

Mainichi Shinbunsha (1965) *Nihon no dōbutsuki* [Record of the animals of Japan]. Tokyo: Mainichi Shinbunsha.

Mano, T. and J. Moll (1999) 'Status and management of the Hokkaido brown bear in Japan'. In C. Servheen, S. Herrero and B. Peyton (eds) *Bears.* Gland, Switzerland: IUCN, pp. 128–131.

Marvin, G. (2000) 'The problem of foxes: legitimate and illegitimate killing in the English countryside'. In J. Knight (ed.) *Natural Enemies: People–wildlife Conflicts in Anthropological Perspective.* London: Routledge, pp. 189–211.

Nishiura, S. (1989) 'Tamba-miyama shuryō hiwa' [The secret hunting history of Tamba-miyama]. In Sanson Minzoku no Kai (ed.) *Shuryō (Hunting).* Tokyo: Enterprise, pp. 165–194.

Otake, T. (1987) 'Shishi ryōken to shite no nihonken' [Japanese dogs as boar-hounds]. In Aiken no Tomo (ed.) *Nihonken taikan* [A general survey of Japanese dogs]. Tokyo: Seibundō Shinkōsha, pp. 81–84.

Satō, M. (1979) 'Gyōkai rire jōhō' [Information from the trade]. *Shuryōkai*, vol. 23, no. 4, pp. 111.

Satō, Y. (1979) 'Fuyugomori chokuzen no kuma o karu gokai na ōgata haundo no ryōmi' [The hunting style of exciting large hounds used in bearhunting prior to winter denning]. *Shuryōkai* vol. 23, no. 3, pp. 109–111.

Song, S. H. (2000) 'The Great Pigeon Massacre in a deindustrializing American region'. In J. Knight (ed.) *Natural Enemies: People–wildlife Conflicts in Anthropological Perspective*. London: Routledge, pp. 212–228.

Sutō, I. (1991) *Yama no hyōteki – inoshishi to yamabito no seikatsushi* [The mountain landmark: a record of the way of life of wild boars and mountain people]. Tokyo: Miraisha.

Suzuki, T. (1994) 'Ōmonoryō shotaikenki' [Record of my first big-game hunt]. *Zenryō*, vol. 59, no. 8, pp. 124–127.

Taguchi, H. (1992) *Echigo Miomote yamando ki – Matagi no shizenkan ni narau* [Record of the mountain people of Echigo Miomote: learning from the Matagi view of nature]. Tokyo: Nōsangyoson Bunka Kyōkai.

WKMK (Wakayama-ken Minwa no Kai) (ed.) (1981) *Kumano-Hongū no minwa* [Folktales of Kumano-Hongū]. Gobō: Wakayama-ken Minwa no Kai.

WKRNS (Wakayama-ken Ryōyūkai Nishimurō Shibu) (1979) 'Nanbyō ga chōjū no myōyaku de naotta hanashi' [Tales of wildlife cures for serious illnesses]. *Shuryōkai*, vol. 23, no. 4, pp. 60–62.

Weir, J. (1992) 'The Sweetwater rattlesnake round-up: a case study in environmental ethics'. *Conservation Biology*, vol. 6, no. 1, pp. 116–127.

Yamamoto, H. (1995) 'Inoshishi no kukuriwanaryō' [Wild boar wire-trapping]. *Shuryōkai*, vol. 39, no. 10, pp. 62–65.

Newspapers

Asahi Shinbun (13/12/1985) 'Ryōken ni osowareta shōnen shinu' [Child dies attacked by hunting dog].

—— (18/11/1991) 'Tōzan no josei o inoshishi osou' [Woman hiker attacked by wild boar].

—— (20/11/1991) 'Shuryō chiiki datta jūmin shiteki de kinshi, ichiji wa minka ni nagaretama' [A stray bullet near houses in a former hunting ground closed by residents' pressure]. Ōsaka.

—— (6/12/1993) 'Shika to machigai happō, Hyōgo de sasatori no shufushibō' [Mistaken for a deer and shot, the death of a Hyōgo housewife out collecting bamboo grass].

—— (16/2/1994) 'Teoi no inoshishi ni kamare, shufu kega' [Bitten by a wounded wild boar, housewife injured]. Yamanashi.

—— (18/11/1994) 'Tentō shi sandanju bōhatsu' [A fall and a misfired shotgun]. Gunma.

—— (19/1/1996) 'Nara-Yoshino de taijū 148 kiro no oinoshishi shitomeru' [Catching a giant boar of 148 kg in Yoshino, Nara]. Ōsaka.

MNSC (*Minami Nihon Shinbun Chōkan*) (10/2/1998) 'Oguchi-shi no tsūgakuro de rōken ga jidō o shūgeki' [Assault on child travelling to school by a hound in Oguchi City].

Mainichi Shinbun (18/12/1995) 'Ryōken santo ni osoware, 94 sai no josei ga shibō' [Death of a 94-year-old woman, attacked by three hounds]. Shimane.

—— (18/11/1997) 'Inoshishigari no ryōken santo, gekōchū no jidō osou' [Three boarhounds attack children returning from school]. Fukui.

4

JAPANESE PERCEPTIONS OF WHALES AND DOLPHINS

Arne Kalland

Introduction

In the article 'Why the Japanese are so stubborn about whaling', two American environmentalists residing in Japan suggest that the 'problem' of Japanese whaling can be solved by educating the Japanese: 'Increased education about marine mammals ... will allow Japanese traditions to be altered in positive ways from the inside' (Glass and Englund 1989). However, Japanese attitudes toward whales are more complex than these two environmentalists seem to believe. First, the Japanese do not have less *factual* knowledge about whales than people in Australia, England, Germany and the United States (Freeman and Kellert 1994).[1] What is at issue are different kinds of knowledge, not the amount of knowledge per se. Second, new attitudes to whales do not necessarily replace old ones. As pointed out in a recent publication, Japanese perceptions of nature are not static but are continuously changing with new dimensions or interpretations being added to, rather than replacing, old ones (Kalland and Asquith 1997: 7). This also applies to peoples' perceptions of whales and dolphins. Third, there is not necessarily a contradiction between 'loving' animals and killing them for food. In contrast to the 'no-touch' approach to nature held by many environmentalists in the West, Japanese recognize that it is the nature of things that one organism feeds upon another, creating relations of indebtedness in the process (Kalland 1995a: 246–247).

In this chapter I shall present an outline of Japanese knowledge of whales and indicate the important place of whales in Japanese culture. In the second half of the chapter I shall consider the extent to which this 'traditional' understanding of people–whale relations is changing in the present-day, in the light of the emergence of whale-watching and other forms of cetacean-related tourism.

The 'traditional' relationship

Whales as a natural resource and cherished food

Most Japanese are quick to point out that Japan is a country poor in natural resources and that from early on they had to learn to make the most out of their scarce resources. However, Japan is endowed with rich fishing grounds and marine products that have throughout its history been the most important source of animal protein. For the Japanese, whales have come to symbolize this dependency on the sea. A beached whale was said to bring prosperity to seven villages. A Japanese NGO campaigning for the resumption of whaling in fact calls itself 'Riches of the Sea'. Whales are seen as giving themselves to the hunters, but in return people are morally obliged to make the fullest possible use of the carcasses. Whale oil, which had long been used for oil-lighting, became in the Tokugawa period (1603–1868) an important insecticide used in the ricefields and was thus a significant factor in averting famines (Kalland 1995b: 196). The bones and some of the entrails were used as fertilizer, and there developed an important market for such products. Baleen, whale teeth, jaw bones and sinews became raw materials for various handicraft products. But most important of all, whale meat – including blubber, skin, cartilage, flukes, intestines and genitals – has for centuries been used as food. Japanese people often contrast this near total utilization of the whales with Westerners who 'only take the blubber and throw the meat away' (*abura dake totte niku wa suteru*).[2] This claim seeks not only to legitimize Japanese catches of whales, but also to place the Japanese morally above Westerners, thus contributing to the popular mythology of Japanese uniqueness known as *nihonjinron*. It is a significant ingredient in the Japanese use of nature as ideology (Kalland and Asquith 1997).

The importance of whale meat as food has given rise to a rich culinary tradition. The quality of the meat is finely graded and the various parts are regarded as suitable for different dishes. Whale meat recipes have been included in Japanese cookery books since at least 1489 when it was mentioned as superior food. In 1832 a special whale cookery book, *Geiniku chōmihō*, was published in Hirado, north of Nagasaki, and this divided the whale into no less than seventy named parts, each described in terms of taste and method of cooking. According to this source, roasted red meat (*akami*) can 'taste better than geese and ducks' and *unagi* (the outer side of the upper gums near the baleen) is tender and has a 'noble' taste, whereas the trachea (*nodowami*) is 'given to servants in the countryside' and the duodenum (*akawata*) is eaten by the poor. What part of the whale people ate thus signaled their social position in the community and therefore carried important symbolic significance. In 1989, the only whale meat wholesaler in Arikawa, not far from Hirado, still dealt in sixty items of whale meat, although he had run out of stocks for twelve of these (Kalland 1989: 111).

Regional food preferences have emerged as a result of the history of whaling in particular communities. Such preferences exist both in terms of the species

Figure 4.1 Dolphins landed at a fishing port in Japan

eaten and in the method of cooking. In Arikawa, the most cherished meat in the past was that of right whale, but because this meat is no longer available, the salted blubber of fin whale has become a new favourite. Dried, salted dolphin meat is also regarded as a delicacy, while on Iki Island in the same prefecture dolphin meat is regarded as non-food. In Taiji, on the Kii Peninsula in western Japan, people have developed a special liking for pilot whale, which is often eaten raw as *sashimi*. In Wadaura outside Tokyo, one of the local specialities is dried, marinated slices of Baird's beaked whale (*tare*), while raw red meat (*sashimi*) of minke whales is preferred in Ayukawa and Abashiri in the north. A special New Year's dish in Abashiri is soup boiled with salted blubber. The meat of the sperm whale is the preferred food in some areas of northeastern Japan, whereas in Arikawa it is regarded as edible only in dried form or as fish-pastes (*kamaboko*).

Few things are as symbolically laden as food, and local cuisine is one of the strongest markers of social identity in Japan. The various ways to prepare meat have become important social markers and whalers from different communities never seem to grow tired of discussing local whale cuisine. Beside being a staple – even now some people in these communities will try to have a small piece of whale meat daily 'just to get a taste of it' – whale meat was an indispensable part of all types of community gatherings and celebrations. It used to be extensively served at weddings, funerals, memorials for the ancestors, as well as to celebrate the building of a new house, a child's first day at school, and so on (see Braund *et al.* 1990 for a complete list of such occasions in Ayukawa). It is often the typical food for New Year, and in Arikawa about a fifth of the annual sale of whale products occurs during that season. The other peaks in consumption occur in August, which is the month when Bon (All Souls' Day) is celebrated, and in March/April when people in Arikawa celebrate Girls' Day (*osekku*) in combination with flower-viewing (*hanami*). Special three-layered lunch boxes (*jubako*) are prepared with cherished food: rice on the lower layer, dishes of whale and dolphin in the middle and cakes or snacks on the top layer (Kalland 1989: 112–113).

Taking the term from de Garine (1972), Manderson and Akatsu classify whale meat in Ayukawa as 'super food' due to its double role as being both 'highly valued culturally and a staple' (1993: 210). Moreover, food often figures as 'special products' (*meibutsu*) of localities and whale meat serves that purpose in whaling communities. People will travel to whaling communities in order to eat the 'special product', and whale meat is thus a tourist attraction. Whale and dolphin meat has an important place in local identities. This aspect of cultural diversity is now under threat as the current moratorium on commercial whaling has restricted catches to pilot and Baird's beaked whales and dolphins. No wonder then, that the question of food culture has been one of the main arguments used by the Japanese government for whaling, and a number of the papers presented to the International Whaling Commission (IWC) since 1986 centre on this theme (for example, Braund *et al.* 1990; GOJ 1991a, 1991b, 1992; Ashkenazi 1992).[3]

Three developments stimulated whale consumption at the national level: the use of canned whale meat by the military in the pre-war period; the food shortage during the early post-war years; and the use of whale meat in school lunches. In the post-war years, whale meat accounted for 47 per cent of the animal protein intake of the Japanese people. As a result, many people are convinced that whales saved them from a major famine. Some of the attachment to whales and whale meat in Japan possibly stems from this belief. The use of whale meat in school lunches has also made a lasting impression on many people.

A fourth development that made whale meat an important part of the *national* cuisine was the whaling moratorium imposed in 1987. The anti-whaling campaigns have turned whale meat into a symbol for Japanese culture and eating whale meat has acquired a new meaning: it has become a ritual act through which the partakers express their belonging to the Japanese tribe, not only to a local community as before. Eating whale meat sets the Japanese apart from others; that some other peoples do eat it is simply ignored in this context, as are surveys showing that about 40 per cent of the Japanese do *not* regard whale meat as an acceptable food item (Freeman and Kellert 1994). The Japanese thus become unique, and the whaling issue serves to strengthen the much cherished Japanese myths about national identity.

Whales as fellow beings

Whales have for centuries been hunted and eaten by the Japanese, but whaling has not been conducted in a moral or ethical void. On the contrary, hunting creates – as the Japanese see it – a strong bond of interdependence between man and animal. There is a common belief in Japan that people, animals, plants and even inorganic objects have immortal 'souls', or some innate power, a world view which stresses the interdependence of supernatural, human and animal worlds. Buddhist doctrine takes a holistic view and refers a 'Buddha-nature' in all things, while Shinto refers to *kami* – a supernatural power that resides in anything which gives a person a feeling of awe. Whales are awesome to many people, including the Japanese. According to a widely held belief in Japanese coastal villages, whales are manifestations of Ebisu, the patron deity of fishing. Often Ebisu disguises himself in this way when on festival days he approaches shrines to pray (Sakurada 1973), and in some districts whales are worshipped as 'Ebisu' (Ogura 1973: 143). This belief may combine with another widely held belief that visiting strangers may bring fortunes (Yoshida 1981). The visiting Ebisu may bring riches from the sea in two ways, either directly in the form of a whale carcass or indirectly by chasing fish toward the shore (ibid.).[4]

Whaling activities are intimately bound up with religious beliefs. Whales are gifts from nature and as such they have to be utilized to the fullest. To do otherwise would be an insult to the whale which is believed to have given itself up to human beings so that they can live. To repay the whales for sacrificing their lives, the whalers moreover have to take care of their souls and failure to do this may lead to the souls of whales turning into 'hungry ghosts' which can cause illness, accidents and other misfortunes. It has, therefore, been the practice in many whaling communities to treat the souls of whales similarly to those of deceased human beings. Whales were given posthumous names (*kaimyō*) inscribed on wooden memorial tablets (*ihai*) and included in temples' death registers (*kakochō*). Tombs and memorial stones can be found in at least 48 places, from Hokkaido in the north to Kyushu in the south, and annually at least 25 festivals (*matsuri*), as well as memorial rites (*kuyō*), are held in honour of whales (Akimichi *et al.* 1988: 55–56). A tomb at Koganji (an old temple dedicated to whales in Yamaguchi Prefecture) has been designated a national historical monument. Built in 1962, the tomb marks the burial of 75 foetuses from whales caught before 1868.[5] Every year in late April, the temple is the stage for elaborate memorial ceremonies with Buddhist priests reciting sutras for several days in order to help the souls of deceased whales get reborn in a higher existence. Such services have a number of meanings. The temple priest may perform the memorial service in the belief that the whale will reach enlightenment and thus be released from rebirth into this world and enter Paradise as a 'buddha' (*hotoke*). Some villagers may believe that the whale will be reborn as a human being or as another whale to be hunted. Finally, memorial services are held to ensure that the whalers, and the gunners in particular, are forgiven (*tsukunaru*) for the sin involved in taking life. Memorial services therefore carry special meaning for the gunners who frequently go directly to the temple upon returning home in order to conduct memorial services for the whales they have killed.

Figure 4.2 A whale monument in Taiji, Japan

A number of lesser ceremonies and rituals are performed to repay the whales for their personal sacrifices and thus secure safe voyages and rich catches in the future. The whaling companies used to gather the whalers and their wives for a joint ceremony before the commencement of the season, as well as afterwards, and in some places the wives would go on a pilgrimage to local shrines. Daily religious observances are conducted in front of the family Shinto altar (*kamidana*), praying for the husband's safety and good catches. Similar rituals are performed on the boats, where a piece of the whale's tail may be offered to the Shinto altar. Many rituals serve to connect whalers to each other, to their families and to the whales, thus giving local residents a feeling of common heritage and meaning in their lives. This common cultural heritage is expressed and reinforced in festivals, songs and dances, through which the present is linked with the past. Rituals give the community its distinct character: the set of Shinto deities is unique to each community and the festivals are different as well. But they are all variations on common themes based upon a conception of the whale as a creature with an immortal soul and a world view stressing the interdependence of supernatural, human and animal worlds.

Successes as well as failures are explained in relation to the divine. Accidents may be caused by failure to repay the whale's sacrifice through ritual neglect, but can also be caused by breaking taboos. There are several stories about the malevolent spirits of whales, some of them known throughout Japan. The most famous is the disaster that struck Taiji in 1878 when whalers broke an old taboo and attacked a right whale with calf. More than 100 whalers lost their lives in the following gale (Taiji 1982; Takahashi 1992: 73–74). A similar story is told in Ukushima where, in 1715, a whale appeared in the dream of Yamada Monkurō, the owner of the whaling net, begging not to be hunted until she and her calf had completed their worship at the famous Daihōji (*lit.* 'Great Treasure Temple') in Gotō. Nonetheless, having failed to catch a whale for a long time, the whalers could not restrain themselves and attacked a mother and calf the following day. During the fight, weather conditions changed for the worse and 72 whalers were lost at sea (Kalland and Moeran 1992: 150–1). These accidents, which have become part of the communities' cultural heritage and peculiar identities, have reinforced the validity of the taboo, a taboo which, incidentally, might carry some value in the conservation of whale stocks.

In Japan whales are not 'simply killed and eaten', as claimed by Hoyt (1993: 15). The 'traditional' Japanese relationship with whales is complex. Whales are seen as nourishing food, yet at the same time they are treated with a compassion almost like that shown to one's own ancestors. But at no time have whales been seen as unique. True, they are larger and more awesome than other animals and whale rituals tend to be more elaborate, but this is a question of quantity rather than quality. In temples throughout Japan memorial rites are performed not only for whales but for any life form killed by human beings, even for tools and other objects used and discharged (Asquith 1986, 1990; Reader 1991). Attitudes to whales are therefore firmly rooted in Japanese culture.

A new relationship between whales and people?

In periods of her history, Japan has been eager to import ideas and values from abroad, first from China and Korea, later from Europe and the United States. Today, Japan plays an active role in the new 'global' culture both as a producer and a consumer. This is also true regarding whales and whaling. The Japanese population is not unaware of, or immune to, new perceptions of marine mammals, especially those which have emerged in the Anglo-Saxon world. At the same time, the Japanese authorities, as well as the whaling industry, have embraced the 'global' discourse on 'cultural diversity', arguing that the international campaign against whaling is a case of cultural imperialism which is detrimental to true internationalism.

The Japanese are exposed to Anglo-Saxon images of whales and dolphins in a number of ways, some subtle, others more forceful and even violent. The mass media in Japan closely follows the whaling issue in general and IWC meetings in particular, with frequent comments on and interpretations of Western anti-whaling views. Television programmes, magazine articles and picture books have focused on the aesthetic qualities of whales and dolphins. All over the country, dolphinaria have appeared, where audiences are exposed to dolphins' alleged intelligence and playfulness. Whale-watching has gained considerable popularity since it was first introduced in 1988 and regular tours are conducted from

Okinawa to Hokkaido, with the most successful operations located in Ogasawara (Bonin Islands) and in Kōchi Prefecture (Shikoku).

Western environmental and animal rights organizations, such as World Wide Fund for Nature (WWF), Greenpeace and International Fund for Animal Welfare (IFAW), have established branches in Japan and have at times entertained close relations with local NGOs, such as Elsa Nature Conservancy, the Dolphin and Whale Action Network and Japan Whale Conservation Network. They cooperate during IWC meetings and Japanese organizations have collaborated with foreigners in their attempts to uncover 'illegal' whale meat on the Japanese market, thereby calling into question whether controlled, sustainable whaling is possible at all.

Finally, the Japanese have been witnesses to direct actions performed by foreigners – sometimes with Japanese assistance. Among the early activities that attracted international attention were attempts to stop dolphin drives, which for centuries have been carried out by some Japanese communities. In 1980, an American activist cut the nets which held a school of dolphins captive at Katsumoto, Iki Island. More recently, Greenpeace has tried to interrupt the scientific whaling undertaken by Japan in the Antarctic. On several occasions inflatable boats have positioned themselves between the catcher boats and the whales, and activists have furthermore tried to prevent minke whales being winched up the slipway of the mother ship. Less violent, but probably more important, are the many boycott threats. Although consumer boycotts are unlikely to have much effect,[6] exporters are concerned particularly about a possible American use of the Pelly Amendment to the Fishery Protective Act, which authorizes the US President to prohibit imports from a nation that diminishes the effectiveness of international wildlife conservation programmes.[7]

The Japanese, then, are exposed to a number of images. At one level, whales and dolphins are portrayed as intelligent, playful and caring. Whaling is seen as barbaric and uncivilized, while eating whale meat appears close to cannibalism. At another level, whaling is seen as unnecessary and impossible to control; Japan is accused of hunting protected species and illegally importing whale meat. Finally, an image is drawn of Japan as isolated in the world. Many Japanese perceive their country as under pressure from the international community (Misaki 1993). In a culture that strongly emphasizes and values consensus, such 'external pressure' (*gaiatsu*) is a source of much concern among the Japanese.

How far have these campaigns been successful in changing people's perceptions and attitudes to whales and whaling? Apparently they have had some success. Images of 'cute' whales and dolphins are widespread: at village entrances, bridges, public offices, manholes, parks and playgrounds, *pachinko* parlours, hotel towels, advertisements and so on. In Ayukawa's new 'Whaleland', visitors are invited to listen to 'the messages of the whales' and to imagine themselves 'as though they were whales'.[8] In 1992 the number of Japanese who went whale-watching was close to 20,000 (Hoyt 1993: 16) and continues to increase. Several associations (for example, Geisharen, Whalco, Whale- and

Figure 4.3 The post office in Ayukawa, Japan

Dolphin-Watching Fun Club) have been established to promote whale-watching, and in 1997 there were seventeen registered operators – a figure seemingly on the increase – in eight different prefectures. The revenues are difficult to estimate but local authorities and tourist boards have in some cases supported these operations. Whale-watching has become an important source of income for local people at Ogawasara Island south of Tokyo and in Ōkata-chō in Kōchi Prefecture. Elsewhere, however, some of the operators are fishermen who opportunistically take on board tourists to augment their income from fisheries (Gomez Diaz, personal communication).

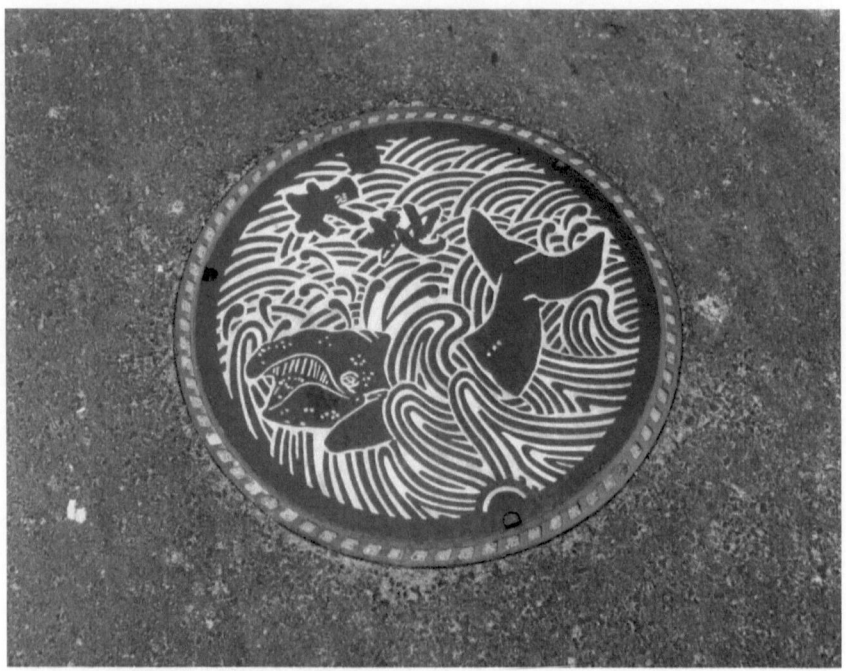

Figure 4.4 A manhole cover in Taiji, Japan

The motives for organizing whale-watching tours are diverse. For many people, whale-watching means business. Tourism is seen as the key industry of the future in many remote communities. But there are other motives at work. When environmental and animal rights organizations like WWF, Greenpeace and the Whale and Dolphin Conservation Society (WDCS) organize or help initiate whale-watching, they see this as an important tool in changing people's attitudes towards whales and dolphins. The 'educational' aspect of whale-watching – that is, to make people feel that whales are 'good' and should not be caught – is repeatedly stressed by these groups. Nature programmes and picture books are used in the same way.

Nevertheless, a survey conducted in 1992 shows that about two-thirds of Japanese support the catching of non-endangered whales for economic and cultural needs (Freeman and Kellert 1994: 309). This is confirmed by several opinion polls conducted by Japanese newspapers in connection with IWC's 1993 meeting in Kyoto. All show a significant majority in favour of whaling, ranging from 54 per cent (35 per cent against) in the *Asahi Shinbun* poll (15 March 1993) to 64 per cent (21 per cent against) in the *Nihon Keizai Shinbun* poll (17 May 1993) and 82 per cent (13 per cent against) in the *Nishi Nihon Shinbun* poll (4 April 1993). According to a more recent survey disclosed by the Prime Minister's Office in 1995, 77 per cent of the Japanese support the resumption of minke whaling (GOJ

1995). In other words, there is little evidence to justify the claim that Japanese attitudes to *whaling* have changed in a negative direction. Nor have people flocked to the anti-whaling organizations, which remain small. Moreover, most of the dolphins and whales adorning bridges, parks and streets are found in former whaling communities (such as Arikawa, Katsumoto, Taiji and Ayukawa). Far from being expressions of a new attitude to whales, they are seen as a means to foster the image of an evolving whaling culture (Takahashi 1987; Kalland and Moeran 1992). Whales and whaling are given new symbolic meanings, which have become important ingredients for the formation and maintenance of local identity. Communities like Ayukawa and Taiji are promoting themselves as 'whale towns' (*kujira-no-machi*) in their attempt to revitalize their communities.

I shall end this chapter with a case study from Katsumoto on Iki Island, in an attempt to illustrate the complexity of popular perceptions of nature in general and whales in particular. Having been the location for direct action in which fishing nets were destroyed in order to release dolphins,[9] Katsumoto is one of the Japanese communities with a first-hand experience of radical Western environmentalism. People were bewildered and had difficulty understanding in the rationale behind the action. However, the drives continued, albeit at a somewhat reduced scale due to dwindling numbers of dolphins appearing at the Iki fishing grounds. The problem became more manageable and a new 'solution' was found when it was discovered that a market existed for live animals. Smaller schools of dolphins were herded into bays where they were kept alive for later sale. Housewives were engaged by the Fishing Cooperative Association to feed the animals, and this seemed to open a new dimension in the human–dolphin relationship in Katsumoto. Several of the women went on to develop emotional ties to the captive dolphins. They wept when 'their' dolphins – ones they had nourished themselves – were sold and were extremely worried during the Gulf War about dolphins sold to Israel.

From selling live dolphins, it was only a short step to the 'Dolphin Park' (*iruka pāku*) concept. When I visited Katsumoto in 1992, the park was not yet completed, but the plan was to keep dolphins in an inlet so that spectators could observe them in their 'natural environment'. The animals were not to be trained and no shows were planned. Everything should be 'natural', that is, people should be allowed to see the dolphins behaving 'naturally' (*shizen ni*). But the visitors should also be allowed to feed and pet them. The park is the main vehicle by which the township is trying to promote itself as the 'Dolphin Town' (*iruka-no-machi*).

The Katsumoto experience highlights the gaps in perception between foreigners – mostly urban Anglo-Americans – and Japanese in two important respects. First, to many Westerners it sounds contradictory to keep dolphins in a concrete arena so that people can observe them in 'nature'. Nature to them stands for the 'wild' – that is, that part of the physical environment not manipulated by human beings. Westerners, including most of those who claim to have an ecocentric as opposed to an anthropocentric world view, see nature as something opposed to culture. There is no such opposition in Japanese culture.

For the Japanese, there is a continuum from raw nature in the wild to tamed nature which merges with culture (Kalland 1995a; Kalland and Asquith 1997). As long as the animals are not trained (or plants not trimmed), they remain closer to the 'raw nature' end of the continuum, whatever their physical surroundings might be, although captive dolphins, simply by being held in captivity, are more tame (cultured) than free-swimming animals. Whether nature is something to be loved or disliked depends on the context. Tamed nature – the tea ceremony, cherry blossoms, flower arrangements, Japanese gardens and trained dolphins – is always cherished. Raw nature might be feared and despised, but is also seen as a source of strength and spirituality.

This brings me to the second point: the compassion that the housewives obviously felt for 'their' dolphins. The dolphins were likened to domestic pets such as dogs. Like pet dogs, fenced-in dolphins came to be seen by the woman as 'cute' (*kawaii*) companions. In the wild, however, they are like wild dogs (that is, wolves), awesome but threatening. Accordingly, if in the future dolphins should again reappear in great numbers off the coast of Katsumoto, they would be regarded by the same women – and the women were very clear about this – as a wild threat competing for scarce marine resources, that is, as pests 'stealing' their fish.

Conclusion

The Japanese are bewildered by what they see as a Western preoccupation with the absolute. Perhaps nowhere is this better expressed than in the anti-whaling rhetoric. To Paul Spong, the man who brought the anti-whaling issue into Greenpeace (Brown and May 1991: 32), whales and dolphins are one-dimensional beings. They are only *positive* (Spong 1992: 25). To the Japanese, however, whales and dolphins are neither good nor bad; they are both, or neither, depending on the context. The Japanese have for centuries seen whales as beneficial beings that give themselves up as prey to human hunters. For this, the whales are duly thanked and honoured in elaborate rituals which rest on a conception of the whale as a creature with an immortal soul, and a world view stressing the interdependence of supernatural, human and animal realms. More recently, fenced-in dolphins have become 'cute' companions, but fully wild dolphins are still regarded as a threat to fish stocks. Nor does compassion rule out consumptive use. While Ayukawa's 'Whaleland' invites people to 'feel like a whale', its souvenir shop sells whale meat and other whale products and local people believe that whale tourism can prosper only in conjunction with whaling (Iwasaki-Goodman 1994: 110). The same contextualization applies to other animals. Keeping turtles as pets and regarding turtles as one of the 'four mythical animals' does not stop people from eating turtles in restaurants.

Many environmentalists and animal welfare advocates see education as an important strategy in changing people's attitudes towards whaling. Some cast doubt on whether they have been successful. It is premature to conclude that

Japanese attitudes are undergoing any radical change. What we are probably seeing are new foreign views being added to, but not replacing, endogenous ones. In this, as pointed out by Masami Iwasaki-Goodman (1994: 193–194), the whaling issue adheres to Kazuko Tsurumi's 'icicle model' of Japanese society, a model which she defines as 'a synchronic cross-sectional view of a society in which the patterns of life of many previous periods coexist with the newest modes' (Tsurumi 1975: 15, in Iwasaki-Goodman 1994: 194).

Notes

1 By factual knowledge I mean knowledge about whale populations and their place in ecosystems, how they are hunted and for what purposes they are used. What the authors of the essay had in mind, however, was probably 'knowledge' about whales' spiritual qualities and intrinsic value.
2 See, for example, Takahashi (ed.) (1990) for such claims made by fifth grade Japanese primary school students. This is largely correct regarding American, British and Dutch premodern whaling, but it should, perhaps, be added that the innovator of modern whaling, the Norwegian Svend Foyn, also saw it as a religious (Christian) duty to make full utilization of the whales (Johnsen 1959).
3 Japan has also stressed the nutritious qualities of whale meat. It has repeatedly been pointed out that whale meat is rich in protein and iron, and has a high percentage of unsaturated fatty acids which lower the cholesterol level and reduce the risk of cardiovascular diseases. Hence, it is 'better' than meat of domesticated terrestrial animals prevailing in the West.
4 The belief that whales chase fish towards shore was also a common belief in Norway from medieval times (Whitaker 1986) until the early twentieth century (Hjort 1902).
5 Elaborate funerals were sometimes performed for embryos (*mizuko*, literally 'water child'), as in Arikawa where the whalers in the days of premodern whaling called the Buddhist priest whenever an embryo was found. A funeral resembling that of a human burial would be performed with all the harpooners formally dressed (Kalland 1989: 109).
6 Despite Greenpeace's claim that Norway lost trade at a value of 450 million *kroner* (about US$80 million) in 1993 due to boycotts against Norwegian fish products and tourism to the country, both fish exports and foreign tourist earnings reached all-time records in that year. Documented losses amounted to about US$1–1.5 million (Bjørndal and Toft 1994).
7 The President has so far refrained from certifying any whaling nations, an act which most likely will be in violation of the WTO (GATT) regulations (McDorman 1991).
8 Unlike its old whaling museum, which focused on the whaling industry, the new 'Whaleland' stresses the aesthetic and symbolic values of whales, values that are alien to the inhabitants of Ayukawa. The locals have mixed reactions to the displays. Some are unhappy about the displays, others see Whaleland as an opportunity for Ayukawa to retain its relationship with whales and whaling (Iwasaki-Goodman 1994: 107–111, 196–7).
9 From the 1960s the dolphin population at the fishing grounds exploded and dolphins emerged as a major competitor for squid and yellowtail. Several measures were tried to get rid of the dolphins until the fishermen decided to drive schools of dolphins towards land where they were killed (Cate 1985).

References

Akimichi T., P. J. Asquith, H. Befu, T. C. Bestor, S. R. Braund, M. M. R. Freeman, H. Hardacre, M. Iwasaki, A. Kalland, L. Manderson, B. D. Moeran and J. Takahashi (1988) *Small-type Coastal Whaling in Japan*. Edmonton: Boreal Institute for Northern Studies, Occasional Paper no. 27.

Ashkenazi, M. (1992) 'Summary of whale meat as a component of the changing Japanese diet in Hokkaido'. IWC Document IWC/44/SEST2, published in *Papers on Japanese Small-type Coastal Whaling Submitted by the Government of Japan to the International Whaling Commission, 1986–1996*. Tokyo: The Government of Japan [1997], pp. 212–219.

Asquith, P. (1986) 'The monkey memorial service of Japanese primatologists'. In W. T. Lebra and T. S. Lebra (eds) *Japanese Culture and Behavior*. Honolulu: University of Hawaii Press, pp. 29–32.

—— (1990) 'The Japanese idea of soul in animals and objects as evidenced by *kuyō* services'. In D. J. Daly and T. T. Sekine (eds) *Discovering Japan*. Toronto: Captus Press, pp. 181–188.

Bjørndal, T. and A. Toft (1994) 'Økonomiske verknader av boikottaksjonar mot norsk næringsliv grunna norsk kvalfangst' [Economic impacts of boycott campaigns against Norwegian industries due to Norwegian whaling]. Report prepared for the Royal Foreign Ministry. Bergen: Norwegian College for Business and Administration.

Braund, S. R., J. Takahashi and M. M. R. Freeman (1990) 'Quantification of local need for minke whale meat for the Ayukawa-based minke whale fishery'. IWC Document TC/42/SEST8, published in *Papers on Japanese Small-type Coastal Whaling Submitted by the Government of Japan to the International Whaling Commission, 1986–1996*. Tokyo: The Government of Japan [1997], pp. 175–190.

Brown, M. and J. May (1991) *The Greenpeace Story*. London: Dorling Kindersley.

Cate, D. L. (1985) 'The island of the dragon'. In P. Singer (ed.) *In Defence of Animals*. Oxford: Blackwell, pp. 148–156.

Freeman, M. M. R. and S. R. Kellert (1994) 'International attitudes to whales, whaling and the use of whale products: a six-country survey'. In M. M. R. Freeman and U. P. Kreuter (eds) *Elephants and Whales: Resources for Whom?* Basel: Gordon and Breach, pp. 293–315.

Garine, I. de (1972) 'The socio-cultural aspects of nutrition'. *Ecology of Food and Nutrition*, vol. 1, pp. 143–163.

Glass, K. and K. Englund (1989) 'Why the Japanese are so stubborn about whaling? A view from the other side'. *Oceanus*, vol. 32, no. 1, pp. 45–51.

GOJ (Government of Japan) (1991a) 'The cultural significance of everyday food use'. IWC Document TC/43/SEST1, published in *Papers on Japanese Small-type Coastal Whaling Submitted by the Government of Japan to the International Whaling Commission, 1986–1996*. Tokyo: The Government of Japan [1997], pp. 195–201.

—— (1991b) 'Age difference in food preference with regard to whale meat. Report of a questionnaire survey in Oshika township'. IWC Document TC/43/SEST4, published in *Papers on Japanese Small-type Coastal Whaling Submitted by the Government of Japan to the International Whaling Commission, 1986–1996*. Tokyo: The Government of Japan [1997], pp. 209–212.

—— (1992) 'The importance of everyday food use'. IWC document IWC/44/SEST4, published in *Papers on Japanese Small-type Coastal Whaling Submitted by the Government of*

Japan to the International Whaling Commission, 1986–1996. Tokyo: The Government of Japan [1997], pp. 223–228.

—— (1995) 'Kujira to shokubunka ni tsuite' [About whales and food culture]. *Monthly Monitor* (October). Prime Minister's Office, pp. 1–49.

Hjort, J. (1902) 'Fiskeri og hvalfangst i det nordlige Norge' [Fisheries and whaling in northern Norway]'. In *Aarsberetning vedkommende Norges Fiskerier for 1902* [The annual report for Norwegian fisheries]. Bergen: Norges Fiskeristyrelse.

Hoyt, E. (1993) *Kujira Watching: Whales and Dolphins – Alive and Being Watched Japanese-style.* Bath, UK: Whale and Dolphin Conservation Society.

Iwasaki-Goodman, M. (1994) *An Analysis of Social and Cultural Change in Ayukawa-hama (Ayukawa Shore Community).* PhD thesis, Department of Anthropology, University of Alberta, Edmonton.

Johnsen, A. O. (1959) *Den moderne hvalfangsts historie* [The History of Modern Whaling], vol. 1. Oslo: H. Aschehough and Co.

Kalland, A. (1989) 'Arikawa and the impact of a declining whaling industry'. In *NIAS Report* (Copenhagen), vol. 1, pp. 94–138.

—— (1995a) 'Culture in Japanese nature'. In O. Bruun and A. Kalland (eds) *Asian Perceptions of Nature: A Critical Approach.* London: Curzon Press, pp. 243–257.

—— (1995b). *Fishing Villages in Tokugawa Japan.* London/Honolulu: Curzon Press/University of Hawaii Press.

Kalland, A. and P. Asquith (1997) 'Japanese perceptions of nature: ideals and illusions'. In P. Asquith and A. Kalland (eds) *Japanese Images of Nature: Cultural Perspectives.* London: Curzon Press, pp. 1–34.

Kalland, A. and B. Moeran (1992) *Japanese Whaling: End of an Era?* London: Curzon Press.

Manderson, L. and H. Akatsu (1993) 'Whale meat in the diet of Ayukawa villagers'. *Ecology of Food and Nutrition*, vol. 30, pp. 207–220.

McDorman, T. L. (1991) 'The GATT consistency of U.S. fish import embargoes to stop driftnet fishing and save whales, dolphins and turtles'. *The George Washington Journal of International Law and Economics*, vol. 24, no. 3, pp. 477–525.

Misaki, S. (1993) 'Japanese world-view on whales and whaling'. In *Whaling Issues and Japan's Whale Research.* Tokyo: The Institute of Cetacean Research, pp. 21–36.

Ogura, M. (1973) 'Drifted deities in the Noto peninsula'. In R. M. Dorson (ed.) *Studies in Japanese Folklore.* Bloomington: Indiana Press, pp. 133–144.

Reader, I. (1991) *Religion in Contemporary Japan.* London: Macmillan.

Sakurada, K. (1973) 'The Ebisu-gami in fishing villages'. In R. M. Dorson (ed.) *Studies in Japanese Folklore.* Bloomington: Indiana Press, pp. 122–132.

Spong, P. (1992) 'Why we love to watch whales'. *Sonar*, vol. 7, pp. 24–25.

Taiji G. (1982) *Kumano-Taiji-ura hogei no hanashi.* Wakayama: Miyai Heiandō.

Takahashi, J. (1987) 'Hogei no machi no chōmin aidenteitii to shinboru no shiyō ni tsuite'. *Minzokugaku kenkyū*, vol. 52, no. 2, pp. 158–167.

—— (ed.) (1990) *A Whale of a Discussion.* Tokyo: Institute of Cetacean Research.

—— (1992) *Kujira to nihon bunkashi – hogei bunka no kōseki o tadoru.* Tokyo: Tankōsha.

Tsurumi, K. (1975). 'Yanagita Kunio's work as a model of endogenous development'. *Japan Quarterly*, vol. 22, no. 3, pp. 223–238.

Whitaker, I. (1986) 'North Atlantic sea-creatures in the *King's Mirror* (Konungs Skuggsjá)'. *Polar Record*, vol. 23, no. 142, pp. 3–13.

Yoshida, T. (1981) 'The stranger as god: the place of the outsider in Japanese folk religion'. *Ethnology*, vol. 2, no. 2, pp. 87–99.

5

CULTURAL UNDERPINNINGS OF THE WILDLIFE TRADE IN SOUTHEAST ASIA

Deanna G. Donovan

Introduction

From all indications, the trade in wildlife in Asia appears to be thriving despite the economic crisis that has plagued the region since 1997. Although statistics are poor, scientists estimate that the illegal trade in wildlife amounts to US$5–10 billion annually, making this illicit trade third in value after the trade in arms and drugs (Lee 1995). Existing records indicate that the United States, Europe and Japan consume 60 per cent of the wild species for which trade figures exist. Statistics for the trade in natural medicines, an estimated 10 per cent of which contain wildlife products, clearly show that China is the major processor and exporter of such products. Anecdotal evidence and market surveys indicate that much of the wildlife and the products derived therefrom that are traded in China originate in Southeast Asia. One survey in Vietnam found the live specimens or products of 23 different species of mammals, more than 36 species of birds (6,000 individuals) and eleven species of reptiles for sale in one market in Ho Chi Minh City (formerly Saigon) (Le 1995, 1998).

Of the 17 countries that, according to scientists, hold more than two-thirds of the planet's biological wealth and diversity, six are in Asia. With four new mammal species having been discovered or rediscovered in Southeast Asia since 1990, the importance of this region should not be underestimated (Rabinowitz 1997; Schaller 1998). However, the unchecked exploitation of forest plants and animals in formerly remote areas of Southeast Asia may threaten not only this valuable biological heritage, but in some cases the very cultures and economies that proper utilization of these resources could help sustain. Apprehensive about losing resources, both biological and financial, governments throughout the region have instituted regulations aimed at controlling this trade. Unfortunately, these efforts to date have been only marginally effective and the drain on important natural resources continues apace (Donovan 1998).

Conservation policies and protocols currently in place have met with only limited success largely because they ignore the complexity of the forces driving the market, both on the demand as well as the supply side. For more effective policy implementation, planners need to understand and address better the cultural and socio-economic aspects of this trade. The purpose of this chapter is to examine in greater depth some of the cultural elements that appear to sustain and stimulate the trade in wild species and their products in this region. The findings and conclusions presented below emerge from the author's research in Southeast Asia, especially Vietnam, one of the key players in this commerce. It is hoped that a better understanding of the cultural underpinnings of the trade will provide the information needed to bring about more effective measures to counteract the illicit trade in rare and endangered species in Southeast Asia.

Historical role of wildlife

Throughout history a great variety of wild animals, large and small, have been collected as curiosities and products derived from them used for many purposes, including clothing, adornment, decoration, charms or trophies, if not for food or medicine.[1] The commerce in wildlife and wild products is not a new phenomenon but rather has played an important role historically in the development of the region. In Southeast Asia, the minority ethnic groups living in rugged mountainous areas at the heart of this region have been key participants in this trade, especially as suppliers but also as traders. It has been argued that such culture-specific specialization, coupled with intercultural exchange, permits more efficient utilization of available resources and in fact represents an intensification of land use (Peterson 1978). Historic ties between upland and lowland groups and between temperate and tropical societies often form the basis of economic and political relationships that exist today.[2]

Archaeological evidence and written records indicate that wild animal and plant products have been major trade items between South and Southeast Asia and China for more than two millennia.[3] Chinese merchants arriving by both land and sea routes have traded with the peoples of Southeast Asia for centuries. As Chinese influence spread with the establishment of trading posts throughout this region, various Chinese customs and technology infiltrated into neighbouring cultures, including Thailand, Vietnam, Tibet and others (cf. Nash 1997, Nguyen 1995). Chinese merchants and other migrants (often from southeastern China) carried with them to foreign shores not only practical skills but their preferences for food, clothing and other consumables. Today the greatest demand for wild animals and their products stems from this cultural heritage based on the traditional use of animal parts and products in culinary and medicinal preparations.

Figure 5.1 A nineteenth-century representation of the butchering of a deer from a temple in Laos

Wildlife as food

Animals have been a key protein source for mankind since the earliest days. Even after the successful domestication of several large animal species in China,[4] hunting continued on a vast scale (Read 1931, Anderson 1988: 20). Wild animals and animal products were consumed domestically but also traded widely. In the Tang era (AD 618–907), international trade, largely via Central Asia, was organized by the government or foreign traders. It was an expensive and complex business regularly interrupted by raiding desert tribes. As a result of these disruptions and an increasing interest in the produce of Southeast Asian forests, there was growing pressure to develop southern trade routes. Subsequently, in the Song era (960–ca. 1260) maritime trade from South China ports developed and essentially anyone who could afford a boat could become a trader. Chinese cuisine benefited as a result of trade expansion into Southeast Asia with a number of new crops introduced and many exotic foods made available in China's markets (Anderson 1988: 60).

Wild game figured prominently in many dishes of this era, as the newly rich urbanities appreciated novel ingredients and sought the display of wealth that the consumption of these rare products provided. Perishable fruits were brought to the capital from great distances, including the area now known as Vietnam (Chang 1977: 155). In the twelfth and thirteenth centuries, some 60 per cent of the more than two thousand people running the emperor's residential quarters were responsible for food and wine and nearly a third of these were specialists in wild game and seafood (Chang 1977: 11). As noted by Marco Polo, who visited the region in the thirteenth century, the markets in Hangzhow, then the capital of China, exhibited an enormous variety of wild game, some imported from great distances.

In 1330, the court physician presented to the emperor a book entitled *Yin-shan Cheng-yao* (Essentials of Dietetics) which included several entries on antelope, bear, various deer, tiger, leopard, marmot, swans, pheasants, cranes and many other animals and birds (Anderson 1988: 75). Although most animals were listed as food, many were included purely for medicinal purposes. One Ming-dynasty (1368–1644) source observed that numerous wild animals, including deer, wild boar, camel, bear, wild goat, foxes, wolves and rodents, among others, were standard fare in the Chinese diet. The hunters, who supplied the market with fresh, dried and salted meat of the highly prized game animals, belonged to a special, revered profession (Chang 1977: 242). Throughout Chinese history the important imperial dynasties have left their mark on the cuisine in predictable ways, namely, with ever more exotic and elaborate dishes.

The Chinese diasporas have carried traditional food and pharmaceutical practices and preferences to nearly every continent. Studies have shown that Chinese emigrants generally retain their dietary habits longer and more faithfully than other ethnic groups. Research indicates that in the process of acculturation traditional foods are abandoned inversely to the order of their cultural importance (Anderson 1988: 210). Thus, the Chinese relinquish first drinks, then snacks and then breakfast, only lastly abandoning the foods associated with specific ethnic traditions and festivals. Worldwide, many cities (e.g. New York, San Francisco and Bangkok) are famous for their 'Chinatown' districts, where residents, many of Chinese descent, and visitors enjoy the varied Chinese cuisine. Historical records indicate that in nineteenth-century Nagasaki (Japan) the Chinese community honoured important guests with elaborate meals, sometimes with as many as sixteen dishes, which included bear paws, deer tails, shark's fin, birds' nests and sea slugs, among other delicacies (Chang 1977: 277). At the start of the century, canned rice birds imported from China were sold in shops in New York City (Simoons 1991: 326). Today, the custom of feasting on rare, exotic beasts has been maintained throughout Southeast Asia not only by the local population, but also, increasingly, by tourists from nearby countries.[5]

Wildlife as pharmacopoeia

Nutritional medicine and culinary art have had a complex relationship throughout Chinese history. The functional value of foods, especially their contribution to health and general well-being beyond purely nutritional aspects, was recognized early on (Weng and Chen 1996). In the Zhou Dynasty (ca. 1030–480 BC), nutritionists served the court as the highest ranking medical personnel (Anderson 1988: 193). Traditional Chinese medicine arose from folk pharmacognosy first recorded between 200 BC and AD 100 during the Han dynasty (Chevalier 1996). Interchange between China and India in the field of pharmacology probably occurred around the first century AD (Hu 1990; Dai and Luo 1996). The first Chinese herbal pharmacopoeia, the *Shen Nung Pen T'sao Ching* (Shen Nong Ben Cao Jing), listed some 365 natural substances, of which from one third to nearly one half were of animal origin (Hsu 1982; Read 1931). Over the following five hundred years the number of natural substances used in Chinese traditional medicine doubled. By the end of the first millennium the total had increased by half again, to just over a thousand. By the end of the sixteenth century, after incorporating information gleaned from the folk healing traditions of Japan and other neighbouring countries, the total number of natural substances noted in the Chinese pharmacopoeia had increased sixfold since first recorded (Hsu 1982: 66).[6]

Numerous reports indicate that from earliest times the Chinese believed rhino horn to have remarkable medicinal and magical powers (Briggs 1951). Tigers, bears and seahorses have also been the source of medicinal ingredients for several hundred years (Mills and Jackson 1994; Mills, Chan and Ishihara 1995; Freese 1998). The medicinal use of bear gall bladder was first noted around the sixth or seventh century AD (Highley and Highley 1998). One of the most famous Chinese materia medica was the *Ben Cao Gang Mu* written by Li Shih-chen (Li Shizhen) during the Ming dynasty around 1597 (Read 1931). Of the 1,892 ingredients listed in the 52 volumes, less than 10 per cent of the total are derived from animals and the majority of these are insects. Of the 63 different animal species (mammals only) listed in the section on drugs of animal origin, however, fully 86 per cent were from wild species.

Despite numerous policies, regulations and international agreements to the contrary, rare and endangered wild animals and/or their products are readily found in markets throughout Asia. Customers range from the least to the most cosmopolitan in society. The enormous variety of animals and plants exploited for this market and the lack of adequate monitoring at virtually all levels make it very difficult to obtain a clear picture of the impact of the current wild species trade. From Southeast Asia, live animals as well as various animal parts and products move east and north by a variety of means, including car, bus, truck, boat, train and porter, to processing plants in Yunnan and Guangxi and then beyond to the richer markets of East Asia, Europe and the United States.[7] China has the dubious distinction of being the largest exporter of processed tiger bone (Mills and Jackson 1994). Trade intermediaries, especially in the larger cities of Thailand and Vietnam as well as Singapore, all strongly influenced by Chinese contacts in the past, are not only significant consumers in their own right but important trans-shipment points.[8] With the support for research on traditional medicines, it is clear that Hong Kong, which already plays a key role in channeling Chinese products onto the international market, is looking for a more significant position in this trade.

Figure 5.2 Shops in China near the border with Vietnam selling a variety of wildlife products (mainly from Vietnam and Laos)

Figure 5.3 Natural products on sale in a National Park in Laos

Cultural influences on market forces

Cultural factors influence the role of different consumables in society as well as the composition of these various elements. The traditions that have developed over the centuries, if not millennia, are shown to persist and influence patterns of consumption today. Similarly, it can be seen that suppliers who have a long history as provisioners may also be influenced by their cultural heritage.

Demand

As one observer has noted: '[C]onsumption ... is not the economist's inscrutable act of shapeless desire. On the contrary, consumption is implicated in identity and is socially communicative as well as technical or material' (Hefner 1998: 25). In China especially, food and medicinal products, both with a long history of royal connections, have shaped the sense of what it is to be a part of Chinese culture. Of all the many uses of wild species and the products derived from them, perhaps the two most important categories are culinary and medicinal preparations. It is widely recognized that food is used not only to satisfy the basic physical needs of nutrition and health but also to meet various human psychological needs for variety, communication and social contact, among other factors (Leininger 1970). Food plays an important role in religion by reinforcing religious ideas; conversely, religion itself then becomes a powerful enforcer of culturally specific food habits and rituals. As in all cultures, the Chinese use food to mark ethnicity, religious and cultural festivals, rites of passage, family events and social transactions. Both ancestors and spirits, believed by many to intercede on behalf of the living to positive or negative effect, are beseeched with elaborate and often expensive offerings of food. Feasts mark special occasions and communicate messages of opulence, status, solidarity and favour. Indeed, one's guest should leave feeling suitably impressed *and* indebted.

The demand for game, or bush meat as it is sometimes called, persists even today at virtually all levels of society. In remote areas where fresh meat is difficult to obtain from domesticated sources, farmers and government workers alike often provision their tables with animals caught in nearby forests, selling or trading any surplus or products therefrom for other goods required (Salter 1993; Anon. 1998). Nowadays, the bulk of demand for wildlife products appears to be driven by largely urban markets, both domestic and foreign. In this context the Chinese market looms large; with an annual population growth rate of 1 per cent, low by Asian standards, China will add the equivalent of two new cities the size of Hong Kong every year. Of a total population of more than 1.2 billion people, an estimated 70 million belong to the class of young, upwardly mobile professionals, most of whom live in urban areas. With disposable income on the rise, these newly prosperous consumers enjoy a growing material culture and conspicuous consumption often involving sumptuous feasts of exotic meats. Wild animals believed to be rare and powerful are especially preferred.

Currently the food industry ranks second among all industries in China. With improving standards of living and the Chinese spending from 30 to 50 per cent of their take-home pay on food, performance in this sector was estimated to rise 18 per cent in 1998 (Brandt 1998). Research has shown that in Hong Kong total expenditure on food increases as a percentage of total income as people get richer. This is contrary to what in economics is known as 'Engels Law', which holds that the proportion of the family budget spent on food declines as family income increases. Clearly in Chinese communities, food remains an important social lubricant and status symbol. Although consumption of wild game worldwide has been tempered somewhat by rising prices due to increasing scarcity, of late it appears to be experiencing something of a revival in China. While renewed interest in cultural heritage may be fueling this demand to some extent, political rapprochement with neighbouring countries as well as trade liberalization has improved the availability of these products. The Chinese are well known for their adventurous consumption; in Guangdong province there is a saying, roughly translated, that 'If it has four legs and it's not a table, we'll eat it' (Brandt 1998). Given the important role that food plays in Chinese culture, the conspicuous consumption of exotic foods by China's newly rich comes as no surprise. Certainly no business deal is complete without a dinner – an indulgence that even the Chinese communists found hard to do away with.

The benefits attributed to the consumption of various animal parts and products derives largely from the user's belief in 'sympathetic magic'. Specifically, this position holds that the characteristics of the animal can be acquired by consuming or wearing something of that animal. It is a belief similar to that espoused in the 'Doctrine of Signatures' associated with plants (Chevalier 1996).[9] With wild animals figuring large in legend, verse and folktales, the socio-biological traits attributed to various wild animals are generally well-known throughout traditional cultures, even in the more remote, pre-literate groups.[10] The fierceness of the tiger, the sexual stamina of the rhinoceros and the timidness of the deer are believed to be inherent in the products derived from these animals. People use preparations made from animal products both to restore health and vigor and to enhance their personal capacities according to the characteristics of the animal consumed.

One of the fundamental beliefs of Taoism, the most ancient of the religions in China, is of a universal energy running through every living and inanimate thing. The disruption of the natural order and smooth energy flow is thought to be the cause of many illnesses. Thus, the basic premise of all Chinese traditional medicine is that disease results from an imbalance within the microsystem of the individual in response to imbalances in the macrosystem within which the individual lives. The aim of treatment then is not only to remove the cause of the illness, but to revitalize the body and reinforce the body's natural resistance (Hsu 1982). Natural substances derived from animals are thought to contain both the chemical constituents and the energy necessary to correct the perceived imbalances and to restore proper energy flows in the patient. This is explained by the anthropologist E. N. Anderson:

> Rare, exotic and unusual foods are considered *pu* [a restorative and stimulant] not just because of cost and strangeness, though these are certainly factors, and conspicuous consumption is a very major part of their use. More basic is the concept of *chi'i* [energy]. In the traditional Chinese worldview, bodily energy, spiritual energy, and the flow of energy in the natural world are all part of one great system. This is true of modern physics too, but the Chinese belief is more extreme, claiming that people can draw on natural energy flow by eating creatures that have a great deal of energy or even by positioning themselves in places that are appropriately located to take advantage of the flow of *chi'i*. The striking appearance of such creatures as pangolins and raccoon-dogs is thought to indicate great energy or unusual energy patterns. Powerful creatures like eagles – to say nothing of the sexually hyperpotent deer – are also obvious sources of energy. The similarity between many of these beliefs and what is espoused as 'holistic medicine' in the West may be the reason why so many consumers in the United States and Europe are being drawn to traditional Chinese medicine.
>
> (Anderson 1988: 193)

Despite the advent of germ theory at the end of the nineteenth century and the influence of modern religions, such as Buddhism and Christianity, many urban dwellers as well as rural residents still cling to the primitive belief that illness is caused by witchcraft or the influence of malevolent supernatural beings, that is, bad spirits (Chevalier 1996). In such societies prescriptions for the prevention and cure of illness may involve invoking the help of more powerful 'spirits', for example the 'spirit' of the rhino or tiger or bear, from which some physical token or talisman is required by the person seeking help. By ingesting or displaying some part of the more powerful animal, the inflicted person believes that he or she can acquire the characteristics and strength of that animal and thus be freed of various afflictions.[11] These beliefs are perceived to be important traditions, elemental to the culture. There is increasing evidence that the resurgence of interest in traditional Chinese medicine is now viewed as an important means of preserving the culture and even as a status symbol among certain groups (Mills and Jackson 1994; Wordsworth 1997).

It is estimated that currently 40 per cent of the population of mainland China – nearly half a billion people – use traditional Chinese medicine on a regular basis (Lee 1999a). The demand for traditional Chinese medicine is increasing not only in Asia but worldwide, with demand in the United States alone reportedly up 280 per cent in the last decade. In a recent survey conducted by TRAFFIC, it was found that more than three-quarters of the Chinese living in the United States have used traditional Chinese medicine at some time and about half use it a few times a year (Lee 1999b). As compared with most Western prescription medicine, traditional Chinese medicine is highly regarded as effective therapeutically yet with minimal side effects. However, the majority of respondents in the TRAFFIC survey readily admitted that they do not know much about the ingredients and assume that these medicines are mostly plant-based. Few users mentally connect traditional medicines with endangered species. Almost a third of respondents surveyed reported that they would continue to use traditional Chinese medicine, even if they knew that it came from endangered species (tiger bone was the ingredient most likely to be used). Willingness to use replacements, however, was strongly tempered by concern for the efficacy of any proposed substitute (Lee 1999a, b).

In the 1990s Southeast Asia has experienced a severe financial crisis, resulting in the devaluation of most of the Southeast and East Asian currencies. As a consequence, consumers find that imported Western medicines are now much more expensive. Governments in China and Thailand openly encourage their citizens to reconsider the use of traditional remedies for less-than-life-threatening illnesses. Substitution of locally produced medicines in place of imported, foreign manufactured pharmaceuticals would mean savings both for the domestic consumer and for governments, most of which are still strapped for foreign exchange. Traditionally in China and more recently in Thailand and Malaysia, governments have encouraged the production of traditional medicine as a means to stimulate the development of local industry and potential exports.[12] This promotion of the substitution of imports by locally produced products has become an additional stimulant to demand. But in some cases the raw materials for the development of the industry may be coming from nearby countries and most often from the scant remaining natural forest.

Supply

The hinterlands of mainland Southeast Asia are home to more than 150 ethnic groups (Lebar, Hickey and Musgrave 1964). For many upland groups, the forest remains an important source of basic needs, including fuel, food, fibre, construction timber and medicinals, in essence both a natural 'supermarket' and an economic 'safety net' in times of need (Lebar et al. 1964; Dang et al. 1993). Much of the forest exploited for wildlife is located in remote areas inhabited by ethnic minorities who still largely depend on swidden cultivation and collection of forest produce for their livelihood. Although much of what is gathered from the forest is consumed by the collector's family, an increasing number of products are brought to the market for sale or exchange.[13] The recent rise in demand coupled with improvement in transportation has hastened the commercialization of traditional commodities and the monetization of rural life.

The sale of forest products, especially wild animals but also plants, has been the traditional means by which groups living in or near the forest were able to generate cash. Whereas previously the government provided many social services and materials free of charge, with the change in economic policies more cash is now needed to pay taxes, school fees, health clinic charges and medicines or to purchase necessities such as salt, iron, cooking oil and rice, after harvest reserves are exhausted. In the area I studied in northern Vietnam, many traditional ethnic groups inhabiting the northern region have been pushed upslope by the influx of both spontaneous and government sponsored migrants and the expansion of plantation tree crops. Under the harsh climatic and topographic conditions of the mountainous areas, a typical farm family is fortunate to produce sufficient food for six to eight months of the year (Donovan *et al.* 1995). For the remaining months rice must be purchased, generally with income earned from the sales of forest products such as bamboo shoots, dried mushrooms, lime, incense sticks, wildlife or other items collected from the forest. With varying skills, knowledge and access to resources, different ethnic groups tend to become associated with specific produce or products.

During the decades of war that dominated this region, people survived in essentially subsistence economies. Having recently rejoined the world economy, these communities, many of which have satellite television, are eager to catch up with their more advantaged neighbours. With the improvement of the road system and encouragement of trade under new economic policies, in Laos, Vietnam and Yunnan, the number of consumer goods reaching rural markets has dramatically increased. Watches, flashlights, lighters, batteries, candy and factory-made clothing are among the most common purchases. Larger consumer durables such as furniture sets, televisions and refrigerators are increasingly appearing, even in seemingly remote areas.

The importance of hunting in each culture varies because distinct groups are associated with particular ecological zones and geographical areas. A prominent upland group, the Hmong[14] immigrated into northern Vietnam, Laos and Thailand beginning in the middle of the nineteenth century. Upland rice has long been their staple crop and opium an important cash crop. Being skilled weapon makers, the Hmong are fiercely independent and enjoy hunting and trapping, often with dogs.[15] With the production of opium discouraged in recent years, the Hmong have increasingly looked to forest products to provide the income needed for store-bought goods, taxes and other cash requirements (Dang *et al.* 1993). Other groups, such as the Dao and Thai who inhabit this region as well, also collect, process and sell forest produce but to a lesser extent. The amount of time that an individual, group or family spends hunting in any given year will vary from family to family, from season to season and from year to year. Recent field reports indicate, however, that with the relatively high prices now paid for certain wild animal species, some collectors spend more of their time pursuing forest products.

Historic trading relationships have been revived, such as the tradition of Chinese merchants living in the valley towns in Laos and Chinese traders traveling from area to area arranging contracts for forest products. As in the past, one sees throughout Southeast Asia Chinese merchants travelling in remote areas

Figure 5.4 A leopard skin for sale in a shop in Sapa, northern Vietnam

to purchase such products. We do not know the extent of these relationships or the strength of their influence in drawing out supplies. It is not uncommon for merchants to extend credit to the farmer-collectors on the promise of delivery of specific forest products. Some scientists suggest, however, that the isolated hunter or specialized hunting group is now largely a myth. Indeed, some say that hunter-gatherer societies could not survive without engaging in some cultivation or trade (Bailey *et al.* 1989, Headland and Reid 1989). In fact, not only does the hunting tradition persist, but it has become ubiquitous in rural areas where resources permit (Le 1995, Trankell 1995, Puri 1997). Even though the amount and types of prey may be constrained by habitat degradation, and destruction and capture largely opportunistic, for many societies hunting remains important in a cultural context. Rituals such as singing, dancing and food distribution often associated with hunting preserve social cohesion and ethnic solidarity.

Anthropological research in Southeast Asia has demonstrated the persistence of the hunter-gatherer lifestyle among ethnic communities, especially in those with regular access to the forest.[16] For many societies hunting is significant as a social activity. Indeed, it may be a key element in defining cultural identity as well as an important means of providing protein and tradable goods.[17] To what extent hunting is and may remain important in the social and cultural fabric of the societies serving the wild species trade is difficult to say. Little specific information exists, either in recent studies or historical records, on the role of hunting in the cultures of the mountainous heartland of Southeast Asia.

Trends and implications

For the majority of the population in Southeast Asia wild animals are powerful as subjects of myth, as symbols, as omens, as well as for meat, medicine and trophy. In the increasingly competitive societies of modern Asia, where conspicuous consumption plays an important cultural role, the demand for wildlife as exotic and expensive consumables will undoubtedly persist for some time. Pressure will grow too for medicinals as the demographic profile adjusts to accommodate an aging population. Accordingly, as the health of the older generation deteriorates, with increased pollution and urbanization hastening the trend, the demand for Chinese traditional medicine with its emphasis on restorative and preventive properties is likely to increase. By contrast, public awareness of the need to protect and preserve wildlife is poor, even in the more advanced economies of the region (Hong 1997).

On the demand side

A recent telephone survey in Hong Kong revealed that 35 per cent of the adult population consult traditional Chinese medicine practitioners and use their products (Lee 1999a). As much as 40 per cent of the population in mainland China is said to use traditional medicine. According to marketing studies the

Chinese consumer is very quality-conscious as well as price-sensitive and brand-loyal (Brandt 1998). Many prefer products from wild sources, believing them to be more potent and therefore more effective (Judy Mills, pers. comm.; Highley and Highley 1998). Some 14 per cent of the traditional medicine users surveyed in a recent Hong Kong study reported that they were willing to pay more for ingredients from wild sources (Lee 1999a). Such attitudes toward wild versus farmed ingredients may be analogous to Western attitudes toward vitamin C derived from rosehips as opposed to laboratory manufactured vitamin C. This may not be entirely unfounded. Increasingly, research shows that some ingredients provide greater than expected benefit due to the particular combination of constituents in the natural product, some of which may not be well studied (Chevalier 1996: 25).

Almost a fifth of the Hong Kong users said they would use traditional medicine even if this were against the law – that is, banned because of using ingredients from protected species (Lee 1999a). Although a sense of political correctness is developing with respect to the way in which the Chinese consumer speaks about domestic brands as opposed to international brands, whether a similar attitude can be developed with regard to the consumption of wildlife is a key question (Wordsworth 1997). Prejudices so strongly linked to cultural tradition are particularly hard to dispel, especially if the pressure for change is seen to be from mainly foreign sources. With greater disposable income, many consumers can now afford to be more selective and to opt for the product and brand that has the highest status.

On the supply side

Many urban as well as rural inhabitants perceive the tropical forest to be a vast storehouse of a great variety of useful products. Rural villagers, however, see many wild animals as pests, even a threat to their own survival in some cases. With hunting traditions strong, attitudes toward conservation are weak, especially for non-tenured resources. As rational peasants, most villagers will scarcely hesitate to take advantage of any opportunity to better their position, financially, materially or socially. Taboos on killing certain species, although widespread, are not universally observed but may be sufficiently respected to have conservation as a secondary effect. By and large, however, hunting practices are designed to maximize efficiency and so methods exist to appease offended spirits when taboos are broken (Hames 1991: 181). It is difficult to dissuade the hunter from capturing or killing a specific wild animal – even the hunter's group totem – if the bounty on that animal (for example, a tiger, a bear or a turtle) is greater than the farmer's or hunter's projected earnings over several months. With scant understanding of market structure (much less the underlying forces), uncertain as to future government policy, handicapped by a lack of information (both technical and commercial), systematically excluded from the higher echelons of trade and subject to the arbitrary application of regulations, villagers readily take advantage of any economic opportunity presented, including the trade in wildlife. Thus, both need and greed are motivating factors.

Rural people's attitude toward resource exploitation, including over-harvesting, is in general one of seeming ambivalence. In interviews in different areas in Vietnam, farmers reported that there were virtually no living animals left in local forests – only rats and the occasional passing bird. They do not seem to consider the loss of certain species as a problem per se. Indeed, farmers perceive the lack of wildlife as only a very local and temporary phenomenon, expecting the area to be re-inhabited by animals migrating into the area from neighbouring regions – just as weeds migrate back into the agricultural field crops and the forest reclaims the swidden fallows. However, the seeming equanimity with which farmers report these conditions belies their concern about their future. Although farmers see an increase in the number of rodents raiding their fields and granaries, they believe that over the next hill there are more wild animals and that eventually these will find their way into the local ecological niche. As most villagers never travel far from their village, they are unaware of the true extent of forest conversion, habitat destruction and the pressure of mushrooming human population on wild places. Species extinction, if understood, is not perceived to be a problem. A similar situation existed in the American West in the nineteenth century with the loss and near loss of several species including the passenger pigeon and the bison respectively.

Domestication would seem to be the logical solution both to relieve the pressure on wild populations and to take advantage of the economic opportunity presented. Some wild animals have been domesticated or semi-domesticated. Game ranching on any significant scale has met with only partial success and is limited to only a few species. Although *bona fide* wildlife farms do exist, many are stocked with specimens captured from the wild, which suggests that the existence of such facilities in the absence of effective regulation of the wildlife trade may actually exacerbate the problem of species extinction (Anon.1996a, Highley and Highley 1998). Remedial measures, such as restocking of depopulated areas, cannot begin until successful breeding programs are developed. Application of advanced technology, such as tissue culture to produce planting stock (as with rare orchids in Vietnam), is for the most part still at an experimental stage. Certainly, if such new technology fails to benefit local communities or enable them to better meet their own needs, or if new technology overturns traditional trading relationships, it may result in accelerating environmental exploitation and degradation as disadvantaged groups are forced by economic necessity to look to other resources. The tendency to collect wildlings for horticultural cultivation and a strong tradition of animal husbandry would suggest that a group such as the Hmong in northern Vietnam could be successful in adapting such new technologies. Experience in northern Thailand and Laos indicates that the Hmong are willing to take the necessary risks to become successful entrepreneurs, adapting well to a role as merchants (Barney 1967, Kunstadter and Kunstadter 1983). However, with economic conditions, resource values and access to technology changing, household economic relationships will be altered. Such shifts might threaten social and cultural cohesion with consequent effects on resource exploitation and environmental conditions.

The moral and philosophical context

Resource use exists within a cultural context that includes a philosophical worldview. Most of East and Southeast Asia where the subject trade thrives is officially Buddhist.[18] Buddhism provides the basis for people to identify with nature. As such, it precludes the despoiling of nature for transitory human advantage. In theory, Buddhist peoples should exist in harmony with nature. From observation of the cultures of mainland Southeast Asia, all claiming to be essentially Buddhist, the belief in harmony with nature at the philosophical level is no blueprint for creating and maintaining such harmony on a day-to-day level. Similarly, it has been observed among hunter-gatherer societies that taboos and totem association with an animal will not necessarily prevent it being killed and eaten (Geddes 1976: 15). Clearly, the resource use of many Buddhist countries is not in harmony with that religion's basic tenets and philosophies. Hunting or game capture becomes 'accidental' or opportunistic rather than a purposeful organized act, so as not to conflict with Buddhist ideas (Trankell 1995: 87). Such modification of societal 'rules' is commonplace. On top of this, radical individualism has also fueled the current environmental crisis, setting the stage to see who can use the Earth's resources faster and more completely (Ferkiss 1993: 225). Conservation policies and programs that rely upon benefits accruing to a generalized group may be doomed to fail in the absence of kinship or other close bonds between members of the beneficiary group. As Ridley (1998) observes '[t]he virtuous are virtuous for no other reason than it enables them to join forces with others who are virtuous, to mutual benefit'. Very conscious of social status and standing, the Chinese consumer may be susceptible to a public campaign in favour of strict wildlife protection.

Conclusion

Wild animals have played an important role in human culture and development throughout history in Asia. The range of variability and historical mutability in human relationships with and sensibilities toward wild animals is very wide – from totem to trade item. The most dramatic change in environmental conditions occurs when resource procurement ceases to be subsistence-oriented and instead becomes market-driven. This impact is magnified when commercial expansion is accompanied by the decline in local institutions of resource management and the breakdown in the regulatory authority.

The current demand for wild animals or their products has been stimulated by a number of factors, including rising incomes in China and the substitution of expensive, imported pharmaceuticals by homegrown remedies as well as an increasing interest in 'alternative' medicines and natural products in the West. A revival of ancient cultural traditions, especially as related to food consumption and medicinal preparations, underpins this demand, especially in Chinese-influenced cultures. Economic liberalization and infrastructure development have strengthened the supply chain so that nowadays products from even the most remote areas can be in the capital city in a few days. This combination of factors sustains a strong level of demand that by all indications is growing. Both domesticated and wild populations feel the pressure of this increasing demand.

On the supply side, availability and access to resources limit supplies. Although the demand for traditional Chinese medicine emanates primarily from East Asia, supplies are increasingly sought from Southeast Asia, the former Soviet Union, North and South America and Africa. Thus the impact of this problem is not limited to Asia. Non-timber forest products, a category that includes wildlife, has always been a source of income for people in remote areas. Whereas in the past (and even today in some more remote areas) wild products may have been bartered, the monetization of the traditional trade has become increasingly common as rural people require cash to pay various fees and services and to purchase the consumer goods flooding in on newly refurbished road systems in this region. Government policy changes with regard to foreign relations, commerce and infrastructure development have dramatically altered the commercial environment. Legislation restricting trade in rare and endangered species, directed largely at the control of trafficking, has had minimal impact in the face of strong demand and the legions of eager suppliers.

Despite several international agreements and much prohibitive legislation, current policies to stem the traffic in wild species are ineffective because of the present level of trade. Given the scope of the problem, the pressure of the burgeoning market and the slow, tortuous path to success of most development projects, it is clear that insufficient resources are being directed to this problem and prospects of preventing the extinction of certain species appear bleak. With demand influenced by cultural factors and supply complicated by social and economic issues, including access to resources, forest conservation and exploitation and direction of upland development, the wildlife trade is a particularly intractable problem. Under the present conditions of virtually unregulated exploitation, the wild species trade in Southeast and East Asia seriously threatens biodiversity.

The pressure of population growth, improved access to remote areas and continued uncertainty in the economic arena are not conducive to conservation. Indeed, these forces may work synergistically to put the remaining forest in jeopardy. In the present state of flux, given the rapid pace of social and economic change, the need for well-considered, socially sound and scientifically based conservation policies supported by strong political commitment and sufficient funds has probably never been stronger. Proceeding with the current focus on interdiction is unlikely to succeed in the face of countervailing pressures, often culturally entrenched, that exist to thwart these efforts. The continuation of present trends will probably lead to the extirpation of several species in this region. For those societies where hunting has played a key role, the destruction of habitat, coupled with over-hunting, threaten not only livelihood but also culture. The dissolution of ethnic minority groups, with their assimilation into the larger population, often results in the loss of their unique knowledge about their environment. To be effective, concern for both biological and cultural loss must be backed by a significant commitment of resources to address both the supply and the demand side of the trade with all its social, economic and cultural complexity.

Acknowledgements

The author would like to thank the World Resources Institute (WRI), Washington, DC, which provided support through its REPSI initiative, a programme funded by the Danish government's Danish Cooperation for Environment and Development (DANCED) and the Swedish International Development Cooperation Agency (SIDA). In addition, the author would like to gratefully acknowledge the support of the John D. and Catherine T. MacArthur Foundation in conducting field research and from the Keidanren Nature Fund of Japan. The Center for Natural Resources and Environmental Studies of Vietnam University at Hanoi provided administrative and logistical support for fieldwork in Vietnam. Personal thanks go to Dr Raj Puri for enlightening discussions on hunting and to Dr John Knight, Dr Jeff Fox, Dr Terry Rambo and Dr Le Trong Cuc for their steadfast support and encouragement. I would also like to thank Ms Le Thai Binh for her excellent assistance both in field research in Vietnam as well as bibliographic work in Hawaii.

Notes

1 Bruemmer (1998); Salter (1993); McNeely and Wachtel (1991).
2 Lebar, Hickey and Musgrave (1964); Schrock *et al.* (1966); Condominas (1972: 203, 206); Pei (1988); Engelhardt (1989); Dang, Chu and Lau (1993); Evers (1985; 1990; 1997).
3 Ma Huan (1433); Briggs (1951); Wolters (1967); Li (1979), Hall (1985); Haellquist (1991); Wicks (1992); Cushman (1993); Reid (1993); Momoki (1998).
4 Archaeological evidence indicates that deer may have been domesticated as early as the Shang dynasty, ca. 1500 BC.
5 Stewart (1998); Highley and Highley (1998); Waldman (1998).
6 It should be noted that a seventeenth-century London pharmacopoeia carried prescriptions with numerous ingredients from wild animal sources, including lion fat, leopard fat, rhino horn, bear gall, bear fat, bear lungs, deer antler velvet, deer blood, deer kidneys, moose antlers, gall of the hare, of the hedgehog and of the mole and wolf's feces (Read 1931).
7 AFP (1998); Anon. (1996a, b); Li, Fuller and Sung (1996).
8 Salter (1993); Baird (1994); Gittings (1998).
9 In medieval Europe the 'Doctrine of Signatures' (i.e., God's signature) maintained that the physical appearance of a plant gave an indication of the ailments it could cure. For

instance, the mottled leaf of *Pulmonaria* sp. was thought to resemble the lung and thus assumed to be useful in treating respiratory ailments, and hence the English name 'lungwort'. Similar beliefs prevail in Africa (Ody 1993).

10 Graham (1954); McNeely and Wachtel (1991); Puri (1997).

11 Such beliefs, known as 'sympathetic magic', are not limited to Chinese culture. In the West different types of meat have been associated with the qualities associated with the animal from which it comes; thus, the prominent role of beef in the diet of the British soldier has been regarded as a significant factor contributing to his valour (Ritvo 1997: 194). Or as noted by Condominas (1972: 204), the Mnong Gar of Vietnam believed that the power of the animal sacrificed (generally water buffalo) would be transferred to the man who sacrificed it. Conversely, the timid nature of the deer is believed by the Penan in Borneo to be contagious and therefore to be avoided by warriors (Puri 1997: 462).

12 Chevalier (1996); Kaur (1998); Taravanit (1998).

13 Westing and Westing (1981); Nguyen XT (1993); Le (1995); Nguyen VC (1996); Compton (1998).

14 Also known as the Meo or Miao.

15 Savina (1930); Mottin (1980); Lebar *et al.* (1964).

16 Rai (1990); Sellato (1993); Brosius (1991); Kaskija (1995); Puri (1997).

17 Lebar *et al.* (1964); Puri (1997); Ingold (1989); Dwyer (1985); Pookajorn (1985).

18 Except for Indonesia and the Philippines, both largely beyond the scope of this study, which are respectively Muslim and Christian, (especially Catholic).

References

Agence France Presse (AFP) (1998) 'Three tonnes of trafficked wild animals returned to Vietnam forest'. 13 July 1998. (via ClariNet).

Anderson, E. N. (1988) *The Food of China*. New Haven: Yale University Press.

Anon. (1996a). 'Bear-faced gall: trade in animals'. *The Economist*. 9 November 1996, 341(7991): 44.

Anon. (1996b). 'Railway smuggling boom'. *Viet Nam News*, Hanoi.

Anon. (1998) 'Wolves, prospectors and land mines threaten wild camels with extinction'. *Current Science* (Middletown) (16 Jan 1998) (via ClariNet).

Bailey, R. C., G. Head, M. Jenike, B. Owen, R. Rechtman and E. Zechenter (1989) 'Hunting and gathering in the tropical rain forest: is it possible?' *American Anthropologist*, vol. 91: 59–82.

Baird, I. (1994) 'The trade in soft-shelled turtles (Trionychidae) between Southern Lao PDR and Vietnam'. Unpublished. TRAFFIC-Southeast Asia, Kuala Lumpur.

Barney, G. L. (1967) 'The Meo of Xieng Khouang Province, Laos'. In Peter Kunstadter (ed.) *Southeast Asian Tribes, Minorities and Nations*, Princeton: Princeton University Press, pp. 271–294.

Brandt, L. (1998) 'The China challenge: the Chinese marketplace offers many opportunities for American food companies'. *Food Formulating*.

Breummer, F. (1998) 'Of monstrous moles and unicorn horns'. *International Wildlife* (May/June), pp. 30–37.

Briggs, L. P. (1951) *The Ancient Khmer empire. (Transactions of the American Philosophical Society)*, Philadelphia: American Philosophical Society.

Brosius, J. P. (1991) 'Foraging in tropical rain forests: the case of the Penan of Sarawak, East Malaysia (Borneo)'. *Human Ecology*, vol. 19, no. 2, pp. 123–150.

Chang, K. C. (ed.) (1977) *Food in Chinese Culture: Anthropological and Historical Perspectives*. New Haven: Yale University Press.

Chevalier, Andrew (1996) *The Encyclopedia of Medicinal Plants*. New York: DK Publishers.

Compton, J. (1998) *Borderline: An Assessment of Wildlife Trade*. Worldwide Fund for Nature (WWF), Indochina Programme, Hanoi, Vietnam.

Condominas, G. (1972) 'Aspects of economics among the Mnong Gar of Vietnam: multiple money and the middlemen'. *Ethnology*, vol. 11, no. 3, pp. 202–219.

Cushman, J. W. (1993) *Fields from the Sea: Chinese Junk Trade with Siam during the Late Eighteenth and Early Nineteenth Centuries*. Ithaca: Southeast Asia Program, Cornell University.

Dai, Y. and X. Luo (1996) 'Functional food in China'. *Nutritional Reviews*, vol. 54, no. 11, pp. S21–S23.

Dang Nghiem Van, Chu Thai Sou and Lau Hung (1993) *Ethnic Minorities in Vietnam*. Hanoi: Gioi Publishers.

Donovan, D. G. (1998) 'The effects of market liberalization on forests and people in northern Vietnam, Laos and Yunnan PRC'. Paper at International Conference of Asian Scholars. June 23–25, 1998, Nordwijkerhout, Netherlands.

Donovan, D. G., A. T. Rambo, J. M. Fox, Le Trong Cuc, Tran Duc Vien (eds) (1995) *Development Trends in Northern Vietnam*. 2 vols. Center for Natural Resources and Environmental Studies, Vietnam University, Hanoi and East-West Center, Hanoi.

Dwyer, P. D. (1985) 'A hunt in New Guinea: some difficulties for optimal foraging theory'. *Man* (N.S.), vol. 20, pp. 243–253.

Engelhardt, R. A., Jr (1989) 'Forest gatherers and strand-loopers: economic specialization in Thailand'. In *Culture and Environment in Thailand*. Bangkok: The Siam Society, pp. 125–142.

Evers, II. (1985) 'Traditional trading networks of Southeast Asia'. *Working paper* no. 67, University of Bielefeld, Federal Republic of Germany.

—— (1989) *Trade and State: Social and Political Consequences of Market Integration in Southeast Asia*. Working paper 127, University of Bielefeld, Federal Republic of Germany.

—— (1990) *Trading Minorities in Southeast Asia: A Critical Summary of Research Findings*. Working Paper 139, University of Bielefeld, Federal Republic of Germany.

Ferkiss, V. (1993) *Nature, Technology and Society: Cultural Roots of the Current Environmental Crisis*. New York: New York University Press.

Freese, C. H. (1998) *Wild Species as Commodities: Managing Markets and Ecosystems for Sustainability*. World Wide Fund for Nature, Washington, DC.

Fox, M. (1998) 'Endangered creatures could save us, if we could save them'. *Reuters* (29 April 1998) (via ClariNet).

Geddes, W. R. (1976) *Migrants of the Mountains: The Cultural Ecology of the Blue Miao (Hmong Njua) of the Thailand*. Oxford: Oxford University Press.

Gittings, J. (1998) 'Smuggled monkeys saved from Asian taste for brains'. *The Guardian* (Manchester) (23 Oct 1998) (via ClariNet).

Graham, D. C. (1954) *Songs and Stories of the Ch'uan Miao*. Smithsonian Miscellaneous Collections, 123(1).

Haellquist, K. R. (ed.) (1991) *Asian Trade Routes, Continental and Maritime*. Studies on Asian Topics no. 13. Copenhagen: Nordic Institute of Asian Studies.

Hall, K. R. (1985) *Maritime Trade and State Development in Early Southeast Asia*. Honolulu: University of Hawaii Press.

Hames, R. (1991) 'Wildlife conservation in tribal societies'. In M. L. Oldfield and J. B. Alcorn (eds) *Biodiversity: Culture, Conservation and Ecodevelopment*. Boulder: Westview Press, pp. 172–199.

Headland, T. N. and L. A. Reid (1989) 'Hunter-gatherers and their neighbors from prehistory to present'. *Current Anthropology*, vol. 30, no. 1, pp. 43–66.

Hefner, R. (1998) *Market Cultures: Society and Morality in the New Asian Capitalisms*. Boulder: Westview Press.

Highley, K. and S. C. Highley (1998) 'Bear farming and trade in China and Taiwan'. *Earthtrust Taiwan*. <www.earthtrust.org/bear.html>.

Hong, C. (1997) 'Public awareness of need to protect wildlife still lacking'. In *New Strait Times* (Malaysia). 14 June 1997, p. 2.

Hoover, Craig (1998) *The US Role in the International Live Reptile Trade*. TRAFFIC-North America. Washington, D.C.

Hsu, H. (1982) *Chinese Herbal Medicine*. Long Beach: Oriental Arts Institute.

Hu, S. Y. (1990) 'History of the introduction of exotic elements into traditional Chinese medicines'. *Journal of Arnold Arboretum*, vol. 28, pp. 487–526.

Ingold, T. (1987) *The Appropriation of Nature: Essays on Human Ecology and Social Relations*. Manchester: Manchester University Press.

Kaskija, L. (1995) *Punan Malinau: The Persistence of an Unstable Culture*. Master's thesis, Department of Anthropology, University of Uppsala, Sweden.

Kaur, J. (1998) 'Ignoring wealth before our eyes'. *New Straits Times* (15/09/98).

Kunstadter, P. and S. Lennington Kunstadter (1983) 'Hmong (Meo) highlander merchants in lowland and Thai markets: spontaneous development of highland-lowland interactions'. *Mountain Research and Development*, vol. 3, no. 4: 363–371.

Le, Dien Duc (1998) 'Wildlife Trade in Vietnam'. In D. Donovan (ed.) *Policy Issues of Transboundary Trade in Forest Products in northern Vietnam, Lao PDR and Yunnan PRC, Vol. 2: Country Reports*. East-West Center, Honolulu and World Resources Institute, Washington, DC.

Le, Dong Phuong (1995) 'Resources management in a mountainous district in northern Vietnam'. Program on Environment, East–West Center, Honolulu (Unpublished manuscript).

Lebar, F. M., G. C. Hickey and J. K. Musgrave (1964) *Ethnic Groups of Mainland Southeast Asia*. New Haven: Human Relations Area Files (HRAF) Press.

Lee, J. (1995) 'Poachers, tigers and bears. Oh my! Asia's illegal wildlife trade'. *Northwestern Journal of International Law and Business*, vol. 16, pp. 497–515.

Lee, S. K. H. (1996) 'A world apart? Attitudes toward traditional Chinese medicine and endangered species in Hong Kong and the United States'. TRAFFIC-East Asia, Hong Kong. (via Internet).

—— (1999). *Attitudes of HK Chinese Towards Wildlife Conservation and the Use of Wildlife as Medicine and Food*. Species in Danger Series. TRAFFIC-Hong Kong (via Internet).

Leininger, M. (1970) 'Some cross-cultural universal and non-universal functions, beliefs and practices of food'. In J. Dupont (ed.) *Dimensions of Nutrition*. Denver: Colorado. Assoc. University Press, pp. 153–179.

Li, W., T. Fuller and W. Sung (1996) 'A survey of wildlife trade in Guangxi and Guangdong, China'. *Traffic Bulletin*, vol. 16, no. 1, pp. 9–16.

Ma Huan (1433) *Ying-yai sheng-lan (The Overall Survey of the Ocean's Shores)*. Translated with introduction, notes and appendices by J. V. G. Mills. Reprint 1997. Bangkok: White Lotus Press.

Marco Polo (1298) *The Travels of Marco Polo*. Reprint 1985. London: Penguin.

McNeely, J. A. and P. Spencer Wachtel (1991) *Soul of the Tiger*. Singapore: Oxford University Press.

Mills, J. A., S. Chan and A. Ishihara (1995) *The Bear Facts: The East-Asian Market for Bear Gall Bladder.* Species in Danger Series. TRAFFIC International, Cambridge, UK.

Mills, J. A. and P. Jackson (1994) *Killed for a Cure: A Review of the Worldwide Trade in Tiger Bones.* Species in Danger Series. TRAFFIC International. Cambridge, UK.

Momoki, S. (1998) 'Was Champa a pure maritime polity?' Paper in seminar on eco-history and the rise/demise of the dry areas in Southeast Asia, 13–16 October, Kyoto, Japan, Center for Southeast Asian Studies, Kyoto University.

Mottin, J. (1980) *History of the Hmong.* Bangkok: Odeon Store, Ltd.

Nash, S. V. (ed.) (1997) *Fin, Feather, Scale and Skin: Observations on the Wildlife Trade in Lao PDR and Vietnam.* TRAFFIC-Southeast Asia, Kuala Lumpur.

Nguyen, Van Thanh (1995) 'The Hmong and Dzao People in Vietnam: impact of traditional socioeconomic and cultural factors on the protection and development of forest resources'. In A. T. Rambo *et al.* (eds) *The Challenges of Highland Development in Vietnam.* Program on Environment, East–West Center, Honolulu, pp. 101–120.

Nguyen, Viet Chien (1996) 'Increase in smuggling of rare animals'. *Thanh Nien Newspaper,* Hanoi.

Nguyen, Xuan Thu (1993) 'A glimpse of the traditional medicines of animal origins'. In *Vietnamese Traditional Medicine.* Hanoi: The Gioi Publishing House, pp. 144 –156.

Ody, P. (1993) *The Complete Herbal.* London: Doris Kindersley Ltd.

Pei S. (1988) 'Plant products and ethnicity in the markets of Xishuangbana, Yunnan Province, China'. In A. T. Rambo, K. Gillogly and K. L. Hutterer (eds) *Ethnic Diversity and the Control of Natural Resources in Southeast Asia.* Program on Environment, East-West Center, Honolulu, Hawaii.

Peterson, J. T. (1978) 'Hunter-gatherer/farmer exchange'. *American Anthropologist,* vol. 80, pp. 335–351.

Pham, Quang Hoan (1995) *The Role of Traditional Social Institutions in Community Management of Resources among the Hmong of Vietnam.* East–West Center Working Papers, Indochina series. No. 5. East–West Center, Honolulu.

Pookajorn, S. (1985) 'Ethnoarchaeology with the Phi Tong Luang (Mlabrai): forest hunters of Northern Thailand'. *World Archaeology,* vol. 17, no. 2, pp. 206–220.

Puri, Rajindra K. (1997) *Hunting Knowledge and the Penan Benalui of East Kalimantan, Indonesia.* Dissertation. Dept. of Anthropology, University of Hawaii.

Rabinowitz, A. (1997) 'Lost world of the Annamites'. *Natural History Journal,* vol. 4, pp. 14–18.

Rai, N. K. (1990) *Living in a Lean-to: Philippine Negrito Foragers in Transition.* Ann Arbor: Museum of Anthropology, University of Michigan. Anthropological Papers No. 80.

Read, B. E. (1931) *Chinese Materica Medica: Animal Drugs.* From Pen Ts'ao Kang Mu by Li Shih-Chen, 1597. Peiping (China): Peking Natural History Bulletin.

Reid, A. (1993) *Southeast Asia in the Age of Commerce, 1450–1680.* Vol. II: Expansion and crisis. New Haven: Yale University Press.

Reuters (1998) 'Alleged exotic reptile traffickers face U.S. trial'.

Ridley, M. (1996) *The Origins of Virtue.* London: Penguin Books.

Ritvo, H. (1997) *The Platypus and the Mermaid and Other Figments of the Classifying Imagination.* Cambridge: Harvard University Press.

Salter, R. E. (1993) *Wildlife in Lao PDR: a status report.* IUCN, Vientiane.

Savina, F. M. (1930) *Histoire des Miao.* Hong Kong: Societé des Missions Étrangeres.

Schaller, G. (1998) 'On the trail of new species'. *International Wildlife* (July/August), pp. 37–43.

Schrock, J. L., W. Stockton Jr, E. M. Murphey and M. Fromme (1966) *Minority Groups in the Republic of Vietnam*. Ethnographic Study Series. Dept. of Army Pamphlets. No 550-105. Washington, DC: Department of the Army.

Sellato, B. J. L. (1993) 'The Punan question and the reconstruction of Borneo's culture history'. In V. H. Sutlive, Jr (ed.) *Change and Development in Borneo*. Williamsburg: Borneo Research Council, Inc.

Simoons, F. J. (1991) *Food in China: A Cultural and Historical Inquiry*. New Haven: Yale University Press.

Stewart, A. (1998) 'Wildlife tempting educated gourmets'. *South China Morning Post*. 30 November 1998. (via ClariNet).

Taravanit, P. (1998) *Medicinal Plants and Options under IMF-led Economy*. Thai Farmers Research Center Co. Ltd.

Trankell, I.B. (1995) 'Cooking, care and domestication: a culinary ethnography of the Tai Yong, Northern Thailand'. *Uppsala Studies in Cultural Anthropology*, vol. 21. Uppsala: Acta Universitatis Uppsaliensis.

Waldman, P. (1998) 'Desperate Indonesians devour country's endangered species'. *Wall Street Journal*. 26 Oct., 1998.

Weng, W. and J. Chen (1996) 'The eastern perspective on functional foods based on traditional Chinese medicine'. *Nutrition Reviews*, vol. 54, no. 11, S11–S16.

Westing, A. H. and C. E. Westing (1981) 'Endangered species and habitats of Vietnam'. *Environmental Conservation*, vol. 8, no. 1, pp. 59–62.

Wicks, R. S. (1992) *Money, Markets and Trade in Early Southeast Asia: The Development of Indigenous Monetary Systems to AD 1400*. Ithaca: Southeast Asia Program, Cornell University.

Wolters, O. W. (1967) *Early Indonesian Commerce*. Ithaca: Cornell University Press.

Wordsworth, C. (1997) 'China: the "X" generation'. In *Advertising and Marketing Guide to China*. Shanghai: J. Walter Thompson.

6

COCONUT-PICKING MACAQUES IN SOUTHERN THAILAND

Economic, cultural and ecological aspects

Leslie E. Sponsel, Poranee Natadecha-Sponsel and Nukul Ruttanadakul

Introduction

In a fascinating novel, *That You Shall Know Them* (Vercors 1953), a journalist discovers human-like primates being forced to work in plantations in New Guinea. He believes that this is slavery because of their closeness to humans. He brings a female back to London where they have a child through artificial insemination. The infant is recognized as human by the state and church through registering the birth and baptism. Next he kills the child with strychnine poison, and then insists on being tried for murder to force a test case. In court the lawyers and the jury debate whether or not the infant he killed was human or a sub-human animal. Although fiction, the book is inspired by some facts. In parts of South and Southeast Asia macaque monkeys are trained to harvest tree crops. Whether or not this is humane might be debated by some of those concerned with animal rights, and Vercors was a pioneer in that arena.

In this chapter we attempt to integrate analyses of ecological relationships and cultural representations of monkeys in southern Thailand. We argue that the explanation of this phenomenon may be found in a combination of factors: economic opportunity; macaques as 'weed species' (as explained below); the appropriateness of macaques for arboricultural labour; and the cultural meaning of monkeys, especially in the local syncretic religions. The result is an evolutionary shift in types of relationships between humans and macaques, from an emphasis on monkey competition (crop raiding) and, in response, human predation on the monkeys, to a form of cooperation among the two species.

Curiously, this phenomenon has never been researched systematically in any detail; so far we have found only five brief anecdotal articles (La Rue 1919, Gudger 1923, Sitwell and Freeman 1988, Peffer 1989, Sirorattanakul 1997) and one more detailed article focused on the training of the monkeys (Bertrand 1967). This subject does not seem to have been taken very seriously, but just dismissed as monkey business! However, coconut production is an important component of local subsistence and economy in southern Thailand and elsewhere in South and Southeast Asia, and in many of these areas monkeys are an integral part of the production. This is one of only three examples of primates which may have been semi-domesticated, as far as we have been able to determine, and thus it appears to be an extremely rare phenomenon.[1] Furthermore, this phenomenon may have significant implications for the future evolution and conservation of macaques as well as for understanding animal domestication.

This chapter draws on information from three sources: our field observations of macaque monkeys trained to pick tree crops; our interviews with the owners and trainers of the monkeys in Muslim and Buddhist villages; and relevant background and supplementary literature. The chapter is organized as follows: first, we sketch the context of coconut plantations; second, we discuss aspects of the religions and cultures relevant to this use of monkeys; third, we directly describe the monkeys; and fourth, we develop our main argument regarding ecology and conservation.

Coconuts

The coconut palm (*Cocos nucifera*) has been recognized as the 'tree of life' in many coastal areas of islands and continents throughout the humid lowland tropics. For some cultures the coconut palm is also a sacred tree, and the coconut is still used as a religious offering in some parts of Southeast Asia. The coconut palm is a reliable source of food and water when other sources fail, and it can provide low income households with a dependable and quick source of cash as needed, such as from copra (dried coconut meat). Almost every part of the palm has many uses. In addition to nutritional food and drink, it may provide material for building houses, furniture and fences, charcoal for fuel and dozens of other products for local and external consumption. Thus, this palm is an integral and important part of the subsistence, economy and culture of many local communities (Persley 1992).

There are two types of coconut palms, tall and dwarf. The tall type grows up to 30 metres in height; begins to produce nuts after five to seven years; is most productive at fifteen to twenty years; and can produce for up to sixty years. Under favourable conditions it will produce sixty to seventy nuts per year that mature within twelve months of pollination. Coconut palms grow throughout most of Thailand, but nearly half are in the southern peninsula, especially along its eastern coast, where conditions are optimal. The palms may be grown in home gardens, along rice paddy dikes, in orchards and in monocrop or polycrop plantations in areas ranging from a few rai to 50–100 rai (1 hectare equals 6.25 rai). Intercropping may include cocoa, coffee, pineapple, spices, fruit trees and medicinal plants. Animals may be raised under the palms: cattle and other livestock, bees in hives and catfish or prawn in ponds (see Klodpeng 1990 and Rattanapruk 1991). Although quite variable, we estimate that on average

5–10 per cent of the land of a village may be devoted to coconut palms. A typical plantation is about 15 rai in size with about 25 trees per rai. Ripe coconuts are harvested from each tree almost monthly. In Thailand production is for household subsistence and for the domestic market, but relatively little is exported abroad. Coconut is used extensively in Thai cooking and provides about half of the fat in the diet (Donner 1982: 100–103, 492–493).[2]

The development of palm plantations is part of the process of agricultural diversification and regional specialization encouraged by government policies in Thailand in recent decades (Wilson 1983: 109). Railroads built in the first two decades of the twentieth century and highways and roads since the third decade and especially from the 1960s have greatly facilitated transport and trade which was previously mainly by sea on ships from ports like Pattani. Rubber (*Hevea brasiliensis*) was introduced into Thailand in 1901 as a new cash crop and has rapidly transformed the south into a market-oriented cash economy (Stifel 1973, Wilson 1983: 108). Also, monocrop plantations of rubber trees of a high-yield variety have almost completely replaced the previous mixed crops of rubber and fruit trees since the government started promoting the export of rubber in 1961. The expansion of monocrop plantations of rubber as well as coconut palm and oil palm (*Elaeis guineensis*) has brought about the conversion of vast areas of various kinds of natural forest into a mosaic of diverse agroecosystems. For example, between 1930 and 1990, the area of rubber plantations expanded by 16.4 times and that of coconut plantations nine times (Wilson 1983: 118–120, National Statistical Office 1991: 45). During this same period forest cover in the south declined from about 75 per cent to 15 per cent, reflecting a tremendous loss of habitat for numerous wildlife species including primates (see Leungaramsri and Rajesh 1992, Boulbet 1995, Sponsel 1998).

The expansion of coconut palm plantations in southern Thailand may be related to the decimation of palms in the central and northeast regions during the nineteenth century by infestations of the black beetle (*Oryctes rhinoceros*) and the red weevil (*Thynochophora*) (Donner 1988: 100). This expansion is also related to the search by Muslims for new economic opportunities as the fisheries yields have declined in recent decades.[3] However, since 1981 coconut production has declined along with yields and market values in comparison to rubber and oil palm. In some areas that formerly specialized in coconut production, such as the island of Ko Samui, this industry is declining as the economic focus has shifted to tourism (Prachuabmoh 1992: 138–139, 174–176).

Culture

Southern Thailand is a fascinating region because of the sharp contrasts in cultures (Thai Buddhists and Thai Muslims of Malay heritage) and environments (coastal to inland forest biomes), all existing in close proximity because of the long and narrow shape of the peninsula. There is a large 'Malay' population, as a result of the arbitrary delineation of the border between Thailand and Malaysia in 1909 by the British. While many Muslims speak Thai and consider themselves to be Thai, their culture and language (Pattani dialect) are derived from Malay, and most call themselves Jawi. They are 'people of the Book', the Koran, and their religious beliefs and practices are those of orthodox Muslims: five daily prayers in the direction of Mecca, Friday sermons in the local mosque, observance of Islamic holy days and so on. However, their religion is actually syncretic, as it includes a mixture of elements from the previous substrata of religion – animism, Hinduism and Buddhism, as well as Islam.[4] The distinctive combination of domestic animals is one way to recognize the ethnicity of a village. A Buddhist community will have dogs and pigs, whereas a Muslim community will have goats and pet birds (especially doves used in song contests), often monkeys, but no dogs or pigs. In general, dogs and monkeys are incompatible in the same village.

It appears to us that the use of monkeys in southern Thailand is mainly a pragmatic matter of economics. Furthermore, we argue that the cultural meaning of monkeys facilitates their economic use (or more accurately, exploitation) and, in general, their relatively humane treatment in captivity. The Ramayana, the Hindu epic from India, remains popular to this day in shadow plays for both Muslims and Buddhists in Thailand, as well as for many people elsewhere in Southeast Asia. Among other things, this legend expresses attitudes about the desirable relationships between humans and animals, and it may facilitate the custom in question here. There are four principal characters: Rama, an incarnation of the god Vishnu; his wife Sita; the demon-king Rawana (Thotsagun in the Thai version); and Hanuman, the general of the monkeys. Hanuman becomes a hero by collaborating with Rama in rescuing Sita from Rawana (see Ludvik 1994). Coomaraswamy (1965) highlights the importance of Hanuman:

> It may be questioned whether there is in the whole of literature another apotheosis of loyalty and self-surrender like that of Hanuman. He is the Hindu ideal of the perfect servant, the servant who finds full realization of manhood, of faithfulness, in his obedience – the subordinate whose glory is in his own inferiority.
>
> (In McNeely and Sochaczewski 1995: 61)[5]

This respect for monkeys is also reflected in Buddhism, because in one of his incarnations the Buddha was a monkey. Also, the Buddha used an allegory about an elephant, monkey, rabbit and partridge to illustrate mutual respect among animals (see Byles 1967: 167–168). Moreover, Buddhism holds, in principle, that humans are part of nature and should interact with it in a humble, compassionate, non-violent and beneficial manner (for example, Buri 1989: 52, Chapple 1993). This attitude may be reflected in one handler's observation that, in contrast to harvesting coconuts on a plantation, his monkey does not have to work as hard when putting on a show for tourists (Sirorattanakul 1997).

At the same time, monkeys are definitely considered to be inferior to humans. In Islam, for example, men can be transformed into monkeys (or pigs) as a punishment by God (Masri 1989: 3). Furthermore, one of the most degrading insults in Islam is to liken another human to a monkey (Vire 1986: 131–132). This cultural prejudice rendering the monkey as 'other' and a lesser species also facilitates their exploitation. Nevertheless, monkeys are still recognized for their exceptional positive qualities among animals, such as their cleanliness (perhaps following observations of the social grooming of their hair) and their considerable ability for imitation. Recognition of the obvious physical resemblance of monkeys to humans may have contributed to the Islamic prohibition on eating monkey flesh, since it would be close to cannibalism (Vire 1986: 131–133). Our Thai Muslim informants believe that, like humans, the monkey has some kind of spirit or soul. More generally, the ideal principles of Islam view animals as part of Allah's creation, thus they should be treated with appropriate reverence, non-violence, compassion and loving care, even though (and this could be contradictory) humans may also use them when necessary to satisfy their basic needs and for convenience. Thus, animals may be considered to have rights by virtue of being part of creation.[6] Furthermore, it is the responsibility of humans to recognize all of this and to treat animals ethically because as spiritual and moral beings humans are considered superior to animals, even if animals may be superior physically.[7]

Islam, Hinduism and Buddhism were profoundly influenced by the earlier religion of Jainism (Chapple 1993: 9, 26–27, 43, 79, 112). While there are certainly significant differences among these four religions, there are also remarkable similarities. The quintessential similarity is recognition of the unity and identity of all life, and the corresponding respect and reverence for all life, at least in principle. The concern for non-violence (*ahimsa*) is central to Jainism, where to harm other beings is to harm oneself. This follows from the Jain belief that all beings have a life force (*jiva*) and that all beings are reincarnated after death into new beings. Jainism is unparalleled as a religion focused on the daily practice of non-violence to all beings (Chapple 1993: 4, 10, 21). (Hindus, however, may practise some forms of animal sacrifice.) In both Jainism and Buddhism, non-injury to animals is an important concern. As Chapple explains:

> Animals are regarded to be none other than our very selves. The underlying assumption is that each and every human being has experienced a wide variety of animal births in prior incarnations and that if one makes a mistake of significant proportions during human birth one will again be born as an animal in punishment for wrongdoing. One of the great acts of wrongdoing that will cause rebirth as an animal is undue harm to an animal or human, both of which are to be seen as akin to oneself.
>
> (Chapple 1993: 43)

It would seem that this common attitude toward animals in these different religions would be mutually reinforcing. At the same time, it must be cautioned that, as in the case of all religions and societies, practice may not always approximate to ideals.

Monkeys are clearly a multivalent symbol for Thais, which allows for latitude in their interpretation, including rationalization of the use of monkeys for arboricultural labour. For example, Thai Buddhists draw a strong overt distinction between human (*khon*) and animal (*sat*) (Wijeyewardene 1968: 76, Tambiah 1969: 443, 446). Referring to another person as an animal, especially a dog (*ai ma*) or monkey (*ai wok*), is a form of verbal abuse. Likening someone to a monkey is to suggest that they are somehow less than human (Wijeyewardene 1968: 79, 87, Tambiah 1969: 435). But the custom of using monkeys to harvest tree crops poses a challenge to the binary oppositions of animal–human, wild–domestic, and nature–culture. The monkeys are borderline cases of domestic and domesticated.

On the one hand, they are given human names; they are kept in the village adjacent to the house of the handler and during a storm brought onto the front porch of the house or into a shelter; they are regularly provisioned with food and water; they are taken to an irrigation canal or other body of water to swim or bathe; they are groomed by their handler using a comb (and the monkey may groom the owner in turn); and they are given a ride on the back portion of the bicycle, motorcycle or cart of the handler en route between house and plantation. If a monkey dies it is buried, although not near human graves and without any ritual. On the other hand, monkeys are always tethered on a long rope or chain (around the neck) and are never allowed to roam freely. Usually, there is no attempt to breed monkeys and no artificial selection for genetic traits to improve the stock. When monkeys become too old to be productive, they are generally released back into the forest, though some are kept as pets. Indeed, the Thai classification from forest to field to village reflects a continuum of landscape ecology from wild to domestic, rendering such distinctions relative rather than absolute (see Gillogly 1988: 60).

Macaques

Given their arboreal habit, locomotor agility, manual dexterity and intelligence, primates appear to be the only mammals which might be considered biologically pre-adapted for harvesting tree crops. Among the many species of primates in tropical Asia, macaques seem to be the most appropriate for use in harvesting tree crops. Macaques are widespread geographically, diverse ecologically, flexible in arboreal and terrestrial locomotion and omnivorous in diet. Some species are large in body size for monkeys. Their habitat ranges from tropical to temperate forests, and also includes mangrove forests, grasslands, and dry scrub and cactus (Napier and Napier 1967: 207).[8]

There are fragmentary and mostly anecdotal reports scattered in the literature regarding the use of the pig-tailed macaque (*M. nemestrina*) for picking coconuts in several parts of Southeast Asia.[9] India, Thailand, Malaysia and Indonesia (especially Sumatra and Borneo) are among the countries mentioned, although often the reference is only to Southeast Asia in general and no countries or species are specified.[10] The full geographic range and antiquity of this custom is not known. The custom probably extends back to at least the early nineteenth century in Sumatra and Borneo (Gudger 1923). It may have spread from Sumatra into Malaysia and then Thailand along with Islam, the latter in the fourteenth century. We have found no historical documentation for the custom's antiquity in Thailand, but an informant in his seventies claimed that the monkeys had been used in this manner as far back as the generation of his grandparents. That places the custom at least back into the mid-nineteenth century. The use of monkeys probably increased markedly with the expansion of coconut plantations during the twentieth century.

In southern Thailand both pig-tailed macaques from inland rain forests and crab-eating macaques (*M. fascicularis*) from coastal mangrove forests are used to harvest tree crops, especially coconut palms in plantations mostly along and near the coast, but also several species of fruit in orchards and forests, some further inland. The pig-tailed macaques are more likely to be used for climbing coconut palms. The crab-eating macaques are used for harvesting various kinds of fruit trees where they have the advantages of lighter weight and a longer tail with some grasping function. Other local tree crops harvested by monkeys include *ma-toom* (Bengal quince, *Aegle marmelos*), *sa-to* (parkia, *Leguminosae*) and *sadao* (neem, *Melia sp.*). Also, monkeys shake tree branches to harvest mangoes (*Mangifera indica*) and tamarind (*Tamarindus indica*).

In southern Thailand, coconut plantations are an important component of the local economy of villagers, and monkeys are usually an integral part of production. For example, in one village surrounded by coconut plantations, almost every one of the 200 households had its own monkey. However, in a second village of 200 households, only seven had monkeys, while in a third, six out of 29 households had plantations but there were only two monkeys. These variations reflect different sizes of plantations and different degrees of villager involvement in harvesting coconuts on plantations. On small plantations of less than 100 trees, the harvesting is usually done by young boys. But on larger plantations men use their specially trained monkeys. An efficient monkey can pick 500–1,000 coconuts a day and some informants say even up to 2,000. Most monkeys work with their handler from about 8 am until 5 pm, except for an hour of rest at noon. A handler working on the plantation owned by another person may keep about one-fifth of the nuts he collects. Some coconuts are used for household consumption, but most are sold. With a well-trained monkey working daily, a handler can earn 5,000 baht (US$200) a month or more harvesting coconuts and other tree crops. For some families this may provide up to half of their income. Although monkeys can harvest tree crops much more productively, cheaply, and safely than humans, the coconut tree may harbour hazards. The thick fur of the monkey protects it from most arachnids and insects such as scorpions, bees and ants, but sometimes monkeys die from the bite of a poisonous snake or break an arm or worse in a fall from a tree.

Monkeys are usually trapped in the wild; some informants, however, report that an infant monkey may be taken from its mother after she has been killed by a hunter. Informants say that monkeys do not breed well in captivity, even though one 74-year-old man had made a career of breeding and training monkeys. He kept them in a group of three males and two females. The usual lack of success in breeding monkeys in captivity may reflect the trauma of their capture and separation from their natural social and biological environments and other deprivations. Males are preferred over females because they are more productive in harvesting coconuts due to their greater size and strength. A freshly captured male monkey of one to two months old may be sold for 1,000–1,500 baht (US$40–60), and a female of the same age would be about 500–700 baht ($20–28). A trained four-year-old pig-tailed macaque can sell for 1,500–10,000 baht ($60–400), depending on its attributes which include how tame, smart, strong and hard working it is. The crab-eating macaque is less expensive at 800–1,000 baht ($32–40).

The training of the monkey may start as early as seven to eight months. But the best age to train a monkey is when it is between two and five years old (even though maternally deprived infants are likely to be traumatized and terrified). Usually only about two to three weeks are required to train a monkey to the minimum standard. They learn to follow at least six verbal commands, select coconuts in different stages of ripeness, and twist the stem to remove the nut. Training is based on applying or withholding physical punishment, such as hitting or beating with a stick, and on food rewards for appropriate behaviour, such as a drink of coconut milk. Although the trainer and subsequent handler dominate the monkey with fear of punishment, informants claim that an affectionate social bond is developed between them.[11] Furthermore, the monkeys owned by different handlers are kept separately rather than released into a group in an enclosure, although individuals may be in sight of one another. The monkey is attached by a chain and rope to a heavy pole about five metres in length which leans against a tree. This allows the monkey to move around on the ground, up and down along the pole and into the branches of the tree. Three times each day the monkey is given food, including rice, chicken or duck eggs, bananas and other local fruits and coconut juice. The cost to maintain the monkey amounts to only about 300 baht (US$12) per month. Obviously the monkey is an extremely cheap labour source, once the costs of the initial purchase and training are subtracted.

Ecology

In an article on the evolutionary implications of macaque feeding ecology, Richard et al. (1989) have distinguished between weed and non-weed species. They tentatively classify the crab-eating macaque as a weed species, but not the pig-tailed macaque – (the latter frequents primary rain forest more than secondary growth). They explain the characteristics of weed species in the following way:

Although the metaphor may not be totally apt, there are some striking parallels between weed plants and weed macaques. Weed plants thrive where people leave their mark on the land; they are plants which spread where people travel and settle down. They depend on people and, in fields, they compete with people. The weed macaques often live in and alongside towns and villages, and they exploit the fields of the farmers and secondary growth nearby. Indeed, they often depend directly or indirectly on human activities for a substantial portion of their diet. In short, like weed plants, weed macaques can be construed as human camp followers that may even occupy some habitats only because human disturbance is present.

The distinction between weed and non-weed macaques is a working hypothesis. We are not implying that non-weeds never raid crops or that weed species cannot survive without access to human resources. There are occasional reports of crop-raiding by all of the macaque species and of weed macaques in primary forest, far from human habitation. However, we suggest that there are marked differences in the frequency and success with which species exploit human resources.

(Richard *et al.* 1989: 573)

They go on to assert that this relationship between weed macaques and human farming is an adaptive strategy through which the monkeys respond effectively to the disturbance of their natural habitat (ibid. 1989: 570). Weed macaques are comparable to pioneer species in the early stage of ecological succession (plant community development). Weed macaques are also comparable to ecotone species which specialize in taking advantage of the adaptive opportunities afforded by the edge or transition zone between two different environments, such as forest and field (Richard *et al.* 1989: 575). Some species, especially the crab-eating macaque, seem to be just as well adapted to a mosaic of disturbed and secondary forests and agroecosystems, maybe even more so, than to primary forests (Bishop *et al.* 1981: 154, 158, Johns and Skorupa 1987: 175). A moderate degree of disturbance in a forest can increase food availability and in turn increase the primate population size and density (Wilson and Johns 1982: 206). Richard *et al.* (1989: 586–588) use their model to explain the coincidence between the distribution of agriculture and weed macaques and the widespread occurrence of the weed macaques compared to non-weed species. We extend this argument further within an evolutionary framework of shifts in the type of symbiosis between macaques and humans (Table 6.1).

Perhaps the entire range of different types of symbiosis generally recognized by ecologists is represented in one circumstance or another in the widespread, diverse and adaptable group of more than a dozen species of macaques. However, we hypothesize the following evolutionary sequence in the main type of symbiosis in southern Thailand, and by implication, possibly elsewhere (Table 6.1). As the agricultural frontier expands into the forest, the macaques

Table 6.1 Shifts in the cultural ecology of macaques

Habitat	Forest	Field	Village
Processes (ecological and economic)	deforestation	agricultural expansion	coconut commercialization
Ecological role monkey	omnivore	competitor	cooperation
Economic role	prey →	pest → (crop raider)	coconut-picking
Symbolic role	strange → wild meat nature	→	familiar domestic pet culture

raid crops (for example, Eudey 1986). The human response is to try to eliminate or reduce this competitor (crop pest) by predation. Because of the religious prohibition on eating monkeys, the hunter is less likely to gain energy and nutrients from the kill, and the costs of energy expenditure in predation may outweigh benefits or at least be disadvantageous. Recall that Hinduism, Buddhism and Islam all discourage eating monkey meat, and that elements of all three variously permeate the syncretic religion of the populace in southern Thailand. Thus, one would expect that the prohibition would be more likely than not to be honoured, even though exceptions occur. Under these circumstances, eventually it would be to the advantage of both the macaques and humans to develop a relationship of cooperation, one in which both species benefit. This is what seems to happen when the monkeys are used for harvesting tree crops. However, the monkeys do not enter this relationship voluntarily (though that is not a prerequisite for the application of the term cooperation in a biological sense). The human handlers provide the monkeys with water, food, shelter, protection, care and affection. In turn, the macaques are an economic asset to the handlers since they can pick coconuts faster than humans, thus increasing efficiency and production.

The weed species model and this evolutionary sequence of types of symbiosis have a broader implication – they focus more attention on domestication as an ecological process. That is, anthropogenic environmental change becomes a catalyst for domestication, rather than the usual emphasis on a particular species as the target for artificial (human) selection replacing natural selection. They imply that the initial stages of domestication may be part of a more or less natural ecological process, and that this may be largely if not wholly inadvertent rather than involving conscious human agency. That is not to suggest that this evolutionary sequence of types of symbiosis is inevitable and invariant, only that it may fit the specific circumstances of the historical ecology of many of the areas where macaques are used to harvest tree crops.

Conservation

Since the 1960s numerous national parks and wildlife sanctuaries have been established by the central government in Thailand, yet most are inadequately administered (Blockhus *et al.* 1992: 75–78). Outside of these 'protected' areas, forest conversion to monocrop plantations and other agroecosystems is rampant; thus, the destruction of the habitat of primate species remains very serious. On the other hand, the commercial export of all macaque species has been banned since 1976 (Jintanugool *et al.* 1982: 41). Also capture of primates for the domestic trade at town markets seems to be very limited (Round 1990: 38, Robinson 1994: 120). Our survey of two markets in Pattani throughout 1988 did not record any monkeys. However, some local restaurants which specialize in wildlife meat, especially for Thai Buddhists from the northeast who work on fishing boats, occasionally include macaques and langurs on their menu. A few of our informants reported that they sometimes shoot monkeys while hunting. This paucity of primates in the markets may reflect their depletion in the forests more than any adherence to government laws and regulations (for example, Dearden 1995).

Today, for Thailand, conservationists list all species of monkeys as threatened and all species of gibbons as endangered (Humphreys and Bain 1990: 239–266). Thus, the variety of cooperative relationship between humans and macaques described above assumes further importance – it may actually be of some significance in inadvertently conserving these two macaque species in spite of the marked decline in their forest habitats. Indeed, the main habitat of the crab-eating macaques, the mangrove forest, may be nearly completely destroyed within a decade (Katesombun 1992: 104). By several different means we estimate, very roughly, that there may be several thousand monkeys maintained for harvesting coconuts in southern Thailand alone.

This custom, however, appears to be on the decline for several reasons. The market competitiveness of the coconut is declining. Young Thais are less likely to join the occupation of monkey handlers because through Western education, the media and other influences they are attracted to more prestigious and higher paying jobs. In addition, in recent years in some locations where tourism has developed, monkey handlers and their monkeys have shifted from picking coconuts to putting on shows for tourists. During the peak of the tourist season (January–May) on Samui Island (Ko Samui), one monkey owner earns up to 600–700 baht per day (US$24–28), much more than he could earn by putting the monkey to work on plantations (Sirorattanakul 1997). Consequently, if present trends continue, the macaques may have only a temporary escape from the destruction of their natural forest habitats and they may be headed toward extinction.

Many of those who are especially sensitive to the problem of speciesism and environmental ethics and animal rights would likely argue that there is cruelty to animals in this custom because they are trapped in the wild; because physical punishment is meted out to them in training; because they are required to work hard; and because sometimes monkeys which are either too aggressive or old are killed. Furthermore, individual monkeys are removed and isolated from their normal life in a social and reproductive group as well as from their forest habitats. On the other hand, it appears to us that, in general, compassion prevails in the handler's treatment of his monkey. Of course, it might be argued that good treatment of monkeys can be accounted for on the pragmatic grounds that the handlers depend, for their livelihoods, on the well-being and cooperation of the monkey. However, we suggest that religious concerns tend also to be involved, as discussed earlier.

As far as we are aware, for Thai people the use of the macaques as arboricultural labour is not controversial. There seems to be essential agreement among villagers from diverse historical, cultural and religious backgrounds on this matter. We are not aware of any criticism or opposition regarding this practice from any environmental or animal rights organizations, either local, national or international. There are much larger concerns for threatened and endangered animal species with respect to habitat degradation and destruction as well as the illegal traffic in wildlife trade.[12] While personally we do not approve of this custom, rather than targeting single species and single issues for conservation, we advocate placing priority on the conservation of biodiversity at the ecosystem level through more effective administration of protected areas by the government (national parks and wildlife sanctuaries) and by the community (community forests and sacred places) (see Sponsel *et al.* 1998).

Conclusion

In parts of Thailand and elsewhere in South and Southeast Asia, coconuts and/or monkeys are an important component of the livelihoods of local human communities. Based on our fieldwork we have documented a fascinating example where macaques are trained and used to harvest coconuts and other tree crops in southern Thailand. We have argued the following propositions: that this traditional practice is facilitated and reinforced by religious and other cultural beliefs, values and attitudes toward the monkeys, which also promote their humane treatment; that the relationship between humans and macaques has changed over time from predation to competition to cooperation with agricultural conversion of their forest habitats and their consequent new status as a weed species; that the trend toward domestication of the macaque is part of the ecological process of anthropogenic disturbance of its habitat; and that to some degree this cooperative relationship may promote macaque conservation. This rare, curious and neglected phenomenon of the semi-domestication of a primate species deserves more systematic and detailed investigation by primatologists, cultural anthropologists, economists and others.

Acknowledgements

Financial support to Leslie Sponsel for this field research on human ecology in southern Thailand was provided by the Wenner-Gren Foundation for Anthropological Research, Fulbright–John F. Kennedy Collaborative Research Grant and the University of Hawai'i Research Council. Personnel and facilities at the Prince of Songkla University in Pattani greatly facilitated our research, including former Dean Preeya Viriyanon. Professor Rawiwan Chaumphruk of PSU kindly provided information on monkeys in Jawai folklore and culture. Vivaswan Verawudh (Chulalongkorn University) helped in locating information in the library at Chulalongkorn University in Bangkok and in translating Thai material. Pierre Le Roux (National Research Centre of France) kindly provided additional information from his own research with Jawi. Michael Dove (Yale University) first alerted us to the article by Richard *et al.*, on weed macaques, which became a central part of our argument. Lucy Wormser (Executive Director, Pacific Primate Sanctuary, 500 A Haloa Road, Haiku, Maui, Hawai'i 96708–9362, http:www.planet-hawaii.com/pps/) provided penetrating comments that stimulated us to substantially revise some statements that suffered from speciesism.

Notes

1 These are the only journal articles we have found on coconut picking monkeys despite a literature search by us and by the Primate Information Center of the University of Washington in Seattle and an inquiry to the Wisconsin Regional Primate Research Center in Madison.

2 For a more detailed description of coconut palms in one Muslim village in southern Thailand, see the classic study of Rusembilan by Fraser (1960: 56–76).

3 See Fraser (1960), Polioudakis (1996), Quarto (1992), Ruohomäki (1996) and Ruyabhorn and Phantumvanit (1988).

4 This ethnographic sketch is based mainly on our field observations. For further information, see Firth (1946) and Fraser (1960).

5 The veneration of Hanuman is at least part of the reason why many cultures in South and Southeast Asia consider monkeys to be sacred and/or taboo. See Aggimarangsee (1992), Bishop, *et al.* (1981), Graham and Round (1994: 101, 115), McNeely and Sochaczewski (1995: 61–62) and Seth and Seth (1986).

6 There is even a famous Islamic treatise on animal rights from the tenth century called *The Case of the Animals versus Man* (Al-Safa 1978).

7 Al-Qaradawi (1960:343–345); Khalid and O'Brien (1992: 14–19, 63–64); and Masri (1989: 1–31).

8 Depending on different classifications, there are 12–19 species of macaques, and all are found only in Asia, except for the so-called Barbary Ape (*Macaca sylvanus*) of northwest Africa. Thailand has five species of macaques, all sympatric in the western portion of the country, as in the Huai Kha Khaeng Wildlife Sanctuary (Eudey 1986, Stewart-Cox 1995: 122). Primate species in Thailand include: slow loris (*Nycticebus coucang*), rhesus (*Macaca mulatta*), crab-eating macaque (*M. fascicularis*), assamese macaque (*M. assamensis*), pig-tailed macaque (*M. nemestrina*), stump tailed macaque (*M. arctoides*), langurs (*Presbytis melalophos, P. phayrei, P. obscura, P. cristata*) and gibbons (*Hylobates lar, H. pileatus, H. agilis*).

9 Child (1964: 131), Donner (1982: 493), Fraser (1966), Gudger (1919; 1923), Heiser (1990: 161), La Rue (1919), Lekagul and McNeely (1988: 281), Medway (1978: 51), Roonwal and Mohnut (1977: 178), and Rowe (1996: 128).
10 Also this custom has been reported to us in personal communications for southern India by Safia Aggarwal, for northeastern India by Mahesh Rangarajan and for Sumatra by Sumastuti Sumukti and Gerard Persoon.
11 See Bertrand (1967: 484) for more detail on training.
12 Brockelman (1987), Graham and Round (1994), Humphrey and Bain (1990), Sponsel (1998), Sponsel and Natadecha-Sponsel (1991; 1998) and Stewart-Cox (1995).

References

Aggimarangee, N. (1992) 'Survey for semi-tame colonies of macaques in Thailand'. *Natural History Bulletin of the Siam Society*, vol. 40, pp. 103–166.

Al-Qaradawi, Y. (1960) *The Lawful and the Prohibited in Islam*. Indianapolis: American Trust Publications.

Al-Safa, I. (1978) *The Case of Animals versus Man Before the King of the Jinn: A Tenth Century Ecological Fable of the Pure Brethren of Basra* (translated by L. V. Goodman). Boston: Twayne Publishers.

Bertrand, M. (1967) 'Training without reward: traditional training of pig-tailed macaques as coconut harvesters'. *Science*, vol. 155, pp. 484–486.

Bishop, N., S. B. Hardy, J. Teas and J. Moore (1981) 'Measures of human influence in habitats of South Asian monkeys'. *International Journal of Primatology*, vol. 2, no. 2, pp. 153–167.

Blockhus, J. M., M. Dillenbeck, J. A. Sayer and P. Wegge (1992) *Conserving Biological Diversity in Managed Tropical Forests*. Gland, Switzerland: International Union for Conservation of Nature Forest Conservation Programme.

Boulbet, J. (1995) *Towards a Sense of the Earth: The Retreat of the Dense Forest in Southern Thailand During the Last Two Decades*. Pattani, Thailand: Prince of Songkla University.

Brockelman, W. (1987) 'Nature conservation'. In A. Arbhabhirama *et al.* (eds) *Thailand Natural Resources Profile*. Bangkok: Thailand Development Research Institute, pp. 91–119.

Buri, R. (1989) 'Wildlife in Thai culture'. In *Culture and Environment in Thailand*. Bangkok: Siam Society, pp. 51–59.

Byles, M. B. (1967) *Footprints of Gautama the Buddha*. Wheaton, IL: The Theosophical Publishing House.

Chapple, C. K. (1993) *Nonviolence to Animals, Earth, and Self in Asian Traditions*. Albany, NY: State University of New York Press.

Child, R. (1964) *Coconuts*. London: Longmans.

Coomaraswamy, A. K. (1965) *History of Indian and Indonesian Art*. New York: Dover Publications, Inc.

Dearden, P. (1995) 'Development, the environment and social differentiation in Northern Thailand'. In J. Rigg (ed.) *Counting the Costs: Economic Growth and Environmental Change in Thailand*. Singapore: Institute of Southeast Asian Studies, pp. 111–130.

Donner, W. (1982) *The Five Faces of Thailand: An Economic Geography*. St. Lucia, Queensland: University of Queensland Press.

Eudey, A. A. (1986) 'Hill tribes peoples and primate conservation in Thailand: a preliminary assessment of the problem of reconciling shifting cultivation with conservation objectives'. In J. G. Else and P. C. Lee (eds) *Primate Ecology and Conservation*. New York: Cambridge University Press, pp. 237–248.

Firth, R. (1946) *Malay Fishermen: Their Peasant Economy.* London: Routledge and Kegan Paul.

Fraser, T. M., Jr. (1960) *Rusembilan: A Malay Fishing Village in Southern Thailand.* Ithaca, NY: Cornell University Press.

Gillogly, K. A. (1988) *The Ethnography of Animal Husbandry in Lowland Thailand: Aspects of Human, Animal, and Crop Interactions.* Honolulu: M.A. University of Hawai'i Thesis in Anthropology.

Graham, M. and P. Round (1994) *Thailand's Vanishing Flora and Fauna.* Bangkok: Finance One Public Co., Ltd.

Gudger, E. W. (1919) 'On monkeys trained to pick coconuts'. *Science,* vol. 49, no. 1258, pp. 146–147.

—— (1923) 'Monkeys trained as harvesters: instances of a practice extending from remote times to the present'. *Natural History,* vol. 23, no. 3, pp. 272–279.

Heiser, C. B. Jr. (1990) *Seed to Civilization: The Story of Food.* Cambridge: Harvard University Press.

Humphrey, S. R. and J. R. Bain (1990) *Endangered Animals of Thailand.* Gainesville, FL: Sandhill Crane Press, Inc.

Jintanugool, J., A. A. Eudey and W. Y. Brockelman (1982) 'Species conservation priorities for Thailand'. In R. A. Mittermeier and W. R. Konstant (eds) *Species Conservation Priorities in the Tropical Forests of Southeast Asia.* Gland, Switzerland: Occasional Papers of the International Union for the Conservation of Nature Species Survival Commission Number 1, pp. 41–51.

Johns, A. D. and J. P. Skorupa (1987) 'Responses of rain forest primates to habitat disturbance: a review'. *International Journal of Primatology,* vol. 8, no. 2, pp. 157–191.

Katesombun, B. (1992) 'Aquaculture promotion: endangering the mangrove forests'. In P. Leungaramsri and N. Rajesh (eds) *The Future of People and Forests in Thailand After the Logging Ban.* Bangkok: Project for Ecological Recovery, pp. 103–122.

Khalid, F. and J. O'Brien (1992) *Islam and Ecology.* London: Cassell Publishers Ltd.

Klodpeng, K. (1990) 'Coconut based farming systems in Thailand'. In Sumith de Silva (ed.) *Coconut Based Farming Systems.* Manila: Asian and Pacific Coconut Community, pp. 198–201.

La Rue, C. D. (1919) 'Monkeys as coconut pickers'. *Science,* vol. 50, no. 1286, p. 187.

Lekagul, B. and J. A. McNeely (1988) *Mammals of Thailand.* (2nd edition) Bangkok: Sha Karn Bhaet Co.

Leungaramsri, P. and N. Rajesh (eds) (1992) *The Future of People and Forests in Thailand After the Logging Ban.* Bangkok: Project for Ecological Recovery.

Ludvik, C. (1994) *Hanuman.* Delhi: Motilal Banarsidass Publishers.

Masri, A. B. A. (1989) *Animals in Islam.* The Athene Trust.

McNeely, J. A. and P. Spencer Sochaczewski (1995) *Soul of the Tiger: Searching for Nature's Answers in Southeast Asia.* Honolulu: University of Hawai'i Press.

Medway, L. (1978) *The Wild Mammals of Malaya: Peninsular Malaysia and Singapore.* New York: Oxford University Press.

Napier, J. R. and P. H. Napier (1967) *A Handbook of Living Primates.* New York: Academic Press.

National Statistical Office (1991) *Quarterly Bulletin of Statistics,* vol. 39, no. 2. Bangkok, Thailand: National Statistical Office, Statistical Data Bank and Information Dissemination Division, pp. 45, 72–73, 111.

Persley, G. J. (1992) *Replanting The Tree of Life: Towards an International Agenda for Coconut Palm Research.* Oxford: C-A-B International.

126

Pfeffer, R. (1989) 'On Malay peninsula picking coconuts is monkey business'. *Smithsonian Magazine*, vol. 19, no. 10, pp. 111–118.

Polioudakis, E. J. (1996) 'Resource management, social class and the state at a fishing village in Southern Thailand for 100 Years'. In *Traditions and Changes at Local/Regional Levels* (Volume I of the Proceedings of the 6th International Conference on Thai Studies). Chiang Mai: Chiang Mai University, pp. 377–389.

Prachuabmoh, C. (1992) *Socio-Cultural Change In Southern Thailand 1950–1990: A Documentary Study*. Bangkok: Thammasat University/The Thailand Development Research Institute Foundation.

Quarto, A. (1992) 'Fishers among the mangroves'. *Cultural Survival Quarterly*, vol. 16, no. 4, pp. 12–15.

Rattanapruk, M. (1991) 'Small scale processing of coconut products in Thailand'. In B. B. Pangahas (ed.) *Small Scale Processing of Coconut Products*. Manila: Asian and Pacific Coconut Community, pp. 169–173.

Richard, A. F., S. J. Goldstein and R. E. Dewar (1989) 'Weed macaques: the evolutionary implications of macaque feeding ecology'. *International Journal of Primatology*, vol. 10, no. 6, pp. 569–594.

Robinson, M. F. (1994) 'Observation on the wildlife trade at the daily market in Chiang Khan, Northeast Thailand'. *Natural History Bulletin of the Siam Society*, vol. 42, no. 1, pp. 117–120.

Roonwal, M. L. and S. M. Mohnot (1977) *Primates of South Asia: Ecology, Sociobiology, and Behavior*. Cambridge: Harvard University Press.

Round, P. D. (1990) 'Bangkok Bird Club survey of the bird and mammal trade in the Bangkok weekend market'. *Natural History Bulletin of the Siam Society*, vol. 38, no. 1, pp. 1–43.

Rowe, N. (1996) *The Pictorial Guide to the Living Primates*. East Hampton, NY: Pogonias Press.

Ruohomäki, O. (1996) 'Resource conflict and resolution in a southern Thai fishery'. In *Cultural Crisis and the Thai Capitalist Transformation* (Volume II of the *Proceedings of the 6th International Conference on Thai Studies*). Chiang Mai: Chiang Mai University, pp. 285–296.

Ruyabhorn, P. and D. Phantumvanit (1988) 'Coastal and marine resources of Thailand: emerging issues facing an industrializing country'. *Ambio*, vol. 17, no. 3, pp. 229–232.

Seth, P. K. and S. Seth (1986) 'Ecology and behaviour of rhesus monkeys in India'. In J. G. Else and P. C. Lee (eds) *Primate Ecology and Conservation*. New York: Cambridge University Press, pp. 89–103.

Sirorattanakul, T. (15/3/1997) 'Monkey see, monkey do'. *Bangkok Post*.

Sitwell, N. and M. Freeman (1988) 'Monkey see, monkey pick'. *International Wildlife*, vol. 18, no. 3, pp. 18–23.

Sponsel, L. E. (1998) 'The historical ecology of Thailand: some explorations of thresholds of human environmental impact from prehistory to the present'. In W. A. Balee (ed.) *Advances in Historical Ecology*. New York: Columbia University Press, pp. 376–404.

Sponsel, L. E. and P. Natadecha-Sponsel (1991) 'A comparison of the cultural ecology of adjacent Muslim and Buddhist villages in southern Thailand'. *Journal of the National Research Council of Thailand*, vol. 23, no. 2, pp. 31–42.

Sponsel, L. E., P. Natadecha-Sponsel and N. Ruttanadakul (1998) 'Sacred and/or secular approaches to biodiversity conservation in Thailand'. *Worldviews: Environment, Culture, Religion*, vol. 2, pp. 155–167.

Stewart-Cox, B. (1995) *Wild Thailand*. Cambridge: MIT Press.

Stifel, L. D. (1973) 'The growth of the rubber economy of Southern Thailand'. *Journal of Southeast Asian Studies*, vol. 4, no. 1, pp. 107–132.

Tambiah, S. J. (1969) 'Animals are good to think and good to prohibit'. *Ethnology*, vol. 8, no. 4, pp. 423–459.

Vercors, J. B. (1953) *That You Shall Know Them*. New York: Popular Library.

Vire, F. (1986) *Kird*, in The Encyclopedia of Islam, C. E. Bosworth, *et al.*, Leiden, The Netherlands: E. J. Brill, vol. 5, pp. 131–134.

Wijeyewardene, G. (1968) 'Address, abuse and animal categories in Northern Thailand'. *Man* N.S., vol. 3, pp. 76–93.

Wilson, C. M. (1983) *Thailand: A Handbook of Historical Statistics*. Boston: G. K. Hall and Co.

Wilson, W. L. and A. D. Johns (1982) 'Diversity and abundance of selected animal species in undisturbed forest, selectively logged forest and plantations in Eastern Kalimantan, Indonesia'. *Biological Conservation*, vol. 24, pp. 205–218.

Part II

WILDLIFE PESTS AND PREDATORS

7

WILDLIFE DEPREDATIONS IN JIGME DORJI NATIONAL PARK, BHUTAN

Klaus Seeland

Introduction

The Kingdom of Bhutan is situated on the southern slopes of the Eastern Himalayas, covering an area of 40,076 sq. km and spanning various climatic zones, from subtropical to alpine and arctic. Southern Bhutan lies at an altitude of about 200 metres (above sea level), rising to an altitude of 7,500 metres in the north where it constitutes a part of the Great Himalayan Range. The country is renowned as one of the Eastern Himalayan hotspots of biodiversity. Its environment has remained largely intact, in contrast to the major degradation which has already severely affected other areas within the Hindu Kush–Himalaya–Karakoram region. In 1996 Bhutan had a per capita income of US$478, which places it among the least developed countries in the world. It has a Five Year Plan economy which enables those who administer the country's protected areas to develop nature conservation and sustainable protection issues as a strategic development policy. Bhutan consists mostly of forest, which covers 72.5 per cent of the land area, whereas arable land under cultivation makes up only 7.8 per cent.

The vast majority of the Bhutanese population of 640,000 are Vajrayana Buddhists. Buddhist values emphasize a respectful attitude towards all living beings and their protection. Despite this prevailing cultural attitude, hunting and poaching still occur in some remote parts of the country, even in protected areas. The Buddhist monarchy was established in its present form in 1907, when the first hereditary king, Ugyen Wangchuck, was elected. The monarchy is held in high esteem by the population and has, together with the monasteries, an all-pervasive influence in Bhutanese society. The position of the monarchy in Bhutan allows the central administration to enact an effective nature conservation policy, which is highly appreciated by the foreign donor countries. However, no official research has so far been undertaken concerning the perception and attitude of the rural population towards nature conservation and its impact on them. An increasing number of problems are emerging from the fact that a policy of nature conservation, as a modern Western legacy, has been imposed on a culture consisting of predominantly self-sustaining rural communities. This chapter analyses the socio-cultural impact of wildlife depredation in the context of Bhutan's national conservation policy.

In the remote areas of the world, such as in the Himalayan Kingdom of Bhutan, the rural population, as well as flora and fauna, appear to have been spared many of the environmental problems which increasingly threaten the South Asian region. Yet it is becoming increasingly apparent that the thinly populated regions in the middle of a Shangri-La-like landscape have to face severe problems of survival which are to a certain extent caused by mostly internationally assisted efforts to protect local wildlife. In 1966 Bhutan began to demarcate protected areas predominantly for the conservation of wildlife (Seeland 1998). For more than 25 years the area of what is now the Jigme Dorji National Park (JDNP) has been declared, first, a wildlife sanctuary and, later on, a national park, to protect rare and endangered and (to a remarkable extent) endemic fauna and flora species. In particular, hunting and the starting of forest fires have been banned in these areas. This nature conservation policy has been largely successful in matters of bio-conservation. However, the policy has had negative effects on the livelihoods of the local inhabitants in the park. Local people suffer from food scarcity caused by crop and livestock depredations of the wild animals now protected by the park.

In 1999 I carried out a survey to assess the present-day living conditions of people dwelling in a national park. One of the research objectives of this survey was to discover whether the phenomenon of wildlife damage had increased since the formal establishment of the protected area, and if so, in what way. Other questions included the relationship of local people to wildlife, the perceptions of the situation among affected people, their attitudes towards likely relief measures and towards compensation for wildlife-related losses. Although this discussion is directed to Bhutan, the issues raised have a more general bearing on the phenomenon of people–wildlife conflict, especially where this is related to conservation policies.

Wildlife depredations in a Buddhist cultural context

The livelihoods of rural Bhutanese farmers at this altitude are typically based on growing a range of different crops and keeping a variety of animals such as horses, bulls, oxen (to plough their fields), cows, pigs and chickens. Their diet consists of maize, barley, wheat, millet and a variety of vegetables. There are hardly any other regular sources of cash income and few off-farm employment opportunities. But this livelihood is threatened by wildlife. Wild animals come to feed on their crops and wild predators prey on their livestock. This phenomenon of wildlife pestilence is, of course, widespread and farmers tend to respond with

Figure 7.1 A farming settlement in Bhutan

an assortment of defence measures. Bhutanese villagers make efforts to defend their fields and animals from park wildlife. Some households guard their fields overnight. This field-guarding, performed by men, takes up a great many hours and, as an extra burden after a hard day's labour in the fields, is exhausting for those who do it. Fields are also protected by constructing dry stone walls and by other barriers covered with thorny plants. Whenever they spot animals in their fields, they try to chase them away by shouting or throwing stones at them. But over time, as the animals get used to people doing this, these actions prove less and less successful in keeping the animals away. What is striking about rural Bhutan, however, is that the rural population is constrained in certain fundamental ways from effectively protecting their livelihoods from the wildlife threat.

One source of constraint arises from religion. Mahayana or Vajrayana Buddhism plays a dominant role in the religious life and culture of Bhutan. Life as such is highly valued by orthodox Buddhism and the killing of any living being is considered a sin. Whereas farmers elsewhere trap and hunt problem wildlife, in rural Bhutan these lethal measures cannot be taken. Wildlife depredations are often endured with religious fatalism, as all living beings are generally perceived as equally valuable forms of life. In this sense it is only natural that animals take whatever they can get to feed on. From a Buddhist perspective, an animal is a being that has not attained the higher spiritual state of human consciousness. To be incarnated as a human being is a privilege that obliges man to accept that animals act according to their instincts to satisfy hunger. Thus animals cannot be expected to behave otherwise and are not to be punished for preying on people's livestock, feeding on their crops or destroying their fields.

Political aspects of wildlife legislation

Apart from the religious aspect, there is a legal ban on killing wildlife according to the Forest and Nature Conservation Act (RGOB 1995), except when an animal raids one's own fields or attacks human life (ibid., Chap. VII, Art. 22 b). As a result of these restrictions, life in rural areas is exposed to economic loss and danger. Poaching and hunting still occur to a certain extent, but those who get caught are drastically fined. In these circumstances wildlife has become a serious threat to livelihood in several parts of Bhutan. The fact that the Royal Government of Bhutan has declared more than 26 per cent of its territory to be 'protected areas' has aggravated the situation. For the people who live near the park there are no legal or other provisions to compensate for their losses due to wildlife depredations. Although the government was to a large extent financially assisted by foreign donors in establishing the protected areas, wildlife problems and related compensation schemes have never been on the political agenda, either by the donors or the government. Affected households are left alone to cope with their problems, but their ability to take measures is minimal. Yet the rising number of wildlife and the increase in human and livestock populations are leading to a situation where either wildlife or livestock numbers will have to be limited.

Bhutan's neighbours, India and Nepal, have similar problems (Schultz 1986, Singh 1991). The much larger human populations in these countries have created an even more precarious situation. Some years ago there was a lively debate in India about the notion of joint management of protected areas in order to harmonize local people's interests with the goal of biosphere conservation (Sarkar *et al.* 1995). Religious feelings and taboos matter in India and Nepal (both predominantly Hindu), just as they matter in Bhutan. But the presence in India of forest-dwelling indigenous people (Adivasis), and in Nepal of ethnic groups which, though Hinduized, hunt and poach, makes the situation different. In Bhutan the density of wildlife populations is extremely high in some areas, but there are practically no legal means or measures which are culturally and religiously appropriate by which the farmers could regulate and balance such populations by themselves. Another problem arises from the social control in Bhutanese villages. This is so strong that even if a farmer wanted to violently respond to wildlife pests, he would be reluctant do so for fear that neighbours might notice.

The research setting

In March 1999 I conducted a household survey in Damji, located in Khamey Geog ('block'), at the southern border of Gaza Dzongkhag, one of the four settlements in this largest and northernmost district of the Himalayan Kingdom of Bhutan (see Figure 7.2). The total area of the district is 4,410 sq. km, about 10 per cent of the Bhutanese territory. It ranges between 1,200 metres and more than 7,500 metres above sea level and 50 per cent of the area is between 4,200–5,400 metres above sea level. Barren land (rock and areas covered with snow) makes up about 62 per cent of the area and forest covers 33 per cent, the rest being pasture (5.3 per cent) and agricultural land (0.2 per cent) (RGOB/LUPS 1997). Most of this area belongs to Jigme Dorji National Park, which includes some areas of Paro, Thimphu and Punakha Dzongkhag, amounting to an overall 4,200 sq. km, including a buffer zone. Damji is located on the upper western valley flank of the Mo Chu river at an elevation between ca. 2,400 metres and 2,700 metres; it comprises thirty households.

According to statistical survey figures of the mid-1990s, there were about 6,500 permanent local inhabitants living in 1,000 households within Jigme Dorji National Park (UNDP/GEF 1997: 25). As there are no official figures available, I estimated from my own calculations that there are approximately 210 people living in Damji's 30 households. The settlement pattern of Damji is dispersed: houses are situated at various levels on terraced grounds in the middle of the agricultural land, and the village is stretched out over about half a kilometre in a north–south orientation on both sides of the main trail to Gaza Dzong, the district headquarters. There are neither roads nor electricity in the district, although water is supplied at a few public places along the trail through water pipes with taps. One small shop stands in the centre of the main settlement, which comprises the wealthiest and oldest houses and a primary school further up on the left side of the trail. Almost 15 per cent of Khamey Geog rely on animal products for their income; 75 per cent of the local income is from agriculture and 10 per cent is earned from wage labour.

The survey

Methodology

According to informants from the Forestry Service Division, the whole settlement is affected by wildlife depredations. A random sample of one-third of the households was taken. In my house-to-house survey, a senior member of each of the ten households was interviewed with the help of a structured questionnaire. During earlier visits to Bhutan, had discussed the design and range of questions with well-informed officials of the Nature Conservation Section in the Forestry Service Division, as well as other people outside the government. The survey was carried out in the middle of March, the season of ploughing, when manure is spread out on the fields and preparations are made for sowing.

I carried out the interviews with the help of an interpreter, Mr Sangay, of the Forestry Service Division, and all the people we approached agreed to be interviewed. During an official meeting, which happened to take place at the

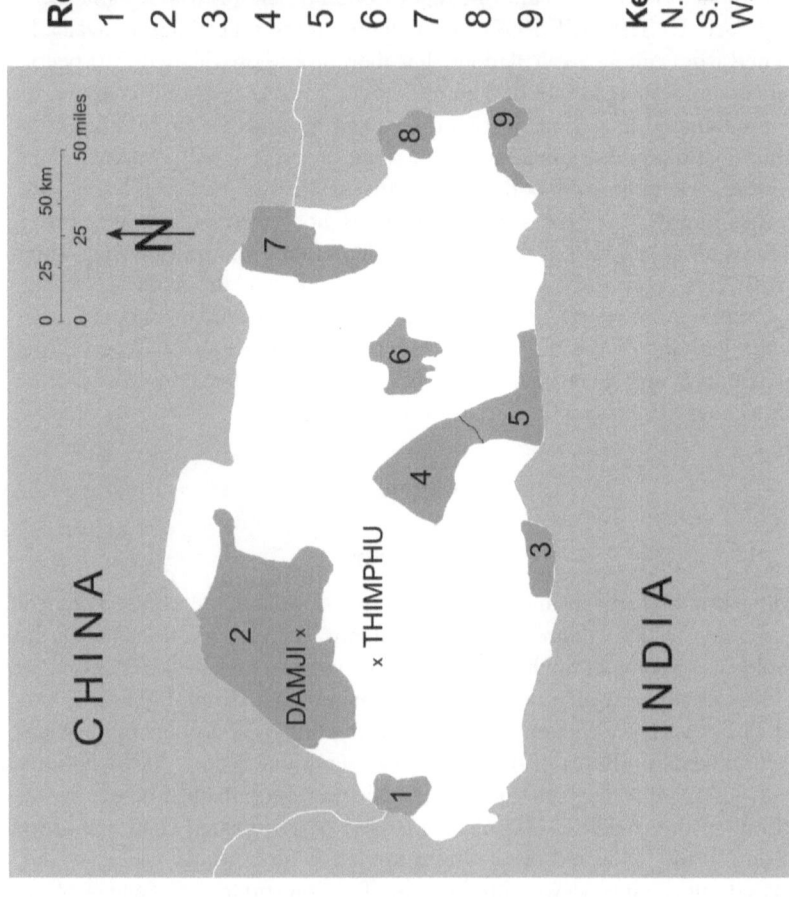

Reference

1 Torsa (S.N.R)
2 Jigmi Dorji (N.P.)
3 Phibsoo (W.S.)
4 Black Mountain (N.P.)
5 Royal Manas (N.P.)
6 Thrumshing-la (N.P.)
7 Kulong Chu (W.S.)
8 Sakteng (W.S.)
9 Khaling/Neoli (W.S.)

Key:
N.P. = National Park
S.N.R. = Strict Nature Reserve
W.S. = Wildlife Sanctuary

Figure 7.2 Map of the Himalayan Kingdom of Bhutan

inauguration of a newly built kitchen near the primary school, the district governor (*Dzongdag*)[1] spoke to local people and encouraged them to participate in the survey. A statement on the topic of wildlife depredations was translated to the audience (more than 200 people of both sexes and almost every age group) and then discussed briefly in public. The questionnaire consisted of fifteen questions, some of them divided into sub-questions. We interviewed three women in their thirties and seven men between the ages of 29 and 69, usually at home and only twice outside their houses. The general atmosphere was open-minded and people showed an awareness that wildlife depredations were an urgent problem that needed to be solved.

Survey results

The surveyed households own between 1.7 and 7.7 acres of land and range in size between two and twelve persons, with an average of almost seven people. Most of the households are settlements which are inhabited throughout the year; the exception is one household, which is the winter domicile of someone who lives in Gaza in summer. Figure 7.3 shows the household size, the amount of land owned and the distribution of livestock among the surveyed households, as well as the unequal economic wealth of local households.

All of the respondents had been affected by wildlife raids between one and sixteen times over the past five years. Most damage had been caused by wild boar (*Sus scrofus*), sambar (*Servus unicolor*), bear (*Selenarctos thibetanus*), barking deer (*Elaphodus* sp.), wild dog (*Cuon alpinus*), tiger (*Panthera tigris*), monkey (*Presbytis*

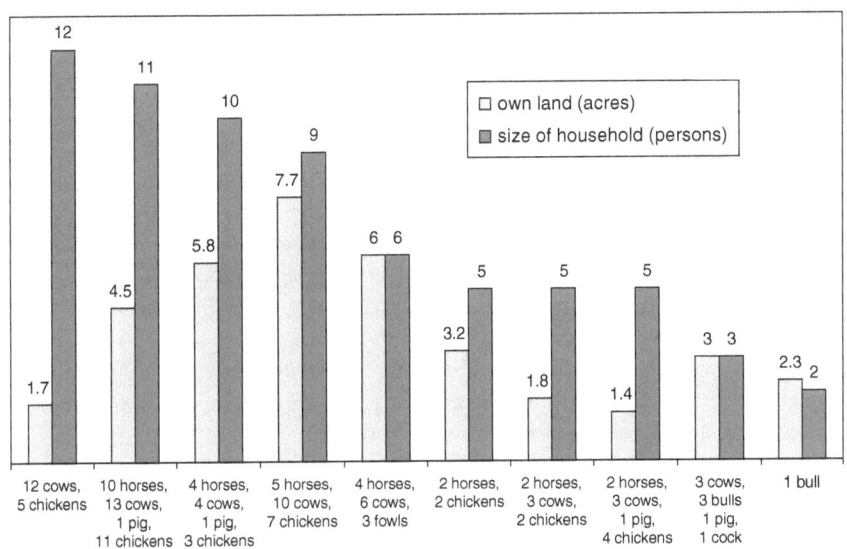

Figure 7.3 Size of households, agricultural land and livestock

entellus) and leopard (*Panthera pardus*). As for the security of people and livestock, half of the respondents considered bear, tiger and leopard to be the most dangerous animals. The greatest threats to livestock property, such as chickens, piglets and puppies, are eagles, hawks and bears. Sixty per cent of respondents say that the fields are mainly affected by wild boar and the same amount of respondents say that bear, wild boar and barking deer together are the major cause of crop loss. On average, the villagers of Damji complain that they suffer about a 50 per cent loss of their annual means of subsistence from wildlife depredations. If it were not for work as porters or as labourers in road construction, their livelihood would be seriously threatened.

All respondents would like to be compensated for their loss. However, 20 per cent have no idea how and by whom this could be done. Half of the respondents perceive difficulties, either because there would be too many villages to be compensated or because it would be too difficult to prove their loss. Moreover, 20 per cent do not like the idea of becoming dependent on the compensating agency, that is, the government. On the other hand, 30 per cent see no difficulties in the process of being compensated. The answers to the question 'To what extent should those people be compensated who suffer from loss?' show great variations. More than one-third of the respondents have no idea about it. But 18 per cent think that it should be 50 per cent of the loss, while another 18 per cent say that 30–40 per cent of the loss would be adequate compensation. Only 27 per cent would ask for a full compensation of their loss caused by wildlife damage. Assuming that there would be a possibility to provide the affected villagers with compensation, half of the respondents came up with ideas about how they could be compensated, while the others have no idea about it. One-quarter believe that fencing material provided by the government would be the best means to compensate them and to put them in a position to help themselves, and 15 per cent think that cash or kind is best.

In an overall assessment of how to overcome the abundance of wildlife and the damage it causes, only one-fifth of the respondents say that it is necessary to reduce the number of wildlife in this part of the national park. Forty per cent are in favour of compensation and 20 per cent argue for protection measures. Nobody suggested removing the legal ban on hunting. Nevertheless, two out of ten people would be prepared to kill animals if they had harmed their household, crops or fields – referring predominantly to wild boar, wild dogs and tigers. Half of the interviewees are convinced that it is their own responsibility to tackle the problem of wildlife depredations. Only one out of ten holds the government responsible for doing something about it. Forty per cent either did not answer or thought that it was nobody's responsibility and attributed it to 'fate', in the sense of an unavoidable situation in which they find themselves. They attribute this situation to their personal karma.

The respondent's perception of whether the numbers of wildlife have generally increased, remained the same or decreased is not clear and to a certain extent even contradictory. For instance, 30 per cent of respondents claim that wildlife has definitely increased. However, while one-third of the interviewees claim that the number of wild boars has increased, a similar percentage claims that it has decreased. A possible explanation for this could be that the number of those animals which prey on the wild boar, such as wild dogs, leopards and tigers, has increased, thus reducing their number for at least some time. 20 per cent believe that the amount of barking deer, bears and wild dogs have increased, while three respondents out of ten say that the number of sambar has increased. Only the tiger seems to have become more rare in the area. Altogether, these figures show that wildlife damage has had a major impact on rural livelihoods. The responses to wildlife depredations include double cropping due to seed loss (to crop-raiders), premature harvesting (to avoid pre-harvest crop-raids) and involuntary fallowing of scarce agricultural land. These countermeasures impose a heavy burden on the farmers of Damji. So far the Forest and Nature Conservation Act has succeeded in protecting the forest and wildlife, but this has been in an unbalanced way which has led to unbearable living conditions for the local human population.

Local perceptions of wildlife

The rural population in northern and central Bhutan is exclusively Buddhist and generally has a peaceful attitude towards other living beings. In accord with the 'great tradition' of the Buddha's teachings, they consider it to be the right of creatures other than man to feed on whatever is at their disposal. In everyday life villagers worship several deities such as those of heaven (*lha*), of the mountains (*tsen*), of the land (*sadag*) and the spirits of the underworld (*lu*). Throughout my interviews in Damji and informal talks with rural people in other parts of Bhutan, it was evident that people strongly believe in the core teaching of Buddhism, that human beings, because of their superior karmic status, should be compassionate towards other forms of life. Animism, or the so-called 'little tradition' of local religious cults, is closely interrelated with monastic Buddhism. Lay priests (*gomchen*) in almost every village perform smaller rituals and look after the village temple (*lhakhang*). Monks and particularly high ranking lamas (reincarnations of important religious teachers, abbots, etc.) are often invited to perform rituals in individual households and are highly venerated authorities on questions of Buddhist morals and codes of conduct.

What this means for farmers is that they should protect their fields and harvests from wildlife pests in a passive or non-violent way, such as chasing the animals away, rather than by killing, trapping or otherwise harming the animals. Farmers tolerate animal feeding to some extent, believing that animals can take their share of what human labour produces. The villagers are prepared to suffer rather than to take measures against living beings that may influence their personal karma in a negative way. The increase in wildlife damage to their property and sometimes their lives is viewed as a repercussion of sinful behaviour in one's previous existences and accepted as inevitable punishment. Although discussions with villagers revealed a strong self-esteem and trust in their own capacities to cope with these problems, almost all of those interviewed were convinced that they would have to live with increasing wildlife nuisance in the future. A growing number of living beings, whether humans or animals, is generally perceived as the core element of life and not something to be opposed by selfishly obstructing their existence. Another constraint on reacting to wildlife damage is the local belief that wild animals are representatives of one's ancestors or local deities. Such animals are not to be harmed, because of the possibility of revenge and because of pious feelings towards them and living beings generally. In the interviews some people argued that ancestors or local deities threaten them intentionally just to demonstrate their power and demand respect.

For most of the farmers interviewed, 'wildlife' (*re dag sin chen*) meant those animals that harm them or their property. As they do not cause damage, takin (*Budorcas taxicolor*) and muntjac (*Muntiacus muntjac*) are not perceived by the farmers as 'wildlife'. However, some farmers claim that all animals living in the wilderness, irrespective of whether they are harmful to humans or not, are 'wildlife'. This categorization is made by observation and experience and is thus rather subjective. Those animals that attack human beings, such as bear, tiger and leopard, are generally perceived as dangerous by the majority of the interviewees. Other animals, even though they threaten the survival of the local families, by creating a food shortage and, therefore, despair, are not 'dangerous' in the sense of attacking other living beings.

Forest policy impact on rural social life

The impact of forest policy and wildlife conservation on rural social life is remarkably strong in rural areas and even stronger in protected areas. The proportion of people who know that there is a Forest and Nature Conservation Act since 1995 has doubled from 20 to 40 per cent, as compared to those who knew about the 1969 Forest Act. 60 per cent of the interviewees still do not know about the 1995 Act. Nonetheless, the ban on hunting and starting forest fires, which are the core elements in both Acts, has an adverse effect on the local subsistence economy. In the government's food security report (which covers 18 districts), Gaza Dzongkhag is the most affected district, with a total food crop loss of 37 per cent (Choden and Namgay 1996: 4). By contrast, the Bhutanese average is 14.5 per cent. In a survey of wild boar crop damage covering five districts of Bhutan (in which 386 households were selected, a sample of 13 per cent), 60 per cent of the farmers interviewed held the strict implementation of the Forest Act responsible for the increased number of wild boars (ibid.: 30). The survey's conclusion is that the local farmers' inability to solve their wildlife problem has a demoralizing effect on them, such that they view their future prospects with despair.

Perhaps it is due to a rather negative perception of the Forest Act's impact on local lifestyles that forest offences are quite frequent. Wangchuck (1997: 62) gives a detailed review of all reported cases from 1991–96, showing that, out of 256 reported cases, illegal hunting was the most frequent offence – making up 33 per cent of the detected cases. The second most numerous category of offences was illegal felling of trees (30 per cent) followed by illegal fishing and illegal appropriation of forest products (14 per cent each). Sixty per cent of the offences happened in rural areas and 63 per cent involved farmers. Approximately 30 per cent of the offences happened within parks and protected areas, amounting to some 50 per cent of the fines (personal communication S. Wangchuck, November 1997).

When laws such as the Forest Act end up favouring wildlife and nature protection and threatening human survival, they must be revised. Yet the Bhutanese Government has taken no measures to tackle the wildlife problem or to come up with a compensation scheme for the affected farmers. One possible interpretation of this state of affairs is that the Government is trying to discourage farmers in protected areas from staying on and is, in effect, encouraging them to migrate to other, more favourable and safer places. However, the formal policy of the National Park Management regime affirms that no resettlement of people residing in the park areas can take place. One policy objective holds that 'Park Management will provide assistance to those who may wish to move closer to places of employment outside the park'. But no new settlements within the multiple-use zone and the enclave or buffer zones will be allowed. Still, given the annual population growth of 3.1 per cent, the area of cultivation will sooner or later have to be expanded, if local people are to continue their traditional agricultural lifestyle.

Compensation

One urgent problem is that of compensation. Compensation payments have not yet been discussed. Nor has there been any policy decision to offer practical relief from the wildlife threat to the park and protected area dwellers. It is just too politically sensitive and delicate for the Bhutanese Government because it might open the floodgates for an unaccountable wave of financial demands and subsequent litigation. At the moment, the Nature Conservation section lacks an adequate number of sufficiently trained personnel to cope with this problem. There is an unofficial awareness that this will be a major obstacle to wildlife and biodiversity conservation and good relationships with the concerned village communities. At the same time, there is rapid human population growth and, particularly in the Northern Wildlife Circle, a growth in the numbers of livestock. In Jigme Dorji National Park overgrazing has already taken place and the livestock of nomads compete with wildlife for the dwindling fodder resources. Therefore, an Enclave and Buffer Zone Management Policy, a Multiple-use Zone Policy and a Seasonal Grazing Zone Policy have been included in the Management Plan (UNDP/GEF 1997: 41–44).

The forest policy situation in Bhutan is characterized by central legislation, backed by international conservation policy agendas and respective financial assistance by foreign donors. From a national perspective, the economic loss suffered by farm households in Damji is politically negligible, as in many other remote pockets of Bhutan, at least for the time being. Because of the wide range of wildlife species affecting the farmer's fields, political and economic compensation is not automatic. Only wild boar damage, common across Bhutan, is recognized as a matter of national interest.

Nature conservation areas such as those of the Jigme Dorji National Park and the various defined zones will, sooner or later, have to be categorized into several zones with respective classes of wildlife density. In areas prone to wildlife damage due to excessive numbers of animals, the *Dzongkhag* administration will have to design a 'compensate-the-most-needy-farmers' scheme to maintain social equity between households. The Enclave and Buffer Zone, the Multiple-use Zone Policy and Seasonal Grazing Zone must be supplemented with a compensation scheme and appropriate hunting regulations. As a guideline, the percentage of loss in a farm household could be based on the average annual income standard of the most recent five years, taking into account changing parameters such as the number of household members and livestock. Detailed wildlife damage assessments and recommendations for the whole of Bhutan have been made (Wollenhaupt 1991, Choden and Namgay 1996), but they have not been taken up politically. No larger investigation, however, has inquired into the living conditions of people inside protected areas.

The Buddhist cultural heritage and the influences of the international conservationist movement have created a unique situation which cannot be easily compared with similar situations in India and Nepal (Nepal and Weber 1993, Saharia 1984). The political will at the *Dzongkhag* and the central Government level is necessary to realize the importance of structural relief measures, before the situation in the protected areas reaches a state of crisis. It is not desirable – nor realistic in the long run – for the government to exclude environmental management from the protected areas. Funds to finance the compensation measures could come from ecotourism in Jigme Dorji National Park (and other protected areas) through entrance fees or from fringe benefits for the local population, or they will have to be integrated into the park management plan account under recurring costs. Local participation and cooperation would be enhanced as a result. Whether a more positive perception of the Forestry Service will emerge remains to be seen.

An essential prerequisite for a compensation regime in Bhutan is a social impact assessment of protected area programmes, carried out on a regular basis, which frames the objective of protecting nature in terms of the cultural context of the area in question. The above interview data show the people's readiness for more protective measures, a high degree of local self-organization and a willingness to solve these problems on their own, as well as a tremendous patience with the troubles caused by wildlife. This cultural capital would be of immense value for the cooperative (if not joint) management of this protected area. An effective compensation regime also depends on ensuring that there is social equity in the assessment of damage and in administering compensation. To achieve this social equity, a local institution is to be established at *Dzongkhag* level, consisting of the park manager of JDNP, the Divisional Forest Officer, *chimi* (member of the National Assembly), *gup* (headman of a block) and respected village elders. Institutionalization will be a necessary step for a responsible body at the regional level to regulate wildlife damage in a competent and less bureaucratic way. An institution democratically legitimized and sanctioned by the central government may be in a position to overcome the adverse effects of a strict wildlife conservation policy on the local population.

The politics of compensation

International nature conservation policies have an immense social impact on the lifestyle and prospects of survival of local village communities, particularly in remote and thinly populated areas. Precarious situations which have not primarily evolved from local conditions – such as in the case of Damji – need outside assistance when problems arise that cannot be solved by the local people themselves. An amendment to existing legislation would have to acknowledge and reflect its shortcomings arising from negative developments not foreseen at the time the legislation was originally drafted. That the wildlife situation in JDNP is unbearable for the local farmers and intolerable for an enlightened nature conservation policy has been shown by larger surveys in Bhutan (Wollenhaupt 1991, Choden and Namgay 1996) and by my own interview findings referred to above. Compensation through measures of self-regulation in areas prone to wildlife damage, combined with a degree of decentralization, will ultimately lead to the building of institutions with the participation of all relevant officials and concerned people at the local or regional level. The interdependence of park management goals, wildlife conservation objectives and local livelihood interests will have to be recognized through a process of negotiation and compromise.

It is presently an open question whether international NGOs, such as the World Wide Fund for Nature (WWF) or the Bhutan Trust Fund, are going to put these matters on their agenda. More than a decade after nature conservation planning was instituted for northern Bhutan (Blower 1989), in which neither wildlife depredations nor compensation was considered, it is high time to discuss this issue at a structural level for the whole of Bhutan. It will be up to a task force to organize a survey in every park or protected area to obtain a reliable country-wide account of the extent of wildlife damage. The establishment of institutions could be a joint initiative of the *Dzongkhag* administrations and the Nature Conservation Section under the Forestry Services Division. Whether these institutions will be able to mitigate the economic pressure of wildlife damage on rural households will depend on the establishment of cooperative relations between the local peasants and government administration.

143

In order to put the example described above into a theoretical context of political economy, it is necessary to consider the general framework of Bhutan's links with international politics and NGOs, soliciting nature conservation objectives in countries such as Bhutan, where they can be sure to achieve best results. Politically, Bhutan is interested in having high profile relationships with the politically influential Western industrial countries because of its geopolitical location. Encapsulated between two superpowers, its basic aim is to secure its survival as an independent state. The recent history of the neighbouring Sikkim, which had been an independent monarchy but was then merged with the Indian Union in 1975, shows how easily a small state's sovereignty can be lost (Singh 1988: 262–276). At the time of its unification with India, Sikkim had no major relations with the international community, nor was it a member of the United Nations.

Since the late 1990s there have been political problems on Bhutan's southern border with India, largely due to agitation and a guerrilla war waged by the ULFA (United Liberation Front of Assam) – comprising the Bodo (an Assamese tribe) – with the Indian Army and Bhutanese security forces. The guerrillas cross the border to-and-fro and hide partly in the forests of the Royal Manas National Park. Bodo guerrillas largely finance themselves through poaching – using the money they make from poaching and selling wildlife trophies to buy food from the local villagers and to procure arms and medicines. Wherever possible, they fraternize with and seek the support of the local population. This situation threatens to have an adverse effect on wildlife conservation, for it could encourage poaching by national park villagers, on top of that carried out by the guerrillas. When wildlife depredations are not compensated by the state administration, villagers may well be tempted to poach animal species that fetch a high price on the black market. In this way, they would both remove wild animals that threaten their livelihoods and earn money. They would be able to conceal their actions by claiming that it was done by the Bodos. In the Southern Wildlife Circle, the Bodos pose the problem. In the Northern Wildlife Circle, on the other hand, the threat to the park comes from medicinal plant collectors entering from Tibet, and over-grazing by wildlife, along with the growing number of yaks (Seeland 1998). The denial of compensation to the local victims of protected wildlife may well have the effect of radicalizing them in the long run.

Social equity demands a fair distribution of opportunities and measures to meet natural hazards and particular risks between the various sections of a society. The state should therefore endeavour to establish suitable institutions to regulate wildlife damage in protected areas. Self-organizing and self-regulating institutions would be a form of social capital that could help to start solution-finding processes (Ostrom 1995). Local subsistence livelihoods must be guaranteed by structural measures such as regulating the number of wild animals, or by technical aid to protect fields and livestock from depredation or by means of compensation. A compensation policy would have to define 'area-specific-risk-indicators' (ASRI) in which social, economic, demographic and occupational indicators are used to assess the risks known to occur within a particular zone. According to the risk standard of a respective zone, wildlife damage compensation could be facilitated by the local institutions, to be appointed as described above, but supplemented with a grant-in-aid provided by the central government through the *Dzongkhag* administration.

Conclusion

Bhutan is a small, undeveloped country which has entered a network of international obligations with donor countries that financially assist the country's nature conservation programmes. Over the past thirty years there have developed legal provisions for setting up protected areas and national parks in all parts of the country. However, despite this international incorporation of Bhutan, the views of rural Bhutanese people on wild animals and their depredations, and the way to respond to them, have not changed. As a result, the import of Western conservationist concepts in the form of restrictive legislation has had an adverse effect on food security at the local level. Rural lifestyles in Bhutan have changed little over recent decades, but people have recently been confronted with phenomena to which their norms and values respond in a traditional way, i.e. by acceptance of suffering. Suffering is a major concept in Bhutanese everyday life. People believe that negative karmic influences from their previous existences (Dzongkha, Tibetan, *leh*) have an impact on their present-day lives. Even if no opposition to conservation measures existed on the part of affected villagers, the clash between unprecedentedly large numbers of wild animals and a traditional religious outlook on everyday life poses a challenge for nature conservation policy. Although local people accept their changed living conditions at the moment, the administration needs to find a solution for what could become a political issue in the long run. In the present situation, compensation schemes would seem to be the only feasible way to avoid acute hardships for the rural population.

Note

1 All local terms in brackets and italics are in Dzongkha, the national language of Bhutan.

References

Blower, J. H. (1989) *Nature Conservation in Northern and Central Bhutan*. Rome: FAO.
Choden, Dechen and Kinzang Namgay (1996) *Project For Assessment of Crop Damage by the Wild Boar: Report on the Findings and Recommendations of the Wild Boar Survey*. Thimphu: National Plant Protection Centre, Research Extension and Irrigation Division. Ministry of Agriculture.
Nepal, S. K. and K. E. Weber (1993) *Struggle for Existence: Park-People Conflict in the Royal Chitwan National Park, Nepal*. Bangkok: Asian Institute of Technology.

Ostrom, E. (1995) 'Constituting social capital and collective action'. In R. O. Keohane and E. Ostrom (eds) *Local Commons and Global Interdependence: Heterogeneity and Cooperation in Two Domains*. London: Sage Publications.

Royal Government of Bhutan (RGOB) (1995) *Forest and Nature Conservation Act. Thimphu. RGOB/LUPS (Land Use Planning Section) Atlas of Bhutan, 1:250,000: Land Cover & Area Statistics of 20 Dzongkhags*. Thimphu: Ministry of Agriculture.

Saharia, V. B. (1984) 'Human dimensions in wildlife management: the Indian experience'. In J. A. McNeely and K. R. Miller (eds) *National Parks, Conservation and Development: The Role of Protected Areas in Sustaining Society*. Washington, DC: IUCN and Smithsonian Institution Press.

Sarkar, S., N. Singh, S. Suri and A. Kothari (eds) (1995) *Joint Management of Protected Areas in India: Report of a Workshop*. New Delhi: Indian Institute of Public Administration.

Schultz, B. O. (1986) 'The management of crop damage by animals'. *Indian Forester*, vol. 112, no. 10, pp. 891–899.

Seeland, K. (1998) 'The National Park management regime in Bhutan: historical background and current problems'. *Worldviews: Environment, Culture, Religion*, vol. 2, pp. 139–153.

Singh, A. K. J. (1988) *Himalayan Triangle: A Historical Survey of British India's Relations with Tibet, Sikkim and Bhutan 1765–1950*. London: The British Library.

Singh, R. L. (1991) 'Wildlife conservation and eco-development programme: a case study'. *Indian Forester*, vol. 117, no. 10, pp. 804–811.

UNDP (Global Environmental Facility) (1997) *Integrated Management of Jigme Dorji National Park (JDNP)*. BHU/96/G33/A/1G/99 [BHU/96/008/A/01/99 (UNDP)], draft version. Thimphu.

Wangchuck, S. (1997) 'Local perceptions and indigenous institutions as forms of social performance for sustainable forest management in Bhutan'. Zurich: D. Nat. Sc. Thesis, ETH Zurich, No.12217 (unpublished).

Wollenhaupt, H. (1991) *Wildlife Management Bhutan – Game Damage and its Compensation: Proposals and Recommendations for Matters under Special Consideration Regarding the Wild Boar (*Sus scrofa*)*. Thimphu: RGOB/Ministry of Agriculture/Department of Forests/FAO. Field Document No. 17, FO:DP/BHU/85/016.

FARMING THE FOREST EDGE

Perceptions of wildlife among the Kerinci of Sumatra

Jet Bakels

Introduction

Anthropologists have long viewed hunter-gatherers and farmers in mutually exclusive terms. They have contrasted the 'relations of sacred companionship between men and animals' of hunter-gatherers with the dualistic view of nature among farmers (Sinclair 1977: 20). As Sinclair argues, in the shift from hunter to farmer, '[t]he psychology of the forest man gave way to the psychology of the field man, as timber retreated before the ax' (ibid.). Nurit Bird-David states that cultivators 'see themselves as living not in [the forest] or by it, only despite it ... opposing it with fear, mistrust and occasional hate' (Bird-David 1990: 190). According to Peter Boomgaard, 'peasants and farmers throughout history, in whatever part of the world ... feared and often hated 'wild nature' (Boomgaard *et al.* 1997: 17).

To what extent is this true? In this chapter I shall examine the view of wildlife among a group of forest-edge cultivators in the Kerenci district of Central Sumatra. I show that for these cultivators the view of 'wild nature' is rather more complex than is suggested above. These farmers, cultivating rice, vegetables and cash crops, routinely suffer from wildlife pestilence and in response hunt these animals – in what might be called defensive hunting. Farmer antagonism with wildlife is therefore clearly evident among the Kerinci. But forest wildlife are not viewed simply as 'pests' that can be straightforwardly removed by hunting and trapping. Rather, the hunting of these crop-raiding animals has to be understood in terms of the larger relationship between these cultivators and the forest around them.

In what follows I shall sketch the conceptual framework in which the forest domain, with its spirits and wild animals, stands in a dynamic relationship with the village domain. I shall describe the forest as a 'cultural space' which is populated by animals and spirits that live according to their own rules of behaviour and in which men are subject to specific regulations. I shall subdivide the 'wild' category into wild prey (such as wild pigs and deer) for human hunters, and dangerous wild predators. Special attention will be paid to the way in which wild prey is treated in connection with ideas of regeneration. Does the Kerinci cultivator feel responsible for the animals he kills, as he does for the crops he cultivates? If this is so, it is, as we shall see, only part of the picture. I hope to show that the reciprocal element in the cultivator's perception of human–environmental relations is essential, and that there is much more to the forest and its creatures than fear alone (see also Pálsson 1996: 71). Additionally, I shall explore the usefulness of absolute distinctions, such as that between hunter-gatherers and agriculturalists with respect to their conceptualization of the forest and its wild animals.[1]

The Kerinci

The Kerinci people are predominantly rice farmers. Hunting is of marginal importance, whether in terms of meat or in terms of ideology (cf. Ellen 1996b: 602). Geographically, Kerinci is a fertile valley (about 70km long and 10km wide) in the middle of a mountainous area (Bukit Barisan) in the mid-western part of Sumatra. As an administrative unit, the Kerinci district comprises almost 300,000 inhabitants. A large lake lies in the centre of the valley, surrounded by wet rice fields (*sawah*) and small villages; the traditional long-house structure is still visible in the layout of the wooden and stone houses. Some villages are situated on the lower slopes and practice dry-rice farming and agroforestry. Behind these forest gardens a lush tropical forest stretches endlessly into the Kerinci Seblat National Park, the largest nature reserve in Sumatra and home to bears, deer, elephants, leopards, rhinoceros and tigers. Local legends speak of a man-like ape, the *orang pendek*, that wanders in the deepest woods, but no scientific proof of its existence has been given so far.[2] Although the Kerinci people are Islamic, pre-Islamic ideas dominate in the realm of animal and wilderness lore, perhaps because Islamic models are either absent or in contradiction with fundamental Kerinci attitudes (see Bakels 1994). The tiger is the animal that attracts most attention in myths and taboos, as seems to be the case wherever the animal exists (Thapar 1992). Thus the most 'wild' animal is the one which is most encapsulated with cultural meaning. Beliefs connected with the animal include the idea that a tiger will never kill a person who has followed the *adat* (customary) laws of behaviour; only trespassers are attacked.[3] The other important mythical animal in Kerinci culture is the python, called *ular sawo* or *ular naga*.[4] This snake appears in myths as a chthonic animal, associated with origins and creation. Some ceremonial flags used by the Kerinci have the shape of a python.

In Kerinci, people enter the forest to collect forest products, to hunt, to clear the ground for a new forest garden and to seek contact with the nameless population of forest-spirits (*mambang*) and spirits of the ancestors.[5] The Kerinci people consider themselves to be immigrants to the area (with the classificatory 'younger' status), and to owe respect to the *mambang*, which are seen as the autochtonous population (with the classificatory 'older' status) and rightful owners of the land. The *mambang* are primordial beings, instrumental in giving the environment its shape. In the distant past they threw rocks at each other, and some rocks hit mountain peaks, giving them their specific appearance today. The *mambang* also keep nature's secrets, such as the knowledge of medicinal plants, and bestow fertility on crops and on men. Most village genealogies start with the marriage of a *mambang* wife with an immigrant man. *Mambang* are honoured during all important rituals with offerings. It is in relation to the presence of the *mambang* that Kerinci people obey specific rules when in the forest.[6]

Tame and wild animals

In the world view of the Kerinci people, there is a categorical division between 'tame' animals (*binatang jinak, ternak*) and 'wild' animals (*binatang buas*, or *ganas*). The tame animals – i.e. water buffaloes, cows, goats, chickens, horses, dogs and cats – are 'under the power' of men (*dikuasai oleh manusia*); it is men who 'own' (*punya*) these animals. As men are the 'herdsmen' (*gembalo*) of the tame animals, the village leaders are the herdsman of the people. All belong to the village domain. The wild animals in the forest are also thought to have a herdsman. But their 'herdsman' guardian is not human, but a kind of *mambang* spirit. This spirt, referred to as the *gembalo*, is generally depicted as a child-size male. Some say he is very handsome, but according to others he is the opposite – shaggy and clothed in rags. He herds and protects the animals. He 'lets them out in the morning and collects them again at night'. The *gembalo* of the wild animals understands their language. He regulates the animal *adat*. For him, it is the wild animals which are tame and, correspondingly, it is the cattle owned by humans which are wild.[7]

To make it clear that a certain animal belongs to the village domain, one has to give it a mark of possession lest the spirit of the forest takes it away with him. This is of special importance when an animal is outside the village: a cat has to wear a string around its neck, a cow or water buffalo a bell around its neck. When a water buffalo is left alone near the forest for an extended period, a symbolic fence is constructed to protect it against the spirit of the forest and against the tiger. The owner makes a square by placing four stalks of the sugar palm on top of poles stuck in the ground. In this way it is clear that the animal is 'legally ours' (*hak kita*) and 'that there is an owner' (*ada yang punya*). The Kerinci people believe that the spirit of the forest keeps his animals together without a string or fence. This 'fencing off' of the village domain *vis-à-vis* that of the forest finds its parallel in the ricefields. The borders of the fields are marked by specific plants that are disliked by the spirits of the forest, in order to discourage them from eating the rice. The background of this anxiety is perhaps that, as stated earlier, the Kerinci people see themselves as immigrants who have taken the land occupied by nature spirits – these spirits might want to take back what once belonged to them. According to a Kerinci *adat* specialist, the fact that Man owns the water buffalo or cow is stressed in a small ceremony that was held (until the beginning of the twentieth century) every time a calf was born. The newly born calf would receive a garland of flowers, which was placed around its neck. The owner of the calf would invite over the children of the village and serve them rice porridge, and village leaders were informed of the birth. In this way the calf was incorporated into the village domain. The reason for inviting children was, according to my informant, that the spirit herdsman of the wild animals was himself the size of a child. But small children are also the ones who tend to look after the animals.[8]

When a family buys or slaughters a water buffalo, cow or goat (a 'four-legged animal', *kaki empat*), the village leaders have to be informed because 'it is the grass of the village domain that is transformed into blood and meat' of the animal. When such animals are slaughtered in the context of village rituals, the meat distribution reflects the social hierarchy of the village. The leaders receive portions of the most prestigious meat, such as the heart, the liver and the brains. As in many other rural societies (see Kreemer 1956), the water buffalo can be seen as the symbol of the village, and its communal consumption can be understood as an expression of the ideal unity of the village and the superior status of village leaders. In Kerinci, a small part of the treasured organs of the slaughtered animals are taken outside the village and offered, in a raw state, to the spirits of the forest and to the tiger. This symbolically expresses that, although men and spirits belong to one moral order and maintain relations with each other, their spatial domains are separate: the village is for humans and the forest is for *mambang*. This overview of ideas about the position of specific categories of animals indicates that there is a preoccupation with marking the borders of village and forest, especially in the liminal zone of fields and gardens that lies between the village and the forest. This *bosrand* (forest-edge) ambiguity (see Baud 1997) is also apparent in the symbolism of wild species for which there exists a tame variety. The forest surrounding the valley contains the mountain goat (*kambing hutan* or *serou*) and the forest chicken (*ayam hutan*). It is considered to be a bad omen when these forest animals come too close to the village – such events presage illness among their domesticated relatives.

The Kerinci grow vegetables and fruits (such as cassava, tomatoes, beans, aubergines, pumpkins, onions, peppers, *durian*, *manggis*) for their own daily use mostly in small gardens (*pelak*) near the village. The larger gardens (*kebun*) are generally located a bit higher up the mountain and are used for the cultivation of cash crops like cinnamon, clove, coffee and vegetables that are sold on the local market. By contrast, rice is grown on the lowest slopes and on the plains. Thus the mountain slopes behind the villages form a mosaic of various crops, mixed with primary and secondary forest. The main threat to the rice is the wild pig, and pig hunts intensify as the rice matures. Several kinds of wild pig, as well as deer and monkeys, pose a constant threat to the garden crops. Families sometimes sleep in the small huts in the ricefields or in the gardens to protect their crops against the animals and, in addition to hunting and trapping, use magic to scare the animals away.

Hunting in the forest[9]

Respect is paid to the spirits of the forest in all Kerinci rituals and myths. Specific pacts have ensured that man has a rightful claim on parts of the land, but, as we have seen, some ambiguity remains, and when new land is brought under cultivation or wild animals are hunted, permission must be sought from the spirits of the forest. In Kerinci, each animal species (deer, wild pigs and so on) has a separate *gembalo*. The only exception is the tiger which has no spiritual *gembalo* but is sometimes considered to be the inhabitant of a tiger village and to fall under the authority of the king (*raja harimau*) of the tiger village. Hunting rituals are directed towards these *gembalo* in the forest who have to give up their 'cattle' to men. Under the influence of Islam, Nabi Suleiman – the Muslim prophet who could talk to the animals – is now addressed before the spirits of the forest are spoken to.

Some hunters claim to have actually seen such a *gembalo*. One had seen a *gembalo* during the 'hot rain' (*hujan panas*, rain while the sun is shining), when the world of the spirits becomes apparent. The *gembalo* was sitting down, crying beside a deer that had been caught in a snare. The animal, which was being used by him as a mount, had been disobedient and had not come home with its *gembalo*. A myth tells of the origin of the hunting rites that were given to men by the *gembalo*. A hunter went out hunting with two dogs and caught all the deer. But one huge animal escaped and flew to the house of the *gembalo*. The hunter followed the animal, but was stopped by the *gembalo* who said to him, 'now that is enough, if you don't stop this I will kill you'. The hunter answered that he would pay compensation (*tebus*) for the animal. The *gembalo* then laid down the conditions under which men could hunt thereafter, that is, the methods of hunting and the accompanying offerings to be made and charms used. Kerinci men hunt individually or in groups. They hunt larger deer, mouse deer and porcupines because these animals destroy crops and because they are good to eat. Wild pigs and monkeys are considered pests and are hunted in order to exterminate them. When hunting in groups, men use *ranjau* – sharp bamboo pikes of one metre in length, which are stuck slanting into the ground. Along with spears and nets, *ranjau* are always used in a drive, in combination with dogs. In what follows I briefly describe the *ranjau* method of hunting.

In my village of residence, Keluru, on the southern border of the lake, deer have not been hunted for some years. They have been almost exterminated in the immediate environment of the village. Although deer are protected by Indonesian law, this legal protection has never, as the men told me, stopped deer hunting. Keluru men simply gave a leg of deer to the nearest police station and everybody was satisfied. Wild pig hunting, by contrast, enjoys an immense popularity. It is a 'sport' for young men. In Keluru (as in most other villages) the pig hunt is organized in a sporting club (PORBI; *Persatuan Ola Raga Berburu Babi*) with paying members. The club generally hunts twice a week, often with other clubs, each taking turns inviting the other to its territory. Groups of between fifty and 200 young men, with a multitude of dogs, then comb the fields, gardens and

Figure 8.1 A pig hunt using dogs and spears

forest for prey. The hunt takes place on a massive scale and is dominated by local youth. It is far removed from the traditional pest control, which is undertaken in terms of an idiom of reciprocity, as we shall see below. In the pig hunt, the old rituals seem to have disappeared, together with the old hunters. Another factor involved in the disappearance of the hunting rituals is perhaps that, under the influence of the Islam, wild pigs are now considered vermin and therefore unworthy of spiritual protection.

Today, only the older men use the *ranjau* method for hunting wild pigs and deer. Younger men prefer to hunt with spears and dogs; they find this more exciting. They associate hunting with *ranjau* with many time-consuming ritual preparations. These rituals begin with the creation of the *ranjau*. Three *ranjau* are important and must be made anew every hunting season. It is the task of the *dukun* ritual specialist to find the right ones. When the hunt starts, the men gather at an open spot in the forest to consecrate the *ranjau*, which are placed in a row on the ground. The *dukun* then collects four types of plants.[10] Three of these are pinched into two cloven bamboo stalks, which are placed on each side of the row of *ranjau*. They are supposed to blind the animals so that they do not see the trap. Once the *dukun* has stuck these into the ground, he recites a charm in the form of a dialogue.

Dukun: Oh guardians of the forest, guardians of the earth
Guardians of the mountains, guardians of the ravines and the rivers;
We are going to terminate our enemy
That destroys our crops.
We ask that it will go without trouble
And that those [animals] do not disturb our crops
And are kept far away [from us].

Then the *dukun* pushes the three main *ranjau* into the ground. For each stalk he pronounces a formula in the form of a dialogue:

Dukun: Oooh Honoured One (*Nenek*) Nabi Suleiman [three times].
Nabi Suleiman: Who calls us?
Dukun: We, who live in a house with a yard
With gardens and ricefields
Who are cultivators.
It is so that our fields
Have been destroyed by your cattle.
Nabi Suleiman: When there are [animals] that have destroyed these,
Kill them
But first inform the *gembalo*.
He is the one who collects them in the evening
And lets them out in the morning
And who has the [task of] disposing of
Illness and confusion.

153

Figure 8.2 A *ranjau* – for use against wild pigs or deer

Dukun:	Ooh Honoured *Gembalo* [three times].
Nabi Suleiman:	What do you want from me?
Dukun:	What we want;
	We who have a house and a yard,
	Gardens and ricefields.
	It is true that our fields
	Have been destroyed by your cattle.
Nabi Suleiman:	If they have destroyed the fields – kill them.
Dukun:	Is it really true that we can kill them?
Nabi Suleiman:	Yes!
Dukun:	When that is so …

The *dukun* then sticks the *ranjau* into the ground and cries, 'When the animals come they will be hit by the *ranjau* and will die'. He repeats this with the two other *ranjau* as well. Thereupon he strikes all the *ranjau* with a bundle of leaves (*daun seleduk*) and commands, 'poison, poisonous!' Once the preparations have been completed, the men collect the *ranjau* and go into the forest to look for a proper spot to set the trap and the hunt begins.

When an animal is killed a 'sign' (*tanda*) has to be left for the *gembalo* 'so he knows that one of his animals has been caught'. Some people stress that this 'sign' is the 'character' or 'kernel' (*sifat*) of the animal which is returned to the *gembalo*. The effect of symbolically restoring to the *gembalo* what is his is that 'the *gembalo* suffers no loss'. Consequently, the hunters can take away the animal's body. Yet people deny that this returned 'kernel' will bring the animal back to life. The kernel consists of fragments of different parts of the animal: a scale of its hoofs and parts of its nose, ears, horns, lips, teeth, hair, meat and tail. All these parts are wrapped in a banana leaf and put in a split branch near the dead animal. As with the slaughtered water buffalo, the meat is distributed according to rank. The head and the largest part of the liver and the heart are given to the *dukun*. The hunter who killed the animal receives a leg. The men say that in their youth all the villagers were given a part of the meat (as is still the case in egalitarian small-scale societies such as the Kubu and the Mentawaians). This is not done anymore, probably due to the growing population of the villages and the erosion of traditions.

Keeping the balance as a cultivator

The notion that wild animals are the possessions of forest spirits is found in different parts of Indonesia (see, for example, Schefold 1988a: 586, for Mentawai). It tends to be associated with hunter-gatherers and their 'animisitic', pre-Islamic belief systems. Here, however, we are presented with a population of Islamic forest-edge cultivators. The question arises as to whether this idea of responsibility for maintaining a balanced relationship with the supernatural keepers of the forest also applies to the cultivator with respect to his hunting of crop-raiding wild animals. When we look at the way the slain animals are 'returned' (in the form of fragments) to the spirits of the forest, we find that there is a difference between cultivating groups, for which hunting is of marginal importance as a source of food (the case in Kerinci and Mentawai), and hunter-gatherers. A prototypical example of this motif of 'keeping the balance' are the pastoral hunters of Siberia, who leave behind in the hunting ground the bones of the big game animals they kill so that they rejoin the animal spirit in order to revive the animal (Friedrich 1964: 205).

This principle of regeneration is also recognizable, to a certain extent, among the Kubu, nomadic hunter-gatherers of Central Sumatra. The bearded pigs (*nangoy*), which are important game animals for the Kubu, are considered to be deities that have temporarily changed their human appearance into that of a wild pig in order to offer themselves to the Kubu as prey. The wild pigs that survive the hunt 'take away with them the remains of their slaughtered fellows, who are resurrected' (Sandbukt 1984: 95). Whether the Kubu themselves play an active role in the regenerative process is not clear from the description provided by Sandbukt. Certainly, such a notion of responsibility for regeneration seems evident among hunter-gatherers in other parts of the world (Friedrich 1964). Among the Nuaulu, a sedentary tribal group in the inlands of Seram (Ceram), hunting is still a central activity, both ideologically and with respect to food supply (Ellen 1996b: 602). When the Nuaulu have killed an animal, they carry out a small ritual in which the body and soul of the dead animal are reunited. One of the aims is to 'return the spirit to the cosmos, and therefore to ensure that finite stocks are not depleted and that hunting prospects remain good' (Ellen 1996a: 117). The neighbouring Huaulu perform a comparable rite in 'leaving the head of the killed animal behind in the forest for the forest deities, the "Lords of the forest" – so that they will not stop providing game' (Valeri 1994: 118).

Could such a wish be expressed by the Kerinci hunters? I do not think so. As we have seen, in Kerinci part of the animal is 'returned'. But this is not seen as an effort to revive the animal or as a means to ensure that the master of the game animals will not withhold his animals, as is the case among hunter-gatherers. Hunter-gatherers consider themselves responsible for the continuous existence of the prey animals on which they depend. But in agricultural and cattle-raising societies, the relation to forest wildlife is different. Wild pigs and deer, although a source of food, are primarily a threat to crops on which the farmer depends. The damage to crops caused by reviving them would outweigh any contribution they would make as meat. We could then ask: Why do the Kerinci 'return' the 'character' of the animal to the spirits of the forest (*mambang*) if it is not to get more meat? Is it simply a memory of a time when hunting was more important? If there is no direct materialistic motivation, perhaps there is an ideological one. Survival of old hunting ideologies can exist within predominantly agricultural societies, for example in the form of rituals or myths.[11] But this is not necessarily the case here.

The restitution of the Kerinci hunters fits well in their 'folk model of nature' in which respectful and reciprocal relations between man and *mambang* stand central. The Kerinci people consider themselves to be immigrants in the area and thus respect is due to the autochtonous population, the forest spirits. Myths recount the marriage ties between *mambang* wives and village founders, thus ensuring a claim on the land and on the mystical powers of the forest spirits. In rituals the help of the spirits is sought. There is a latent but clear concept of an equilibrium – a status quo between the world of men and that of the forest – which is formulated in a primordial, holy oath called *Karang Setia*. According to the tradition of the founding ancestor, the village made a pact (sealed by an oath) with the original inhabitants – the spirits of the forest and the mighty animals such as the tiger – when the village was originally established. Part of this 'deal' with the first inhabitants was that they were appeased and willing to give up part of their territory, and in future they would respect each other's domain. Only when a member of one group intruded on the other's domain would the offended party have the moral right to take action. It is probably for this reason that, in the hunting charms hunters stress to the spirit herdsman that 'his cattle has destroyed the crops'.

This rite puts men and animals in a balanced relationship, as parts of the same moral order. But this mutual respect should not be romanticized. It has nothing to do with 'love' of the animals or respect for them as such. Rather, it should be understood in terms of an overall framework of relations between forest and village, in which the people feel dependent on the powers that inhabit the forest and therefore feel the need to keep up good relations. But this balance only really exists on an ideological level and does not protect the animals from being over-hunted. People feel that, as long as the proper rituals are performed, they can hunt, or rather poach, since deer are now protected in the Kerinci Seblat National Park.

The tiger

There are about 500 tigers left in Sumatra, of which around 70 live in the Kerinci Seblat National Park (Tillson 1994). But the future of tigers looks bleak because the population is too small and too fragmented and suffers from poaching, habitat loss and shortage of prey. The tiger is the only animal for which the 'pact' works (at least until recently). The reason for this is that the tiger is feared for its anger over the loss of its kin. Elsewhere I have gone into the complex symbolism of this animal (Bakels 1994, 2000, see also Boomgaard, this volume). In Kerinci several tiger concepts are acknowledged. There are 'good' tigers, generally called ancestral or spirit tigers (*harimau roh*): such a tiger is the manifestation of an important ancestor, his mount and friend, or a primordial forest-spirit in animal shape. There are also 'bad' tigers from outside Kerinci. They are considered either 'normal' tigers (*harimau biasa*), that is, animals that attack because they are hungry, or weretigers (*cindaku*). The latter are thought to live in a village in the forests of Pasemah, a region south of Kerinci. Inside the village they look like people, but outside they take the shape of a tiger and go out hunting for human hearts, which they have to bring back to their king. These tigers are feared, but the Kerinci people are convinced that the good ancestral tiger generally protects the villagers from attacks by these bad tigers. Every village in Kerinci has its own version of the basic myth, according to which the founder of the village made an alliance with the tiger people and the two sides promised to respect each other's domain: tigers in the forest and people in the village.

In practice, people seem to have a pragmatic attitude towards the tiger. Discovery of a tiger's paw prints in the forest can inspire fear, and some people said they would immediately return home after seeing them. I know of two occasions where men have gone mad from fear after an accidental confrontation with a tiger. But people tend to trust the ancestral goodwill of the animal and can appear amazingly relaxed when a tiger is in the vicinity. On the one occasion when I was with villagers in the forest when fresh tiger-paw prints were found, I was impressed that none of the villagers appeared at all worried.

Judging from my (incomplete) data, I estimate that up to now there has been a tiger incident in Kerinci every two or three years. Tiger incidents usually involve attacks on village livestock. Water buffaloes temporarily left alone in the forest-garden are attacked by tigers. Goats are sometimes victims; there have even been instances of tigers coming into the centre of the village to catch a goat (formerly kept in the village, but now generally kept in barns just outside the village). Humans can also be the victims of tigers. When I arrived in 1991 in Keluru I learnt that in 1990 a tiger had killed somebody in a neighbouring village. That year the same tiger had been seen just outside Keluru. Tiger attacks tend to take place when people are working in their gardens. But this is unusual tiger behaviour; in general tigers do not attack humans or come close to the village. In the villages I studied, tiger attacks on people appear to have occurred every five to ten years.

When tiger attacks do occur, they trigger discussions in Keluru about possible transgressions by villagers of the local laws. In Sumatra it is generally thought that when a tiger appears in the village domain, this is a sign of moral imbalance: someone in the village must have broken the *adat* rules of proper behaviour. For example, a family inheritance has not been divided up justly, or a person has committed adultery. However, when a tiger kills livestock or a person, this is also seen as a brutal intrusion on the village domain, even if this action is triggered by faulty behaviour of the villagers. Then, *but only then*, will Kerinci villagers try to kill the tiger. In Kerinci, the killing of a problem tiger is generally carried out by one of the well-known hunters. After an attack on a livestock animal, the hunter waits at night in a tree above the animal carcass to shoot the tiger when it returns. The help of the authorities is sometimes called on. Nowadays the local police are keen to shoot the animal and sell its skin and bones (illegally) to Jambi or Padang for thousands of dollars. Because the price of tiger skins and tiger bones is so high, a small number of Kerinci men poach tigers (a recent development, as far as I am aware).

Traditionally, it was the local *dukun* who would set up a tiger trap. The trap consists of two sloping boards supported by upright standing poles. Bamboo spears are placed on both sides of the poles. At the end the boards are set slightly apart and bait is deposited, connected with a rope to the supporting poles. When the tiger pulls at the bait, he pulls away the poles, the boards fall and the tiger's stomach is pierced by the bamboo spears placed on both sides and in between the boards.[12] It is believed that only guilty tigers are trapped in this way. Indeed, it is often believed that guilty tigers are sent into the trap by the *raja* of the tiger people himself. The *dukun* takes great care to explain to the '*raja* of the tigers' why the tiger must be caught, that one of the *raja*'s grandchildren has broken the original pact and has 'crossed the gate, broken the fence' (*menyerobot lawang, merompok pagar*). Consequently, the *raja* is asked to send the guilty tiger to meet his punishment, to 'pay blood with blood, liver with liver'.

When a tiger is killed, it is buried with full ceremony, a condition that was part of the original pact. This is called *bangun harimau*. *Bangun* can be translated as rise/revive/wake up, and refers to the paying of blood money. The tiger is put on a bamboo platform as if he were still alive. Then the men (and sometimes also the women) of the village perform a martial dance (*pencak silat*) in front of the dead tiger, a variation called the '*silat* of the tiger' (*silat harimau*). This dance is seen as an expression of honour to the tiger, but at the same time the animal is symbolically revived in the dance. The tiger's body is created anew in the objects the dancers carry: a knife represents the tiger's whiskers, a *pedang* (long sword) his fangs, a spear his tail, a mirror his eyes (spectacles are also used) and a gong his mighty roar. Pieces of black, red and white cotton symbolize his skin. The tiger is believed to copy the movements of the dancers. During the ceremony, the *dukun* explains to the tiger, in ritual speech, why he had to be killed and stresses that the punishment is just.

In the context of the ceremony it seems that both meanings of '*bangun*' are valid. Where a human being was the victim of the tiger, the tiger's death can be seen as the blood money that must be paid in order to pacify the soul of its human victim and, in a sense, to repay his or her family. In this sense the same mechanisms of revenge are at work as in headhunting feuds: blood has to be repaid with blood. But the tiger is also a victim, and the way the dead animal is treated shows similarities to the way a human being who has died a sudden, accidental death (*mati berdarah*) is viewed. People fear that the deceased man will not accept his fate and will haunt those he has left behind. Elements in the mortuary rites as well as the payment of blood money, in the case of murder, must pacify both the family and the soul of the deceased, who otherwise might ask for revenge. Thus the tiger must pacify its human victim with its own death, but in the *bangun* ceremony the spirit of the dead tiger himself and his family need to be appeased as well. In the past the 'character' of the tiger, consisting of the same elements that are described in the hunting rituals, was sometimes buried at the entrance of the village. In this way the spirit of the tiger would be tamed and turned into a guardian of the village.

159

That people in Kerinci are afraid to break the rules of the pact with the tiger is consistent with the fact that tiger poachers in the reserve are usually men from outside the area. However, the possible monetary rewards from tiger-killing are so great that some local men have taken up poaching tigers. Yet fear of the tiger's revenge remains rife. A few years ago a man from Siulak (a Kerinci village) found a dead tiger caught in a rope meant for deer. He so feared the wrath from the tiger's family that he never dared to enter the forest again. He now sells potatoes in Jakarta.[13] Finally, given that the Kerinci are first and foremost cultivators rather than hunters, we might expect that there would be an appreciation of the tiger as a predator of crop-raiding wildlife. Wild predators have often been viewed positively by farmers as a means of controlling the numbers of wildlife pests.[14] The Kerinci also acknowledge this fact, but in my view it does not play an important role in the symbolism of the animal, which is motivated by deeper drives. The meaning of tigers on Sumatra springs from the fact that the animal can kill and is thus feared and respected, as is that other maneater, the crocodile, an animal to which the same sacred status is attributed and which is *not* a crop protector (see Bakels 2000).

Conclusion

The Kerinci farm in a forest clearing. Kerinci cultivators are surrounded by wild animals in the forest. They practically defend their farms through a variety of means, including magical fences and traps. But in so doing, this potentially brings them into conflict with the forest spirits to whom the wild animals in question belong. They attempt to placate these spirits and maintain the balance with the forest through ritual. These forest-edge cultivators are physically separated from the forest – fields are physically cut out of the forest – but in terms of beliefs and rituals they remain joined to the forest. I have sketched several aspects of animal lore in Kerinci and described three kinds of relationships that people maintain with animals: with livestock, with wildlife crop pests and with wild predators. The economic status of the Kerinci people as cultivators has proved influential in the conceptualization of the first two of these categories. It made cows and other domestic animals their possessions and symbols for the village and it determined whether wild animals were considered 'pests'. In the case of the tiger, a mixture of fear and respect inspire its symbolic meaning among the Kerinci people. Perhaps this similarity is due to the fact that in both cases the animal is threatening, and that the forest is considered a metaphysical realm with predominantly benevolent powers.

Perhaps it is surprising that a group of cultivators should attribute the forest such a metaphysical importance, or that there should exist such a dynamic relationship between village and forest. As I mentioned before, some anthropologists have tended to depict cultivators as viewing the forest and its denizens as foes. This is not the case with the Kerinci, who respect the spirits of the forest as the rightful inhabitants of the untouched jungle, instrumental in giving health and fertility to man, animals and crops. Several kinds of treaties, concluded in the mythical past, must ensure man's well-being in the open spot he has cut out of the forest. Likewise they ensure the well-being of certain spirits and animals in their domain. Fear is part of the attitude towards the forest, but people have a pragmatic view of its dangers, such as the wild animals (more likely to be met in forest gardens than in the jungle). In this sense, hunter-gatherers also are afraid (of tigers, for example), yet I never discerned a notion of 'hate', not even in the story of the son whose father was eaten by a tiger. This man responded by bringing offerings to the tiger, because now his father's spirit inhabited the animal.[15]

Nevertheless, hunters kill in the forest. But, at least in theory, they only kill transgressors. They treat them according to specific rites after their death, to ensure a harmonious relationship with the 'culture of the forest' (Schefold 1988b; Howell 1996: 142). Deer are returned to their owners, and the tiger and his family are appeased. Ceremonies accompanying their death contain elements of restitution. However, this is *not* in the sense that people hope the animals will return to the living (as in the case of hunter-gatherers). Kerinci villagers would not want to regenerate animals that are harmful to their crops, to their livestock or to themselves. Rather, their ritual relationship to the forest must be understood in terms of reciprocity with the forest domain. It is against this background of village relatedness to the forest that Kerinci cultivators can be seen as *not* wholly different from neighbouring hunter-gatherers. The Kerinci case suggests that the difference between the worldviews of hunter-gatherers and of cultivators may not be as absolute as is sometimes stated (see Bird-David 1990, 1992; Ingold 1994). In practice, only the fear of the tiger constitutes a full 'balance of power' and is instrumental in making man keep his part of the deal (in Kerinci, but also with hunters and gatherers). In the case of the tiger traditional beliefs have, until recently, provided an effective form of wildlife protection. This leads to the paradox that the most feared aspects of nature are the most likely to be protected, because they are associated with the most awesome powers of retribution.[16]

Notes

1 This distinction is often made; see for example Bird-David (1990; 1992), Oelschalger (1991), Ingold (1994). It seems to me that both the 'trust' and oneness of hunter-gatherers with their environment is overstressed, as is the distrust and separation of agriculturalists with their natural surroundings.

2 Since 1994 an English journalist, Debby Martyr, has camped in the National Park in the hope of establishing definite proof of its existence. She claims to have seen the animal three times and thinks it is either an earthbound gibbon or an orang utan. Plastercasts of its footprints have stirred the interest of primatologists, but there is no decisive answer yet.

3 Elsewhere (Bakels 2000) I have dealt more extensively with the cultural background of the tiger symbolism. It can be interpreted as a means to control fear – as Valeri has

done for the shark in Hawaii (Valeri 1985) – and can also be related to the cultural mechanisms that enforce the social order (Douglas 1966: 13).

4 By applying the name of the Indian holy snake, *naga*, to the python, the Indonesian people seem to have attributed the *naga* concept to their own traditionally holy snake. In India the *naga* is the cobra.

5 Another, exceptional reason to go to the forest appeared to be madness. I have heard of a number of instances in which men took refuge in the woods, in what seemed a wave of panic or madness. Most returned after some days. Madness in Indonesia has an aspect of holiness, of other-worldliness, as part of the forest domain of the spirits. The same parallel between wildness of the mind and the wildness of the forest played a role in the European Middle Ages (see Bernheimer 1952: 12).

6 Upon entering the forest one has to put a leaf behind the ear, just as, men explained, one wears an insignia in the office. Also one cannot talk loudly, use obscene language, mock the tiger, etc.

7 In Europe too the Wild Men of the Middle Ages, a hairy and primitive variety of man, supposed to live in the deepest parts of the forest, were thought to keep wild animals as their cattle, a concept also present in Siberian cultures (Friedrich 1964: 221; Bernheimer 1952).

8 Another reason might be that (important) new life has to be enculturated anyway. Also for newborn children precautions are taken to protect them against the *mambang*, who might want to take them away.

9 I freely use the term 'hunting' for all sorts of methods, including the use of traps, used in the forest or in fields to deliberately kill animals for food or for other reasons. For definitions of hunting, see Ellen (1996b: 601).

10 These are the *ramput ambun* (to *merabun*, that is to blind), the *rumput rantai*, the *daun jeluan* (generally used as a border-marker) and the *daun seleduk* (considered poisonous).

11 In Mentawai there is a mythical episode which tells of how a turtle teaches a man, Pole'le', to eat and revive him by placing the bones in a certain way (Schefold 1988: 597).

12 The tiger capturing method of the Minangkabau is seldom practised by the Kerinci people. In this case a small cage is used, with a live dog or goat used for bait. At night a *dukun* chants for hours, accompanied by the sound of a bull-roarer (see Kartomi 1972) made from the skull of a dead tiger. The sound of the bull-roarer is supposed to confuse the tiger so that he mistakes the bait for his wife/husband and enters the cage.

13 Personal communication, Debbie Martyr.

14 See the chapters by Rangarajan and Knight, this volume. See also Wessing (1986: 5).

15 Perhaps 'hate' is an emotion more fitting for the class of the rulers. Kings have an essentially different attitude towards the forces of the forest than the rural population. They seek to exert their power upon wilderness, and generally limit the need for balance to the minimum. In epic tales of kingdoms we find this attitude (to subdue wild nature) reflected: for example in the Gilgamesh epic, where the guardian of the holy forest is killed and the forest cut down to build the city of Uruk (Schefold 1988, McVey 1993). Perhaps also the killing of tigers in the *rampok macan* at the central Javanese courts has to be understood in this light. Nevertheless many courts have also retained a highly ritualized link with the powers of the forest, such as in the blood offering to Durga as performed by Surakarta court officials (Headly 1979), in the marriage of the Jogjakarta sultan with Lara Kidul, or in the relation between king and *mandala* (Berthe 1965, Bakels 1989).

16 This is for example also the case with other man-eating animals such as crocodiles, which are regarded with the same religious awe as the tiger (Marsden 1991: 501; Kruyt 1935).

References

Bakels, J. (1989) 'Mandala gemeenschappen in West-Java'. *Bijdragen tot de Taal-, Land-, en Volkenkunde*, vol. 145, no. 2–3, pp. 359–364.

—— (1994) 'But his stripes remain: on the symbolism of the tiger in the oral traditions of Kerinci, Sumatra'. In J. Oosten (ed.) *Text and Tales*. Leiden: Research School CNWS, pp. 33–52.

—— (2000) *Het Verbond met de Tijger: Visies op de relatie tussen mensen en mensenetende dieren in Kerinci (Sumatra), Indonesië*. Leiden: Research School CNWS.

Baud, M. (1997) *De filosofie van de bosrand*. Leiden: RUL.

Beek, G. A. van (1987) *The Way of All Flesh*. Dissertation, Leiden University.

Bernheimer, R. (1952) *Wild Men in the Middle Ages: A Study in Art, Sentiment and Demonology*. Cambridge: Harvard University Press.

Berthe, L. (1965) 'Ainés et cadets: l'Alliance et la hierarchie chez les Baduj'. *l'Homme*, vol. 34, pp. 189–223.

Bird-David, N. (1990) 'The giving environment: another perspective on the economic system of gatherer-hunters'. *Current Anthropology*, vol. 31, no. 2, pp. 189–196.

—— (1992) 'Beyond "The Original Affluent Society": a culturalist reformation'. *Current Anthropology*, vol. 33, no. 1, pp. 25–43.

Boomgaard, P., F. Colombijn and D. Henley (eds) (1997) *Paper Landscapes: Explorations in the Environmental History of Indonesia*. Leiden: KITLV Press.

Douglas, M. (1966) *Reinheid en gevaar*. Utrecht: Het Spectrum.

Ellen, R. (1996a) 'The cognitive geometry of nature: a contextual approach'. In P. Descola and G. Pálsson (eds) *Nature and Society: Anthropological Perspectives*. London and New York: Routledge, pp. 103–123.

—— (1996b) 'Individual strategy and cultural regulation in Nuaulu hunting'. In R. Ellen and K. Fukui (eds) *Redefining Nature: Ecology, Culture and Domestication*. Oxford: Berg, pp. 597–635.

Friedrich, A. (1964) 'Die Forschung über das frühzeitliche Jägertum'. In C. A. Schmitz (ed.) *Religionsethnologie*. Frankfurt: Akademische Verlagsgesellschaft.

Headly, S. (1979) 'The ritual lancing of Durga's buffalo in Surakarta and the offering of its blood in the Krendowahono forest'. In *Between People and Statistics. Essays on Modern Indonesian History*. The Hague: M. Nijhoff.

Hell, B. (1996) 'Enraged hunters: the domain of the wild in north-western Europe'. In P. Descola and G. Pálsson (eds) *Nature and Society: Anthropological Perspectives*. London and New York: Routledge, pp. 205–217.

Howell, S. (1996) 'Nature in culture or culture in nature? Chewong ideas of "humans" and other species'. In P. Descola and G. Pálsson (eds) *Nature and Society: Anthropological Perspectives*. London and New York: Routledge, pp. 127–144.

Ingold, T. (1994) 'From trust to domination: an alternative history of human–animal relations'. In A. Manning and J. Serpell (eds) *Animals and Human Society: Changing Perspectives*. London and New York: Routledge, pp. 1–23.

Kaiser, T. (ed.) (1991) *Bärenfest*. Zürich: Völkerkundemuseum der Universität Zürich.

Kartomi, M. (1972) 'Tiger-capturing music in Minangkabau, West Sumatra'. *Berita Kajian Sumatrae*, vol. 2, no. 1, pp. 24–44.

Kreemer, J. De karbouw (1956) Zijn betekenis voor de volkeren van de Indonesische archipel. 's Gravenhage: Van Hoeve.

Kruyt, A. C. (1935) *De krokodil in het leven van de Posoërs*. Oegstgeest: Zendingshuis.

Marsden, W. (1991) *The History of Sumatra*. Kuala Lumpur: Oxford University Press.

McVey, R. (1993) *Redressing the Cosmos: Belief Systems and State Power in Indonesia*. NIAS report no. 14, Copenhagen: NIAS.

Oelschlager, M. (1991) *The Idea of Wilderness*. New Haven: Yale University Press.

Pálsson, G. (1996) 'Human-environmental relations: orientalism, paternalism and communalism'. In P. Descola and G. Pálsson (eds) *Nature and Society: Anthropological Perspectives*. London and New York: Routledge, pp. 63–81.

Persoon, G. (1990) 'The Kubu: survival in a protected jungle'. In R. Schefold, N. de Jonge and V. Dekker (eds) *Indonesia in Focus*. Meppel: Edu'actief.

Sandbukt, Ø. (1984) 'Kubu conceptions of reality'. *Asian Folklore Studies*, vol. 43, pp. 85–98.

Schama, S. (1995) *Landscape and Memory*. London: Harper Collins.

Schefold, R. (1988a) *Lia, das grosse Ritual auf den Mentawai-Inseln* (Indonesiën). Berlin: Dietrich Reimer Verlag.

—— (1988b) 'De wildernis als cultuur van gene zijde: tribale concepten van "natuur" in Indonesië'. In G. Persoon and J. Oosten (eds) *Mens en Milieu*. Special Issue of *Antropologische Verkenningen*, vol. 7, no. 4, pp. 5–23.

Sinclair, A. (1977) *The Savage: A History of Misunderstanding*. London: Weidenfeld and Nicolson.

Thapar, V. (1992) *The Tiger's Destiny*. London: Kyle Cathie Ltd.

Thomas, K. (1983) *Man and the Natural World: Changing Attitudes in England, 1500–1800*. Harmondsworth: Penguin Books Ltd.

Tillson, R. L. (ed.) (1994) *Sumatran Tiger Population and Habitat Viability Analysis Report*. Minnesota: IUCN/SSC Captive Breeding Specialist Group.

Valeri, V. (1985) *Kingship and Sacrifice: Ritual and Society in Ancient Hawaii*. Chicago: University of Chicago Press.

—— (1994) 'Wild victims: hunting as sacrifice and sacrifice as hunting in Huaulu'. *History of Religions*, vol. 43, no. 1, pp. 101–131.

Verrips, J. (1993) 'Het ding "wilde" niet wat ik wilde: enige notities over moderne vormen van animisme in westerse samenlevingen'. *Ethnofoor*, vol. 6, no. 2, pp. 59–80.

Wessing, R. (1986) *The Soul of Ambiguity: The Tiger in Southeast Asia*. Monograph Series on Southeast Asia, Special Report 24. DeKalb: Center for Southeast Asian Studies, Northern Illinois University.

9

PIGS ACROSS ETHNIC BOUNDARIES

Examples from Indonesia and the Philippines

Gerard A. Persoon and Hans H. de Iongh

A poster issued some years ago by the Indonesian Department for Social Affairs shows two pictures (see overleaf). One of them shows two hunters with weapons pulling a dead pig out from the forest. The other picture shows a farmer's house surrounded by chickens and goats. An arrow, which symbolizes 'development', connects the two pictures: from hunting animals to domesticating animals, implying a step from 'primitive' pig-hunter to 'civilized' peasant.

Introduction

The wild pig is a controversial animal in the forested areas of Asia. For forest dwelling hunter-gatherers, wild pigs are by far the most important game animal and the major source of animal protein. But for shifting cultivators and sedentary farmers clearing land in the forest fringe, wild pigs are a mixed blessing. Although they are the most important game animal, they are a nuisance because of the damage done to the agricultural crops like upland rice and corn. There are also what might be 'pig cultures' in various parts of Asia and the Pacific, where pigs are a crucial element in the division of labour, ritual cycles, systems of exchange and land-use patterns. Even though they continue to roam around in the forest environment, these pigs are, to some extent, domesticated, and their prominent position has, in some cases, prevented other ethnic groups from moving into the local area. In a variety of social and cultural contexts in Southeast Asia, pigs mediate relations between different groups of people and even become markers of ethnic identity.

In this chapter we shall discuss the importance of the wild pig in terms of food, cash and culture. We shall look at the different forms of human–pig relations that arise, that is, the pig as a crop pest, the pig as a game animal and the pig as livestock. We shall also explore why the pig is such a controversial animal in different parts of Southeast Asia by examining examples from Central Sumatra,

Figure 9.1 From pig hunter (above) to farmer (below)

from Northern Luzon and from Siberut. Fieldwork in these areas was carried out in various periods from 1980 onwards, but we also draw on secondary literature by scholars working in some of these societies. We shall begin by presenting a brief outline of the animal itself.

Wild pigs in Asia

The wild pig is an important animal in much of Asia, especially China and Southeast Asia. In Asia as a whole, the total amount of consumed pork exceeded twenty million tons per year in the mid-1980s, an amount greater than the total consumption of all other domesticated species put together (ADB 1995). In biological terms the genus *Sus* in Asia comprises seven species and about 22 subspecies which can be divided as follows (Groves and Grubb 1993: 1–3):

1   the Eurasian wild pig, *Sus scrofa* (ca. 17 ssp.)
2   the pygmy hog, *Sus salvanius* (0 ssp.)
3   the Javan warty pig, *Sus verrucosus* (2 spp.)
4   the bearded pig, *Sus barbatus* (3 spp.)
5   the Philippine warty pig, *Sus philippensis* (0 spp.)
6   the Visayan warty pig, *Sus cebifrons* (0 spp.)
7   the Sulawesi pig, *Sus celebensis* (0 spp.).

For this, chapter three species are of particular relevance: the Eurasian wild pig, the bearded pig and the Philippine warty pig.[1] Eurasian wild pigs are found all over Asia (and Europe). They are often hairy and have a brown to blackish colour, and piglets have longitudinal stripes which gradually fade after around six months. 'The Eurasian wild pig occupies a wide variety of habitats from semi-desert to tropical rainforests, temperate woodlands, grasslands and reed jungles; often venturing onto agricultural land to forage' (Oliver *et al.* 1993a: 113). Bearded pigs (*Sus barbatus*) have relatively large bodies, long legs and a 'bearded' jaw, though the rest of the body is not hairy. The bearded pig occurs in archipelagic Southeast Asia and, while still relatively abundant on Borneo, in Malaysia, on Sumatra and on Palawan, it is becoming rare (Caldecott *et al.* 1993: 137). Its natural habitat is tropical evergreen rainforest but within this category it utilizes a wide variety of ecotypes ranging from beaches to upper montane cloud forests. With massive deforestation in Malaysia, Borneo and Sumatra during the past few decades, the bearded pig has lost much of its territory and has been displaced by the invasion of *Sus scrofa* into logged-over and disturbed areas. The most common wild pig is the Philippine warty pig (*Sus philippensis*) with its distribution from Northern Luzon to the island of Mindanao. The natural vegetation of the eastern side of the Philippines, with lowland areas, montane rainforests, mossy forests, coastal zones and beaches, is reported to be the home area of the Philippine warty pig. Massive deforestation throughout the country and expanding agriculture has greatly diminished the territory of the pig.

Palawan Bearded Pig *Sus barbatus ahoenobarbus*
Occurs only on Balabac, Palawan and the Calamian Islands, bu
declining rapidly over most of its limited, remaining range.

Bornean Bearded Pig
Sus barbatus barbatus. Stil
quite widely distributed on mainla
Borneo, where it is avidly hunt
and provides an important source o
protein for many rural people. Mos
of its forest habitat is subjected t
commercial logging, which ma
affect the periodic populatio
fluctuations and migrations of these
animals.

PERAK

SUMATRA

SUNDA
SHELF

BAWEAN

MADURA

JAVA

rage
oated
sular
palm

ig *Sus scrofa vittatus*. A localised variant of
stributed Eurasian wild pig. The only non-endemic
the region, and one of the few wild pigs not
o any degree at the present time.

ene islands/land masses).

Figure 9.2 A section from the poster *Suids of Southeast Asia* produced by the Pigs and Peccaries Specialist Group of the World Conservation Union (IUCN) – Species Survival Commission. (Reproduced with the permission of William Oliver, the designer and illustrator of the poster)

The pig was one of the first animals to be independently domesticated in various parts of Asia. For a long time early domestic pigs continued to be associated with wild pigs and even today this is the case in some areas. In New Guinea a hybrid pig was introduced by humans, and populations of wild and domestic pigs continue to mix. But there is great variation in the methods of domestication and in the interaction between domesticated species and wild populations. Domesticated pigs escape and join herds of wild pigs. In some cases people take piglets from the wild pigs and domesticate them (Van Beek 1987). Consequently, the two populations are not very different genetically. In some areas where wild pigs were – and continue to be – important as a source of animal protein, pig domestication never developed. In Kalimantan no efforts have been made by the local human population to domesticate the wild pig. The Chinese migrants, just as they did in many other places in Southeast Asia, brought along their own preferred species for domestication (MacKinnon *et al.* 1996). But local people did not adopt this custom and continued instead to depend heavily on the wild pig for their meat consumption.

In some places true 'pig cultures' have developed, in which pigs play a central role in the community. The animals are symbols of wealth and status, and extensive and elaborate rituals have developed in connection with their domestication, slaughter and ceremonial exchange. Pigs are used to obtain wives, to promote peace and to compensate enemies and allies; they are the most important means of obtaining personal status and authority. All scarce social goods and benefits derive from the exchange of pigs. It is not so much the production of the pigs for their own sake which has value. Social value only accrues to the producer by exchanging them or by giving them away (Feil 1984: 77). In the New Guinea literature in particular the ceremonial exchange network has received a great deal of attention. But just as the pigs are an important medium for extending social networks, so also can they serve as an efficient boundary. This is particularly the case where 'pig cultures' or groups of hunter-gatherers are close to Muslim groups (see for example, Shanklin 1985: 386; McNeely and Wachtel 1988: 163).

We shall now turn to three socio-cultural contexts in Southeast Asia in which pigs play a central role. In our first example, taken from Central Sumatra, pigs play a crucial role as a marker or symbol of ethnic identity among different ethnic groups, including 'pig lovers' and 'pig haters', to use Marvin Harris's terms. In the second example, from Northeast Luzon, wild pigs are important to all ethnic groups for bush meat. Although in the past a kind of cooperative relationship did exist between two ethnic groups, this has now become a thing of the past. Our third example, from Siberut (West Sumatra), portrays a 'pig culture' in which pigs play a very important role primarily as bearers of social and cultural meanings and as markers of ethnic (and intra-ethnic) identity.

Wild pigs and 'Pig-men' in central Sumatra

In spite of the tremendous recent ecological changes in the central part of Sumatra (the provinces of Riau, Jambi and South Sumatra), there are still a number of hunting and gathering tribes, like the Kubu, Talang Mamak, Sakai, Bonai and Petelangan, who live on game animals, wild tubers and other wild foods from the forest. The most important game animals are wild pigs and deer and, to a lesser extent, monkeys, while in the past elephants and rhinos were hunted as well. Although they did not use powerful weapons, they were able to catch these animals using pitfalls and spears. The bow and arrow and the blowpipe were unknown to them. All of these groups use dogs in hunting. In the past these tribes, consisting of small bands of people living in the forest, have had various degrees of interaction with neighbouring sedentary Muslim farmers. In particular, there has long been some exchange (sometimes in the form of silent barter) of forest products (rattan, resins) with farmers and traders in return for salt, ironware, cloth and tobacco.

In this part of Sumatra there are two kinds of pig: the Eurasian wild pig and the bearded pig. The bearded pig was only 'discovered' on Sumatra by Dr Abbott in 1902 (though the animal was known, from Borneo, long before that time). The Dutch called this animal the *strandvarken* (beach pig) as it was often found in coastal areas. During the fruiting of the durians, however, bearded pigs move in herds into the interior to feed on this tasty forest fruit. They also cross the large rivers in the coastal plains in this part of Sumatra. This was the moment for the hunters to catch the animal, locally known as *nangoi*. Schneider described how they spear the animals just before they reach the bank of the river, giving them little chance to escape (Schneider 1905: 83; 1958: 248). The abundance of meat allowed the people to organize large gatherings and meetings.

The Malay farmers in this area practice an extensive form of shifting cultivation in which they combine annual crops with perennials. Rubber trees have been of particular importance since the beginning of the twentieth century. As a result of this situation, the Malay farmers have transformed the rainforest into a highly diversified area consisting of cultivated upland fields for rice, corn, cassava and bananas as well as fallow fields and rubber gardens. This kind of environment, with its abundance of food resources, is highly favourable for wild pigs. The pigs pose a major threat to the harvest of the Malay farmers, who have no effective means of getting rid of them, as they do not practise pig hunting (though they do possess locally made shotguns). As Muslims, the meat of the wild pig would, in any case, be useless to them. Fencing the extensive fields is deemed to be too labour intensive.

It is against this background that groups of hunter-gatherers have established relatively close relations with the farmers. The hunters guard the fields of the Malay farmers and keep the meat of the wild pigs they kill. They are also employed by the Malay to clear the forests, to weed the fields and to assist in harvesting or rubber tapping. If they succeed in catching a non-*haram* animal alive, they can even sell it to the Malay farmers (to be slaughtered in the prescribed way). However, the hunter-gatherers' attraction to pigs serves to maintain a distance from the Muslim farmers. For example, the Kubu sometimes keep a piglet as a pet, something abhorrent to the farmers. In other words, while wild pigs bring the two groups together in connection with field-guarding, they separate them socially and culturally.

Into this situation, a new set of farmers has entered – transmigrant farmers from Java. In recent decades, large tracts of forest have been converted into other forms of land-use. After large-scale logging and the subsequent process of land clearing, transmigration sites were established as well as plantations for industrial crops like rubber and palm oil. In this situation, pigs are considered to be nothing more than pests. Compared to the land-use practices of the Malay farmers, forest conversion for transmigration sites and plantations implies much more rigorous methods. Instead of clearing relatively small plots for a diversified land-use, the total clearing of the forest land is necessary to make room for the transmigration sites and estates, leaving little room for hunter-gatherers to survive. Transmigration has been going in the provinces of Riau and Jambi for many years. In general, Javanese farmers with a long tradition of wet-rice cultivation were moved to the provinces in the outer islands with different ecological conditions. After clearing the forest floor and after a number of years facing declining soil fertility and rain-fed agriculture, people were forced to change their system of cultivation, and instead began growing sweet potatoes and cassavas as substitute crops. However, these new crops started to attract wild pigs, which rapidly increased to become a major pest to the transmigrant farmers. In a recent article, Simon Rye has described this conflict between the transmigrants and the wild pigs, and the way the newcomers experience it in specific cultural terms. The following remarks draw on this vivid case study (see Rye 2000).

The transmigrants' pattern of land-use is different from that of the Malay farmers. In transmigration sites people are allocated only a few hectares, in accordance with ideas based on permanent and intensive agriculture. No fields should be left fallow. Because of this situation no symbiosis develops between the transmigrants and the hunter-gatherers. There is simply no room for hunter-gatherers to make a living in this kind of environment. There are no fallow fields, there is no forest left and there is little opportunity to provide labour services to the poor migrant farmers. The hunter-gatherers show up in these transmigration sites only to sell medicinal forest plants and roots, or to beg for food at bus terminals. The same holds true for the areas which are turned into large plantations for oil palm, rubber and other industrial crops. Neither the people traditionally inhabiting these areas nor the wildlife itself can survive under these conditions. The absence of food crops makes these plantations relatively barren with respect to wildlife (large mammals in particular).

For the transmigrants and plantation workers from Java, the pig is part of the wilderness. It belongs to the haunted forest. It has become the symbol of all their problems in the new environment. They believe that there is a close association between the wild pig of the forest and the people who live in the forest, and believe in the existence of a 'pig-man', a creature half man and half pig. Both the forest people (the Kubu, Sakai, Talang Mamak and others) and the wild pigs are seen as harmful, polluted and tabooed creatures of the forest (Rye 2000). Another interesting aspect of the situation in Central Sumatra emerges in Errington's (1984) account of a Minangkabau collective wild pig hunt. Although these people are Muslims, from time to time they engage in pig hunting. This pig hunt is more than just an opportunity to kill a religiously prohibited animal or a forest-edge farm pest, but is a traditional male sporting activity which has a religious background. Minangkabau people believe that pigs are former people who were punished for having led evil lives. Muslims believe that evil-doers are judged by Allah and condemned to become pigs. When such people are then killed as wild pigs by hunters and dogs, they go to hell. This makes for a strong deterrent to sinful behaviour (Errington 1984: Ch. 7).

In Central Sumatra only a small part of the population is interested in the wild pig as a source of food or as an important game animal for consumption by the group itself. There is no external market for this kind of bush meat. All other ethnic groups (Malay, Minangkabau, Javanese, Sundanese and so-called spontaneous settlers in the coastal areas like the Buginese and Banjarese) would like to get rid of the animal altogether, but they lack an efficient means to do so. Because the animal is not hunted for food, because of its impressive reproductive capacity and because extensive forest land remains, the Eurasian wild pig is unlikely to be threatened to the point of extinction (even though numbers are certainly declining). The situation of the bearded pig is worse as this animal is more susceptible to ecological degradation and suffers more heavily from hunting pressure because of its seasonal migration in large herds. Moreover, in these deteriorating circumstances, the Eurasian wild pig seems to be pushing the bearded pig out of its declining habitat. But inside and outside protected areas in Riau, Jambi and southern Sumatra, both species have better chances of survival than many other terrestrial mammals because of their adaptability to changing conditions and their reproductive capacity.[2]

Wild pigs and pig bombs in Northeast Luzon

The situation in the Sierra Madre Mountain Range in Northeast Luzon in the Philippines is ecologically comparable to the one in Sumatra, but socially it is totally different. The tribal hunter-gatherers in this forest area are the Agta, who belong to the original Negrito population of the country. The main game animal is the wild pig, in this case the Philippine warty pig and the bearded pig (Mudar 1985), and to a lesser degree the deer. Other large terrestrial mammals are absent, but the smaller game animals include monkeys, wildfowl, fruit bats and monitor lizards. Traditionally, the Agta hunted with bow and arrow and with dogs, but in recent decades they have adopted new hunting technologies such as the shotgun, which is a prestigious possession among young men. The Agta obtain these shotguns from traders, from the guerrillas of the New People's Army

(NPA) or from the logging companies. Hunting is mainly done by men, but women also use bow and arrow and hunt with dogs (Estioko-Griffin 1985: 121).

Apart from hunting for subsistence, the Agta hunt on behalf of the sedentary farmers. These farmers thus obtain their animal protein less from animal husbandry than from exchange relations with the Agta, offering in return a great variety of other products (Headland 1986: 120; Rai 1990: 85). However, due to logging activities and the subsequent influx of migrants from a wide variety of ethnic groups, farmers have moved to the area to start shifting agriculture. As they are either Christians or adherents of tribal religions, wild pig meat is not a forbidden item for any of them; on the contrary, the bush meat of wild pigs is in great demand in Northeast Luzon and in many other parts of the Philippines. In the academic literature there has been much discussion about the significance of increased activities on the forest fringe. In a number of publications Jean Peterson has argued that the edge of the forest is a productive zone, where the number of game animals (especially terrestrial mammals such as deer and wild pigs) increases. Peterson has proposed the 'merits of margins' hypothesis based on the idea that garden hunting contributes to a symbiosis between hunter-gatherers and farmers. Expansion of farming by migrants creates more forest edge, and this is beneficial to terrestrial mammals like wild pig and deer, and indirectly to commercial hunter-gatherers who can more easily satisfy the farmers' demand for bush meat (Peterson 1978, 1981).

Other researchers, in particular Headland (1988), have questioned these merits of the forest edge (at least on a long-term basis) for a number of reasons. In the first place, Peterson's interpretation of the ecotone concept is called into question. Moreover, increased hunting pressure by migrants is thought to have contributed to the long-term depletion of wildlife populations. In addition to hunting by the Agta, migrant groups have also been quick to take up hunting of the highly praised wild game. Together with military activities, the New People's Army and the logging companies, the availability of new hunting technology has put an enormous pressure on wild animals. For the Agta, bush meat became an important source of cash income, in addition to that from wage labour and the harvesting of non-timber forest products. As Headland (1988) reports, the Agta in the Casiguran area gave up hunting as a regular pursuit because the animals became so scarce. Modern firearms, such as home-made shotguns and other professional weapons (loaned or sold to them by loggers, soldiers, local politicians or the NPA), have completely replaced the bow and arrow and therefore increased the efficiency of hunting. The second reason Headland mentions for the decline in game was the desire for wild meat on the part of the new immigrants moving into the area along the newly made logging roads. In the first years, bush meat was turned into a regular cash crop. Large quantities of dried bush meat were even sold to places as far away as Manila, where there was a big market for this product (Headland 1988: 131).

Other means of catching wild pigs include snares, pitfalls, rope traps and so-called pig bombs. Pig bombs are a very effective, though indiscriminate, killer.

The explosive material used in the bombs is phosphorous taken from matches or gunpowder and combined with broken glass or little pieces of metal. The bomb is baited with, and placed within, odorous foods, usually fruits. Once the pig touches the bait, the bomb explodes, killing the animal on the spot. Even if the pig escapes, the 'hunter' will have no trouble in following the blood trail of the injured animal. Some hunters place a large number of bombs along pig paths in the forest. The big advantage of the pig bomb is that it does not require the constant attention of the 'hunter'. But this way of killing pigs has contributed to the rapid decline of the wild pig population and, not surprisingly, has caused regular human injuries and fatalities (Headland 1988, Polet 1991; Leonor 1993, Conelly 1996).

One of the consequences of the reduction of bush meat is that migrant farmers have increasingly taken up animal husbandry as a way of supplying meat for themselves or for the market. Pigs, cows and water buffaloes are used to this end. In general, people do not make use of the offspring of wild pigs; they prefer the much tamer, larger and faster-growing new breeds of pigs that can be obtained through the pig dispersal projects of the government or development organizations. The decline of pigs and other game animals and various non-timber forest products has also contributed to the inclination of the Agta to move closer to their present main source of income-generating activities – that is, the farms of the migrants in the lowlands. Alternatively, they settle down in the logged-over forest to collect fuelwood and sell it in little bundles along the roads (Baldi 1997). Hunting is no longer a major activity for some groups of Agta, and many of them have given up hunting altogether because of its diminishing returns.

A number of new interconnected events point to change in the near future. First of all, most of the official logging concessions have been recently cancelled and illegal logging has been greatly reduced. As a result of this, no more new roads are built and existing logging roads are not maintained and deteriorate rapidly. Consequently, employees of the logging companies have left the area. There is far less trapping going on at the moment. Also, the rate at which new pioneer migrants move into the logged-over forest declines as transport is that much more difficult in the mountainous areas. In addition, there is the establishment of the Northern Sierra Madre Nature Park, which will further hamper human penetration into the area. Although the wild pig population was never a main concern in the establishment of the park – the concern for faunal biodiversity is much more focused on birds and other animals – wild pigs will certainly benefit from the reduced human presence (and human activities) in the park. Perhaps in the future wild pig populations will increase again. If so, the Agta would profit because they are set to receive collective hunting rights as the 'indigenous cultural community'. Outside the park, however, the pressure on wild pigs and other game animals continues to be high. As the supply of bush meat decreases, prices go up, compensating for the diminishing returns from hunting efforts and encouraging continued hunting. As a result, wild pigs have become a scarce resource.

In Northeast Luzon (and in other areas in the country, like Palawan), hunter-gatherers who used to hunt either for subsistence or for commercial purposes increasingly compete with migrant farmers and commercial hunters (see also Eder and Fernandez 1996). In such cases, ethnic differences have no implications for the animals. To all ethnic groups living in the foothills of the Sierra Madre mountains (the Ifugao, Ilocano, Kalinga, Bontoc, Tinggian, Ibanag and many others), the wild pig is equally attractive as potential meat for themselves or as a cash-earning forest product. Members of all ethnic groups are equally quick to adopt new methods of killing the animal, as illustrated by the 'pig bomb'. Rituals employed in hunting among the Agta rapidly declined because of these new hunting technologies (Rai 1991). As a result of increased hunting pressure, most of the original hunters look for alternative activities. The returns from hunting are simply too small. Other non-timber forest products rarely provide an alternative due to deforestation and overexploitation. In most cases, hunter-gatherers become day labourers in the fields of migrant farmers or they become farmers themselves. The wild pig gradually loses its significance and comes to be replaced by domestic pigs and other 'tame' animals.[3] The transition from wild to tame pigs also brings about a change in the use of the animals. While wild pigs were mainly for consumption and for distribution among relatives and friends, domesticated pigs are usually kept as living capital to be sold in the market when the owner is in need of cash.

The 'pig culture' of Siberut

Pigs are of great importance in many tribal societies in Southeast Asia and the Pacific. Apart from being a main source of wealth, pigs play an important role in the ritual cycles, in exchange patterns, in the division of labour and in patterns of land-use.[4] One example of such a pig culture is to be found on the island of Siberut in West Sumatra. Siberut is the largest of the Mentawaian Islands, off the west coast of Sumatra. It is inhabited by about 23,000 Mentawaians and a small number of migrants, predominantly of Minangkabau origin. The Mentawaians are traditionally organized in patrilineal groups of approximately thirty to eighty people living in small settlements, called *uma*, along the banks of rivers. Hunting, fishing and gathering provided most of the daily food. Sago starch, obtained from the sago palm (*Metroxylon sagu*), was and still is the staple food. Stands of wild and planted sago occur in the swampy areas and along the banks of the river. In addition to these food resources, people cultivate root crops, bananas and fruit trees. Pigs are the most important source of wealth on Siberut. Every adult male should have his own herd. They are the most important item during the negotiations about bride-price and pigs are used for the payment of fines in the local justice system. The animals are slaughtered on special occasions and sacrificed in all major rituals. Pork should be offered, for example, to people who have assisted in the building or hauling of a dugout or in the construction of a house.

The wild pig population is the origin of the quasi-domesticated pigs found on Mentawai. It is highly unlikely that traders from Sumatra have ever taken other pigs from the mainland to Siberut, but this cannot be excluded, and the existence of hybrid pigs has been proved in other areas such as Irian Jaya (Oliver *et al.* 1993b: 173). By being fed, free-roaming wild pigs gradually get used to a human presence, while retaining much of their wild character. Only since the beginning of the twentieth century have Chinese traders, Dutch colonial administrators and, later, Japanese soldiers and foreign missionaries imported their preferred species of domestic pigs from the mainland. No doubt some of these animals have escaped, or were given to the local people who released them, and have gradually mixed with the local herds.

Mentawaian pig keeping has a number of salient features. Sows are impregnated by free-roaming boars, give birth in the forest and do not appear until one or two weeks afterwards. It is always a moment of great excitement when a sow first shows up followed by her offspring. While they are still young, the animals are caught in order to make an incision in the ears to indicate ownership of the pig. Mentawaian pig herds comprise animals of all age groups. The largest animals are the castrated boars and they may be surrounded by groups of twenty or thirty sows and young boars. Numerous piglets follow their mothers around or try to find their own food. Pigs are fed with the pith of the sago palm. Thanks to the abundance of sago on the island and the relative ease of providing it to the animals, the labour demands of pig raising are limited. The feeding of the pigs involves felling the tree, transporting it to the field-house along the river, cutting the stem into pieces and then giving it to the animals. The river is the storage room for sago stem (the water prevents it from going bad). Regular feeding of the pigs is necessary to make sure that the animals come back to the field-house, otherwise they may start looking to other places for food, but the animals are free to roam around in the swampy forest. Where fields of root crops are located too near to the pigs, these fields (always very small) have to be fenced in to protect them from damage.[5]

All the pigs end up being slaughtered. But catching a free-ranging, semi-wild pig in a herd is no easy job and has to be done with the help of a strong rattan strap while standing on the verandah of the house at a safe distance from the animals. The strap is placed on the muddy floor and once the pre-selected animal steps into the strap the rattan is quickly pulled. While the animal screams loudly and the other pigs run off into the forest, its mouth and legs are tightly secured to prevent its escape, and it remains in this position until the time of slaughter. Certain ritual measures are taken when pigs are slaughtered. Before taking the animal to the settlement, an offering is made: a little piece of one ear is given to the spirits of the forest so that they will not come after the man who has caught the pig and do him harm. During the ritual slaughter, the *kerei* medicine man utters a spell over the pig's body to invoke the soul of the pig to keep away evil spirits and attract good forces, and to ask the pig for forgiveness and not to be angry. He reminds it that it has always been taken care of, fed and protected, and that that was why it was able to grow to become so big and fat (see Schefold 1980). The animal is killed by pushing a sharp spearhead through its throat. The intestines are taken out, and the *kerei* and other adult men try to foretell the future by 'reading' the animal's heart and lungs, in a kind of fortune telling. The hairs of the skin are burned and the intestines cleaned before the animal is eaten.

Generally, hunting involves primate species and, occasionally, deer. Pigs are not considered game animals because all the pigs around belong to somebody. Even if a pig grows really wild, people rarely take the risk of killing a pig as it may belong to somebody else. If they did, and it became known, they would have to pay a fine (*tulou*) and provide compensation for the killing. The only time when one can kill a pig is if it is destroying somebody's gardens and if the owner, after being warned, does not take the animal away. An example of pig crop-raiding arose in 1994 during a period of extreme drought (lasting five months) when the rivers, usually wide enough to keep the animals away from the settlement, could be easily crossed by the pigs. This gave rise to serious problems in unprotected home gardens. Numerous conflicts occurred and finally people were ordered to confine their animals or risk having them killed.

This close association between pigs and humans, along with the Mentawaian mode of keeping pigs, has been a problem for all administrators on the island. In Dutch colonial times administrators urged the people to build fences for the pigs and to keep the settlements clean, but, after an initial (if minimal) compliance, the Mentawaians refused to continue (Hansen 1915).[6] After Indonesian independence (1949), the new administration ordered the construction of larger villages, spaced at regular intervals along the banks of the river, and pigs were not allowed in these settlements. Instead, people kept their pigs on the other side of the broad river, on small islands in front of the main island or along the seashore at least a few kilometres away from any settlement. The West Sumatran government (in particular the Department of Social Affairs, which is in charge of the development of the 'isolated tribes' in the country) has always considered this way of raising pigs as a 'bad custom', inferior to raising chickens and goats, water buffaloes or cattle. This is because pigs are seen as dirty animals which eat human faeces, spread diseases and cause damage to agricultural crops (a high density of pigs makes the cultivation of annual crops like rice or corn virtually impossible). Another government argument against raising pigs has to do with the taboos – or 'superstition', as the government sees it – connected with their domestication. Mentawaians traditionally believe that the well-being of their herd is to a large extent dependent on the behaviour of the owner and his family. In particular, herd owners and their families should refrain from social and agricultural activities while any of the women family members are in the last phase of pregnancy. From the government's perspective, if pigs were replaced by other animals (such as goats and cattle), these superstitions could be done away with.

Against this background of government pressure, pig keeping has become a site of resistance for Mentawaians. Because of the large number of restrictions and obligations connected to living in the village (school attendance, *gotong royong*, church services, cleaning the yard around the house, the absence of traditional ceremonies), the pig house (and in many cases people even build a traditional *uma*) has became a place of retreat where Mentawaians one can find peace and quiet, where they can carry out traditional ceremonies away from administrators, police officers, civil servants and teachers and where they can escape from the many obligations imposed on them by the state. The 'need to feed the pigs' is always a good excuse not to show up in the village at certain times.

There have been many attempts to replace pigs with other animals, some of which have failed.[7] Interestingly, water buffaloes and cows, which were taken to the island to stimulate modern animal husbandry, were accepted by Mentawaians but raised in basically the same way as pigs. The water buffaloes and cows are not kept at a prepared and fenced pasture area close to the settlement, nor are they used for agricultural purposes like hauling or ploughing.[8] On the contrary, they are taken upstream to the pig house and released to roam around as they please, receiving no attention thereafter. Water buffaloes in particular become wild very quickly when no longer connected to human society (even if only by a rope in the hands of a child or adult). They roam along the swampy banks of the river or in the forest-fields. It is only when the animal is to be killed that it is caught, though this may take a few days of chasing and hunting. Due to the relatively small number of cows and water buffaloes on the island, there seems, as yet, to be no problem regarding knowledge of ownership. This example clearly shows that on Siberut the local 'pig model' of animal husbandry is applied to newly introduced animal species.

The 'bad' aspects of pig raising, along with the state efforts to make people turn to other domesticated animals, form the context in which pigs have become a marker for ethnic identity on Siberut. The raising of pigs in the traditional manner is not just a matter of protein acquisition. It is part of a lifestyle, closely connected with the traditional religion, *sabulungan*, with ceremonies and taboos interwoven with many other cultural elements. People who want to retain that lifestyle refuse to give up their herd and insist on keeping their free-ranging animals. They argue about bride-price, fines and exchange relations in terms of pigs and they honour guests by slaughtering a fat animal. On the other hand, those who want to become modern, and who are no longer attracted to the lifestyle of their ancestors, choose to give up pig raising altogether, considering it to be incompatible with the new, modern lifestyle. But these people still have to find alternatives for pigs in all the social situations in which the pigs are used. In general, it is largely the lifestyle and value patterns of the Minangkabau which set the standards for modernizing Mentawaians. The domestication of pigs is an inter-ethnic marker of lifestyle.

Mentawaian pig raising has helped to prevent Minangkabau migrants from moving into the area (they instead opt for coastal zones). The presence of the pigs makes the Minangkabau reluctant to move inland for agricultural purposes because growing annual crops like rice and corn would be almost impossible with a population of free-ranging pigs nearby. Moreover, the idea of being surrounded by numerous dirty pigs does not appeal to the Muslim Minangkabau migrants. It is a sobering thought, however, that if the Minangkabau migrants did come in, and were minded to do something about these pigs, serious conflict would become inevitable. To the extent that the pigs have kept the Minangkabau at a distance, the animals have been useful in drawing a boundary with a rival ethnic group.

Pig survival, extinction and domestication

The three cases mentioned above can be related to the recent discussions about domesticatory relationships between people and animals and to questions about prospects for extinction or survival of the animals concerned. David Harris (1996: 440), for instance, argues that it is too simple to contrast wild animals with tame or domesticated animals, and to leave out the intermediate forms of human–animal relations. Referring to the earlier work of Zeuner (1963), he differentiates a number of stages in intensity of animal domestication. In his classification, Harris differentiates predation, protection and domestication. He defines hunting as a form of predation where humans compete with other predators for prey. Protection implies intended intervention in order to enhance the reproductive potential of the species by the manipulation of its environment, by offering protection against other predators and by providing additional food. Systems of loose control in semi-wild habitats are often used to that end. Conventionally, the essential criterion that is offered for domestication is the maintenance of a self-perpetuating breeding population of animals genetically isolated from their wild relatives. Gradually, behavioural and phenotypic changes will occur in the domestic stock (D. Harris 1996).

In his categorization, Harris does not mention the kind of hunting that aims not at obtaining game animals for their meat or other products (teeth, skin) but at stopping their ravaging or threatening influence. This is a kind of passive hunting which is spatially limited to areas that need to be protected, rather than the active hunting aimed at obtaining the animal for its meat, skin or teeth, which takes place wherever the animal roams around. This passive form of hunting tends to have limited ecological effects (unless of course destructive methods like poison are being used). Applying these thoughts to the cases presented above, some interesting comparisons can be made. The situations in Central Sumatra and Northeast Luzon seem to offer a direct contrast. In the first region, wild pigs are 'hated' because they are destructive pests; apart from the Kubu, Sakai or Talang Mamak, most people just want to be rid of them. But in Luzon, the second region, the wild pig is in great demand as cheap bush meat. This great demand has never led to any protective measures to secure the animal's future survival; it was, and still is, considered a resource free for all to take. In response to its near extinction, people simply seek alternative sources of wild protein, and other animals become the next target to satisfy the appetite for bush meat. In Central Sumatra this need not be the case. There is only a small market for bush meat anyway and no other game animal has the extremely negative reputation of the wild pig.

Hunting is also less common among the Javanese, Malay and Minangkabau people when compared with the various ethnic groups in Northeast Luzon. On Siberut, as in other 'pig culture' areas, the semi-wild pig is not being threatened at all. People take great care in ensuring the well-being of their herds. Threats come only from outside, for instance from large-scale conversion of the rainforest habitat to other land-uses. As a semi-domesticated animal, the pig can survive without much human attention. Due to the present-day movement of people towards the coast, herds of pigs might increasingly grow wild, but this in itself is not a threat to the survival of the animal.[9]

In Sumatra the killing of pigs (to obtain food by hunter-gatherers and to remove pests) is not followed, once population numbers decrease, by the selection of another animal. The small numbers of hunter-gatherers do not turn to protective measures or even domestication. In Luzon, however, people turn to animal husbandry and livestock raising, including domestic pigs. On Siberut the situation is different. Pig husbandry on the island is a clear example of 'protection' in Harris's classification, that is, of enhancing the reproductive potential of the animal by additional feeding, by spreading the animals over large areas, by regularly reducing the herd through selection of animals to be slaughtered and by castration. The example of the domestication of water buffaloes and cows illustrates that Harris's classification is not necessarily, and in all cases, an evolutionary one. The treatment of livestock depends not simply on the animal itself or its history of domestication. Previous experiences with other animals on Siberut (such as ducks and geese) suggest that the Mentawaian way of pig raising is the locally preferred system of keeping animals. Based on experience, and largely drawn from the pig model, Mentawaians prefer the 'protection' mode for all animals. If the animal turns out not to fit in with this mode of exploitation, Mentawaians are not interested.

Paradoxically, the chances for survival of wild (as opposed to feral) pigs seem to be best in Central Sumatra as long as pockets of rainforests remain, in addition to extensively used fields and other forms of land-use. The absence of a market for bush meat and the defensive motivation for pig hunting (pest control) limit the hunting pressure on the animal. Hybridization of wild pigs with domestic pigs is unlikely to occur as there is virtually no pig husbandry in this part of the island. In Luzon the animal is bound to be brought close to the edge of extinction because of the strong demand for bush meat. Due to hunting pressure, hybridization of wild pigs with free-ranging pigs is limited because the populations are spatially too far apart to meet. Recalling Marvin Harris's discussion of 'pig-loving' and 'pig-hating' cultures around the world, the experience of these two cases prompts the thought that, in some cases at least, it might be better to be 'hated' than 'loved' (M. Harris 1985). On Siberut in particular, the wild pig has in the past been mixed with species of domesticated pigs but only to a limited extent. This population of pigs is unlikely to change in its genetic composition unless major changes are implemented in the near future.

Conclusion

In this chapter we have examined a number of different cultural contexts in which wild and semi-wild pigs are found. We have looked at a number of ethnic interfaces in which pigs are implicated. Pigs can bring different ethnic groups together through trade in wild meat or because one particular group assists another in fighting an agricultural pest. Pigs can also keep different ethnic groups apart: the abundance of pigs may prevent another ethnic group from encroaching on an area because agricultural practices are incompatible with these free-roaming animals. Particularly in areas where Muslim populations are involved, pigs have the potential to exacerbate inter-ethnic conflict. The pig is a symbol of everything that developed, civilized people should not be – that is, dirty, free-roaming and hard to control. In relation to nature conservation efforts in Southeast Asia, it is clear that wild pigs in general are not appreciated for their contribution to the region's biological diversity. In fact, it is only in the Philippines that the various wild pig species have attracted conservationist concern, in the form of campaigns emphasizing that the different species occur 'only in the Philippines' (along with a number of other birds and mammals). The role that these animals play in the forest ecosystem, for instance in the dispersal of seeds, is generally underestimated. In this respect, there seems to be little difference in the basic attitudes of the different ethnic groups in the region. Broadly speaking, wild pigs are animals hunted for their meat or destroyed for the damage they cause. They are not, as yet, the focus of widespread conservation efforts.

Notes

1 For this part of the chapter extensive use was made of the IUCN specialist group report on pigs and peccaries (Oliver 1993), which is a rich source of information on wild pigs.

2 Of some interest in this connection is the position of wild pigs and other forbidden animals in the discourse of Islamic attitudes towards biodiversity and nature conservation. Officially, pigs enjoy the status of 'living beings in their own right, glorifying God and attesting to His Power and wisdom' as differentiated from animals as 'creatures subjected in the service of man and other created beings' (Bagader *et al.* 1994). In practice, however, wild pigs receive little protective attention, both outside as well as inside protected areas. Their role in the forest ecology is often unappreciated.

3 In this context the dog should also be mentioned. Apart from being a pet and a useful animal for field-guarding and hunting, dogs are increasingly eaten and sold for meat.

4 There is an extensive literature on the pig cultures of Papua New Guinea. See, for example, Rappaport (1984), Feil (1984), Rubel and Rusman (1978), and Dwyer (1990).

5 People use living branches of trees, which root and continue to grow, and this living fence will retain its strength for quite some time. These fences are not equipped with snares and pitfalls. Pigs should not be caught in this way.

6 Interestingly, a medical doctor in the colonial service expressed an opposite opinion. He stated that the pigs were extremely useful in cleaning up the rubbish and waste material in the settlements (Van Beukering 1947).

7 Goats were introduced, but they proved a pest, as they eat almost everything, enter the houses and even take laundry away. All of the goats 'disappeared' within a few weeks after dispersal. They were reported to have fallen 'ill'. Usually they were slaughtered and communally eaten.

8 Cows and water buffalos have come to the island through dispersal projects of the government but also in exchange for rattan and *gaharu* (a fragrant wood) during the boom period of this product at the end of the 1980s.

9 Gold and Gujar (1997) provide an interesting perspective on intra-ethnic differences with regard to the pig. It is an area where pigs have become extinct only after a major shift in political power. In the kingdom of Sawar in Rajasthan (India), protection of forests and wildlife was bound up with royal reputation and dharma. The rulers fed, fostered and protected the abundant wild pigs but reserved for themselves the right to hunt them and eat their reputedly savory meat, while the farmers had to cope with ravaged crops. Forests and pigs could only prosper due to brute political power and economic exploitation by the kingdom. The British did not change this situation but added a tax to the pressure on the local people. It was only after the formation of the state of Rajasthan in 1949 that princely power declined and land property was handed down to those who worked it. The subsequent changes in the political economy of the newly created state led to rampant deforestation and the killing of all wildlife and of the pigs in particular. Princely power and wild pigs died together. The striking absence of trees and wildlife in the area is, as stated by Gold, laden with meaning, mainly to be explained by reactions against previous regimes. But this reaction did away with not only the hated wild pigs but also with other valuable forest products (wood, herbs and fruits). Social and political power was not strong enough to differentiate between various kinds of resources and to generate new structures capable of managing these resources in a fruitful manner. After independence the area experienced severe environmental degradation which even now is hard to reverse.

References

Asian Development Bank (Agricultural Division) (1994) *Livestock Sector Study of the Philippines.* Manila: ADB.

Bagadar *et al.* (1994) *Environmental Protection in Islam.* Gland: IUCN.

Baldi, M. (1997) *In Search of a Better Future as Agta: A Study of Changes, Influences and Reactions of Three Agta Groups in the Cagayan Province, Luzon.* Leiden and Cabagan: CVPED Student Report no. 111.

Caldecott, J. O., R. A. Blouch and A. A. Macdonald (1993) 'The bearded pig (*Sus barbatus*)'. In W. L. R. Oliver (ed.) *Pigs, Peccaries, and Hippos: Status Survey and Conservation Action Plan.* Gland: IUCN, pp. 136–145.

Conelly, W. T. (1996) 'Strategies of indigenous resource use among the Tagbanua'. In J. F. Eder and J. O. Fernandez (eds) *Palawan at the Crossroads: Development and the Environment on a Philippine Frontier.* Manila: Ateneo de Manila University Press, pp. 70–84.

Dannenberg, H. D. (1990) *Schwein Haben: Historisches und Histörchen vom Schwein.* Jena: Gustav Fischer Verlag.

Dwyer, P. D. (1990) 'The pigs that ate the garden: a human ecology from Papua New Guinea'. Ann Arbor: The University of Michigan Press.

Eder, J. F. and J. O. Fernandez (eds) (1996) *Palawan at the Crossroads: Development and the Environment on a Philippine Frontier.* Manila: Ateneo de Manila University Press.

Errington, F. K. (1984) *Manners and Meaning in West Sumatra: The Social Context of Consciousness.* New Haven: Yale University Press.

Estioko-Griffin, A. (1985) 'Women as hunters: the case of an eastern Cagayan Agta group'. In P. B. Griffen and A. Estioko-Griffen (eds) *The Agta of Northeastern Luzon: Recent Studies*. Ceby City: University of San Carlos, pp. 18–32.

Feil, D. K. (1984) *Ways of Exchange: The Enga Tee of Papua New Guinea*. St. Lucia: University of Queensland Press.

Fox, J. J. (1977) *Harvest of the Palm: Ecological Change in Eastern Indonesia*. Cambridge: Harvard University Press.

Gold, A. G. and B. R. Gujar (1997) 'Wild pigs and kings: remembered landscapes in Rajasthan'. *American Anthropologist*, vol. 99, no. 1, pp. 70–84.

Groves, C. P. and P. Grubb (1993) 'The suborder suiformers'. In W. L. R. Oliver (ed.) *Pigs, Peccaries, and Hippos: Status Survey and Conservation Action Plan*. Gland: IUCN, pp. 1–4.

Hansen, J. F. K. (1915) 'De groep Noord- en Zuid-Pageh van de Mentawei-eilanden'. *Bijdragen tot de Taal-, Land- en Volkenkunde*, vol. 70, pp. 113–220.

Harris, D. R. (1996) 'Domesticatory relationships of people, plants and animals'. In R. Ellen and K. Fukui (eds) *Redefining Nature: Ecology, Culture and Domestication*. Oxford: Berg, pp. 437–465.

Harris, M. (1985) *The Sacred Cow and the Abominable Pig: Riddles of Food and Culture*. New York: Touchstone.

Headland, T. N. (1986) *Why Foragers Do Not Become Farmers: A Historical Study of a Changing Ecosystem and its Effect on a Negrito Hunter-gatherer Group in the Philippines*. Ann Arbor: UMI.

—— (1988) 'Ecosystemic change in a Philippine tropical rainforest and its effect on a Negrito foraging society'. *Tropical Ecology*, vol. 29, no. 2, pp. 121–135.

Headland, T. N. and J. D. Headland (1997) 'Limitation of human rights, land exclusion, and tribal extinction: the Agta Negrito's of the Philippines'. *Human Organization*, vol. 56, no. 1, pp. 79–90.

Leonor, P. H. (1993) 'Pigs in the wild: overhunting and severe habitat loss are slowly killing off all species of wild pigs here in our country'. *People and Nature*, vol. 2, no. 2, pp. 26–28.

MacKinnon, K., G. Hatta, H. Halim and A. Mangalik (1966) *The Ecology of Kalimantan (Indonesian Borneo)*. The Ecology of Indonesia Series, vol. III. Singapore: Periplus Editions.

Mudar, K. M. (1985) 'Bearded pigs and beardless men: predator-prey relationships between Pigs and Agta in Northeastern Luzon, Philippines'. In P. Bion Griffin and A. Estioko-Griffin (eds) *The Agta of Northeastern Luzon: Recent Studies*. Cebu City: San Carlos Publications, pp. 69–84.

Manning, A. and J. Serpell (1994) *Animals and Human Society: Changing Perspectives*. London: Routledge.

McNeely, J. A. and P. S. Wachtel (1988) 'Soul of the tiger: searching for nature's answers in exotic Southeast Asia'. New York: Doubleday.

Oliver, W. L. R. (ed.) (1993) *Pigs, Peccaries, and Hippos: Status Survey and Conservation Action Plan*. Gland: IUCN.

Oliver, W. L. R., I. Leht Brisbin, Jr. and S. Takahashi (1993a) 'The Eurasian wild pig (*Sus scrofa*)'. In W. L. R. Oliver (ed.) (1993) *Pigs, Peccaries, and Hippos: Status Survey and Conservation Action Plan*. Gland: IUCN, pp. 112–121.

Oliver, W. L. R., C. P. Groves, C. R. Cox and R. Blouch (1993b) 'Origin of domestication and the Pig Culture'. In W. L. R. Oliver (ed.) (1993), *Pigs, Peccaries, and Hippos: Status Survey and Conservation Action Plan*. Gland: IUCN, pp. 171–179.

Peterson, J. (1978) *The Ecology of Social Boundaries: Agta Foragers of the Philippines*. Chicago: University of Illinois Press.

—— (1981) 'Game, farming, and interethnic relations in Northeastern Luzon, Philippines'. *Human Ecology*, vol. 9, no. 1, pp. 1–22.

Persoon, G. A. (1994) *Vluchten of Veranderen: Processen van Verandering en Ontwikkeling bij Tribale Groepen in Indonesië.* Leiden: FSW.

Polet, G. (1991) *Rattan and Bamboo Utilization in a Sierra Madre Community.* Environment and Development Student Report no. 3. Leiden and Cabagan: CVPED.

Rai, N. K. (1990) *Living in a Lean-to: Philippine Negrito Foragers in Transition.* Anthropological papers no. 80. Ann Arbor: University of Michigan.

Rappaport, R. A. (1984) *Pigs for the Ancestors: Ritual in the Ecology of a New Guinea People* (new, enlarged edition). New Haven and London: Yale University Press.

Rubel, G. and Rosman, A. (1978) 'Your own pigs you may not eat: a comparative study of New Guinea societies'. Chicago: University of Chicago.

Rye, S. (2000) 'Wild pigs, "pig-men" and transmigrants in the rainforest of Sumatra'. In J. Knight (ed.) *Natural Enemies: People-Wildlife Conflicts in Anthropological Perspective.* London and New York: Routledge, pp. 104–123.

Sandbukt, Ø. (1988) 'Resource constraints and relations of appropriation among tropical forest foragers: the case of the Sumatran Kubu'. *Research in Economic Anthropology*, vol. 10, pp. 117–156.

Schefold, R. (1980) *The Sakuddei.* Film produced by Granada Television. Boston: Odyssey.

—— (1988) *Lia, das grosse Ritual auf den Mentawai-Inseln.* Berlin: Dietrich Reimer Verlag.

Schneider, G. (1905) *Ergebnisse zoologische Forschungsreizen in Sumatra, I Teil, Säugetiere (Mammalia).* Jena: Gustav Fischer Verlag.

—— (1958) 'Die Orang Mamma auf Sumatra. Reisebericht und ethnologische Studien'. *Vierteljahrschrift der Naturforschenden Gesellschaft in Zürich*, vol. 103, no. 5, pp. 214–286.

Shanklin, E. (1985) 'Sustenance and symbol: anthropological studies of domesticated animals'. *Annual Review of Anthropology*, vol. 14, pp. 375–403.

Van Beek, A. G. (1987) *The Way of All Flesh. Hunting and Ideology of the Bedomuni of the Great Papuan Plateau (Papua New Guinea).* Leiden: PhD dissertation.

Van Beukering, J. A. (1947) *Bijdrage tot de anthropologie der Mentaweiers.* Utrecht: Kemink and Zoon.

Vayda, A. P., A. Leeds and D. B. Smith (1972) 'The place of pigs in Melanesian subsistence'. In P. W. English and R. C. Mayfield (eds) *Man, Space, and Environment: Concept in Contemporary Human Geography.* Oxford and New York: Oxford University Press, pp. 130–136.

WWF (1980) *Saving Siberut: A Conservation Master Plan.* Bogor: WWF.

Zeuner, F. E. (1963) *A History of Domesticated Animals.* London: Hutchinson and Co.

10

'PRIMITIVE' TIGER HUNTERS IN INDONESIA AND MALAYSIA, 1800–1950

Peter Boomgaard

Introduction

Killing tigers is wrong. These words neatly summarize a broadly held consensus on a topic that appeals to increasing numbers of people, as the numbers of tigers steadily drop. In many present-day Western European countries, hunters in general have a bad reputation. In these areas, where wildlife has almost disappeared or is vanishing rapidly, the hunter is seen as an upstart who tries to imitate the gentry of former days, as a backward trigger-happy rural or even as an ecological criminal. The reputation of those who kill tigers is even worse. Such strong feelings are no doubt generated by the near certainty that tigers will become extinct if the tiger killers are not stopped. Also of influence is the popular notion that animals such as wolves, bears, lions and tigers are far less dangerous than people assumed. If these animals started to kill humans, so the argument goes, it was almost always because of human interference (for example, McDougal 1987).

In this chapter, which deals with the nineteenth and the first half of the twentieth centuries, we encounter hunter-gatherers and semi-sedentary peasants living alongside tigers in what are now Indonesia and Malaysia, sometimes called the Malay world.[1] I shall address the following questions. How did people who actually lived next to tigers relate to them? In what circumstances did they kill tigers, and in what ways did they go about it? The language of 'coexistence' is widely used in wildlife conservation circles today. But to what extent did people co-exist with tigers in the past?

Numerous studies of the hunting activities of hunter-gatherers and semi-sedentary peasants are mostly concerned with hunting and trapping as a means to acquire food. This is hardly relevant for tiger hunting, where nutritional considerations were either absent or unimportant. I have recently published elsewhere on hunting in general in Indonesia's past, but there the emphasis was on the ecological impact of hunting, a topic that does not concern us here. Apart from that one article, the history of hunting in Indonesia has received no attention recently.[2] As regards tigers, there is no shortage of articles discussing their present status in the Malay world, but recent studies on the tiger in Indonesian and Malaysian history are rather rare.[3]

Our idea of tiger hunting in Asia in the past is, no doubt, shaped by stories and photographs depicting the massive Indian tiger hunts from the back of an elephant, dating from the nineteenth and early twentieth centuries, organized by local rulers and with many high British colonial dignitaries present. Another image evoked by the phrase 'tiger hunting in Asia' is that of the so-called shikar stories, written down in their hundreds by British hunters who, often alone or with one trusted indigenous guide-cum-assistant (shikari), went after dangerous game, among which the tiger held a place of pride.[4] The reader would be well advised to suppress such images as they are hardly relevant to the subject matter of this chapter. In the first place, tiger hunts by 'tribal' groups had little in common with the ones undertaken by noblemen and Europeans. In the second place, even hunting by aristocrats and Europeans in Indonesia or Malaysia differed from the situation in India.

In the Malay world, rulers and noblemen were often avid hunters, but seldom of tigers. In Sumatra, this may have been related to the absence of the grassy plains so often found in India, and in Java it may have been linked to the absence of elephants. Furthermore, the Javanese rulers had a number of tiger rituals at their disposal, at least from the early seventeenth century onward, which made tiger hunts supervised by king and aristocracy superfluous (Boomgaard 1994). As regards the Europeans, I would like to argue that the Dutch in Indonesia were much less fascinated by and given to tiger hunting than were the British in India. Although part of the explanation might be found in differences in 'national character' between the Dutch and the British, the fact that the British did not hunt many tigers in Malaysia either points in another direction. Even around 1900, when many forests had been cleared and when more climax vegetation than before had been turned into secondary forest, the British hunter Charles Whitney could write that in Malaysia, Sumatra, Lower Burma and Southern Siam 'the jungle is too dense and continuous' to hunt tigers from the back of an elephant. He added: 'In fact, as compared with India, almost no tiger hunting is done in these countries, and that little consists of sitting up over a kill, or, in the dry season, over a water hole' (Whitney 1905: 117, 290). He was clearly referring to European hunting, not to indigenous hunting.

Sedentary peasants

European peasants have, as far back as the documents go, hated and feared animals such as bears and wolves, and continue to do so nowadays even though these animals have become extremely rare. With these notions in mind, we might wish to investigate, before exploring the attitudes of nomadic and semi-sedentary

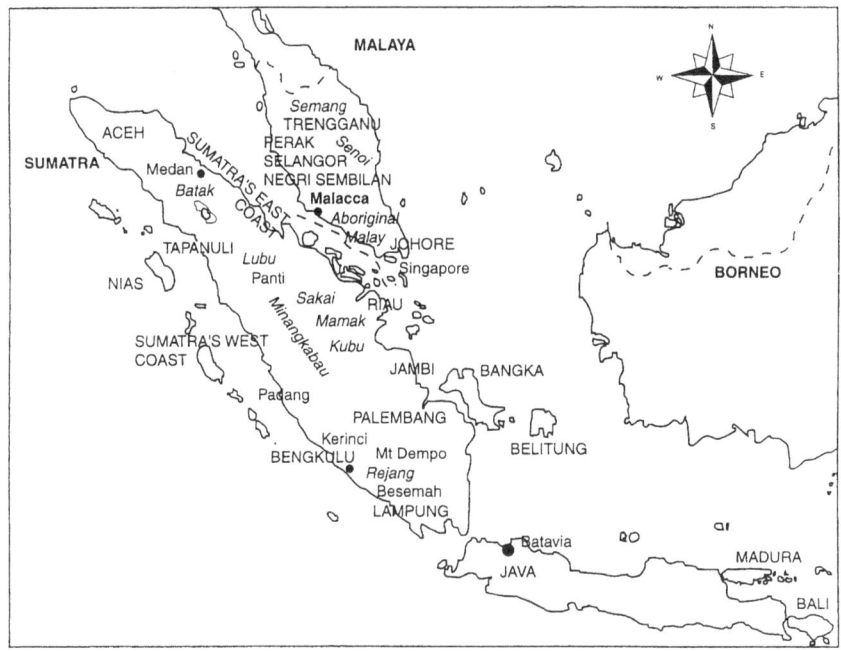

Figure 10.1 Map of the Malay world

groups, the behaviour of the sedentary peasantry of the Malay area. Details are published elsewhere, but the following lines summarize my findings.

Among the sedentary Malay peasants, trapping tigers (and leopards) was more important than hunting them. A specific tiger trap was one of the main instruments in this respect. Poison was also used, but this may have been largely restricted to Java. Java is also the area where collective hunts took care of the remainder of the tigers who had been foolish enough to make the villagers uneasy. In Sumatra these hunts seem to have been rare, and they were not recorded for Malaysia, in both cases perhaps because the natural environment was less conducive to such endeavours. In Sumatra supernatural reasons played an additional role. The feeling that 'blameless' tigers – those that had not killed human beings or livestock – should not be killed, as they could very well be ancestors who meant the villagers no harm, seems to have been fairly general among the populations of Malaysia, Sumatra, Java and Bali. But some groups were apparently immune to such feelings, as is witnessed by the rather cavalier attitude towards tigers of the Achenese, the Batak and the Rawas people in Sumatra, and the populations of northern Banten and Banyuwangi in Java. In three out of these five cases, firearms seem to have played a role.

More mundane reasons for not killing all tigers, such as the fear that this would lead to an invasion of wild boar, were occasionally given in the sources. It

187

may be that such notions were under-represented in the colonial sources, as 'superstition' was a much more appealing, generally accepted explanation among colonial civil servants. This is not to say that local beliefs had nothing to do with indigenous attitudes towards tigers, but these may have been just one factor among many. Whatever the weight of these factors, it could be argued that during the later part of the nineteenth century certain individuals, perhaps mostly orthodox Muslims, were slowly but surely weaned from such animistic beliefs. Spurred on by a generous bounty system and by the increasing availability of more dependable firearms, which they could, moreover, afford to buy with the reward money, they embarked upon a career of professional tiger hunting.

The presence of the colonial state with its bounties and that of European technology in the form of guns and rifles were a stimulus to the killing of tigers by the indigenous peasantry. But the basic mechanisms of hunting and trapping were already in place before the European presence made itself felt. At least in the case of Java, the tiger was on his way out before bounties and improved firearms could make much of a difference. There is no shortage of stories about villagers who lived in constant fear of tiger attacks, or eventually left their villages because of them. Even killed or captured tigers, particularly in Malaysia and Sumatra, were shown deference. Occasionally, when a tiger had been killed by a European or by an indigenous ruler, which would let the villagers off the hook in a spiritual sense, fear would not express itself as deference but as hatred or pent-up rage.

So tigers were said to be both greatly feared and hated in many of these groups and at the same time venerated as the embodiment of ancestors. This has led some writers to argue that the attitude of the indigenous population towards tigers was ambiguous: the tiger as friend *and* foe.[5] My reading of this paradox is different. The fact that many people regarded tigers as ancestors who could have a benign influence certainly does not imply that these tigers were seen as 'friends'. Such a tiger would not hesitate to punish his offspring if they committed an offence against the rules laid down by the ancestors. The ancestor-tiger was a strict disciplinarian, who, like a ruler or a deity, could be benign if he so desired, but who was for the most part greatly feared in his wrath. The tiger is the King or Lord of the Forest, and, as a real king, he is unpredictable, thus keeping people in awe of him. Having dealt with the attitudes of the sedentary Malay peasantry regarding tigers, I shall now turn to the groups that form the main topic of this chapter.

Hunter-gatherers [6]

The only groups that qualify as hunter-gatherers are the Semang in Malaysia and the Kubu in Sumatra. These two groups never numbered more than a few thousand people or in the case of the 'wild' Kubu a few hundred, at least during the period for which we have written sources on these people (that is, the last three centuries).

The Semang

The 'tribe' that I will call the Semang are Negritos inhabiting the forested interior of the northern and central Malay peninsula. They consist of various groups with their own specific names. These groups, in turn, comprise smaller units. This may account for the fact that the information that has come down to us is sometimes contradictory. Of all the 'wild' tribes (*orang asli*) of the Malay peninsula, the Semang spent most of their time hunting and their main source of protein was game and fish. They were and are expert hunters.[7]

> The rain forest is a safe environment for those who know how to deal with its potential dangers. The Batek [Semang] have a realistic appreciation of these dangers; they do not fear the forest, nor do they populate it with imaginary evil spirits as many agricultural peoples do. One source of danger in the forest is wild animals.
>
> (Endicott 1979: 8)

To this, the same author adds: 'The two things the Batek fear most are thunderstorms and tigers' (1979: 67). Another writer agrees with the most important part of this quotation: 'The forest is also the home of two animals that the Semang greatly fear, the tiger and the elephant' (Rambo 1985: 40). Of course, the Semang have always had to come to terms with this fear:

> If tigers or elephants are thought to be nearby, the Batek simply form larger than normal work-groups on the generally correct assumption that there is safety in numbers. They may also take special care to keep their fires smouldering at night, knowing that wild animals avoid smoke and fire.
>
> (Endicott 1979: 8)

How realistic was this fear of tigers? In the 1970s, when the American anthropologist Kirk Michael Endicott reported on his fieldwork, the tiger was no longer much of a threat, but around 1900, when the German ethnographer Rudolf Martin also mentioned the fear of tigers among the Semang, this fear would have been more real than today. Around 1800 the Briton John Anderson reported that beasts of prey (i.e. tigers) were not much of a problem to these groups, and it is certainly possible that in the much denser jungles of the early nineteenth-century tiger attacks were rare.[8]

There are no indications that the Semang systematically attempted to destroy the tigers (and leopards) with whom they shared their habitat. Nor did they hunt tigers (information on leopards is lacking) for food. Although compared to the coastal, sedentary Malays the Semang had few food taboos, all sources agree that they did not eat tiger. Modern sources add that they do not eat elephant, rhino, bear or large snakes either, but Anderson, writing in the early nineteenth century, explicitly mentioned that the Semang ate elephants and rhinos.[9] The reason for not eating these animals, including the tiger, is that 'they were once men', or, as a modern researcher expresses it: 'The Batek say they would be afraid to eat these animals because they might be the bodies of *hala'* who, in new bodies, would take revenge upon them' (Endicott 1979: 64). One supposes that this reasoning applies to the random killing of these animals as well. As regards tigers, the Semang are particularly cautious, because a tiger may very well be the temporary abode of the soul of a dead shaman, who, far from being dangerous, is the guardian of his former camp. The soul of the dead shaman is also the tiger spirit helper or familiar of his successor, guiding his journeying soul during trance sessions.[10] Although the Semang did not kill tigers for food or to rid their environment of dangerous animals, they did kill them nevertheless. The sources do not say under what circumstances tigers were killed by the Semang, but it must be assumed that they did so when threatened by a tiger, or, as almost all other peoples of the Malay world did, in revenge for the death of a human being.

Most authorities agree that the bow and (poisoned) arrow were the preferred weapons of the Semang. Around 1900, however, they seem to have given up this weapon in favour of the blowpipe (or blowgun, *belau*, *sumpitan*) with poisoned darts, the weapon of preference of the Sakai/Senoi, their neighbours. This has puzzled researchers ever since. The abandonment of the bow and arrow has been linked to the arrival of guns and to the difficulty faced in making arrows, but these explanations are, in my opinion, insufficient and partly wrong.[11] It is also puzzling that the spear is not mentioned in the twentieth-century literature, whereas, as will be shown presently, it was the main weapon of the Sumatran hunter-gatherers, the Kubu. It was also the main weapon of the Aboriginal Malay, neighbours of the Semang. In a publication dated 1824, John Anderson named the spear as one of the Semang weapons.[12] So the sequence of events seems to be that the Semang, who possessed the bow, the blowpipe and (possibly) the spear in the early nineteenth century, started to give up the bow and perhaps the spear around 1900, while retaining the blowpipe and occasionally acquiring (rather primitive) guns.

The only reasons that I can think of for giving up the bow around 1900 is an increase in the numbers of small to medium sized animals and at the same time a drop in the numbers of big game. According to research carried out in the Venezuelan Amazon, the number of arrows that one person can take with him is restricted to four, whereas the number of darts is much higher, namely fifty. So although the bow has the advantage of a larger maximum effective range, and perhaps also that of a surer kill if the animal is hit, one has far fewer chances of a successful shot. If the number of targets increases, the blowpipe would be the better choice. Given the fact that by 1900 much more land had been cleared in Malaysia than around 1800, increasing numbers of terrestrial and arboreal mammals, lovers of ecotones or borders between ecological zones, could indeed have ensued, and the Semang could have decided to bet on quantity rather than accuracy and larger killing power. At the same time, it seems that the German ethnographer Paul Schebesta was wrong in his opinion that big game had not become scarce. Rhinos, for instance, had become fairly rare by 1900, and in some areas the numbers of big tuskers among the elephants had been greatly reduced because the Malay had hunted them a good deal during earlier years. Another reason for favouring bows over blowpipes had thus disappeared, and the time-consuming production of arrows could be dropped.[13]

However that may be, the Semang are on record as having used the bow, and possibly the blowpipe, for killing tigers. Both darts and arrows were poisoned, and feathers of the rhinoceros hornbill (*Buceros bicornis*) were used as a magical means for making tiger-arrows even more deadly. The Semang used dogs when hunting, perhaps also when they went after tigers. Schebesta suggests that a Semang person, when alone, 'is a certain victim' when confronted with a tiger, which would imply that the Semang hunted in groups when the tiger was their target. Generally speaking, the Semang do not seem to have been great users of snares. Nor did they use specific, purpose-built tiger traps, so common among the sedentary peasantry of Malaysia, Sumatra and Java.[14] Summing up, it can be said that the Semang, admirably adapted to their rainforest environment, had learned to live with the tiger. They feared the tiger, but they did not attempt to create a tiger-free environment, even though they had the means to do so and would not hesitate to kill a tiger if necessary. That they did not try to kill all tigers may have been partly caused by their belief in supernatural tigers, which were harmless and which even protected 'their' bands. If they killed such tigers, there would certainly be revenge. It was also said that such a spiritual tiger killed game for them, which implies that the Semang shared the spoils of the tiger, as has been reported about similar tribal groups elsewhere. It would have been foolish to kill such a provider of food. Finally, the tiger was also the policeman of their deity Karei, perhaps another reason not to kill tigers unnecessarily.[15]

The Kubu

In terms of ethnicity, the Sumatran (*orang*) Kubu are to be compared to the Jakun and other Aboriginal Malay groups in Malaysia rather than to the (Negrito) Semang. Nowadays it is accepted by most researchers that the Kubu are an example of devolution: they are Malays who, somewhere in the remote past, seem to have given up their agricultural lifestyle and left the coast to start a life as hunter-gatherers in the interior of Sumatra. They are to be found in Dutch sources since the seventeenth century, as 'wild people' living in the interior of Palembang and Jambi, but detailed information pre-dating the nineteenth century is lacking. When more information becomes available around 1870, a distinction is usually made between the 'wild' and the 'tame' or semi-settled Kubu. Unfortunately, we know much more about the semi-settled Kubu than we do about the 'wild' ones.

The Kubu, like the Semang, were reputed to eat practically everything. There were, however, some exceptions, such as elephant and bear. On this, all sources are in agreement. Most sources, particularly the later ones (after 1900), also state that they did not (and do not) eat tigers. However, the German physician Otto Mohnike stated explicitly that they do eat tigers. Mohnike, who during the 1850s spent a considerable period in Sumatra, particularly Palembang, does not claim to have had first-hand experience with the Kubu, so I am inclined to disregard his statement.[16]

As regards killing tigers, the Dutch civil servant J. W. Boers, who visited the 'wild' Kubu in the 1830s, was quite outspoken: '[The Kubu] dares to attack the tiger in the jungle, only armed with a spear'. Apart from a jungle knife (machete, *parang*), the spear, of which they had two types (*tumbak, kujur*), was their only weapon. Since the 1920s only the *kujur* has been mentioned. Iron blades for their spears were acquired from the Malay, in barter for rhino horn and other jungle products. They also had dogs that were used for hunting, but they possessed neither blowpipes nor bows and arrows. Around 1900, the sedentary Kubu relied for their meat to a large extent on traps, nets, snares, including the spring-spear trap (*belantik*) and pitfalls (with sharp sticks, *ranjau*), and in recent times (ca. 1980) the 'wild' Kubu did the same. Anyway, some of the main items on their menu, namely wild boar, deer and monkey, were killed with spears. As spears are more productive and procure a larger weight per kill than blowpipes (as we know from research in Ecuador), there may have been no need for another weapon in addition to the various traps and snares.[17]

Information on the hunting equipment of the Kubu is rather straightforward, but it is much less clear how far it was used for killing tigers. I have quoted Boers, who stated that the Kubu killed tigers with their spears. Almost a century later, the Dutch civil servant W. A. J. M. van Waterschoot van der Gracht repeated this statement, adding that the Kubu climbed into trees in order to regale the tiger with a barrage of spears. G. J. van Dongen, another civil servant, writing in the early 1900s, reported that the Kubu caught tigers in sturdy traps.[18] He gave no details, but he was probably talking about the well-known tiger traps of the sedentary 'Malay' peasantry. However, the most recent source to deal with killing tigers (Persoon 1994: 158), explicitly states that tiger hunts are taboo! This taboo seems to be related to the Kubu conception that the tiger is a kind of deity, or rather, as another recent source has it, that the 'real' tiger is a manifestation of the Tiger Deity. One of the functions of this deity is to accompany the human soul or spirit when it temporarily leaves the body during a shamanistic session (Sandbukt 1982: 3, 5).

This sounds remarkably like the Semang stories on supernatural tigers. The Kubu were also reputed to eat carrion, so they might have revered the Tiger Deity as a provider of meat, as did the Semang. If the tiger was the manifestation of a deity, it must have been rather wrathful, at least if we believe the Dutch civil servant A. W. P. Verkerk Pistorius who visited the 'wild' Kubu in the early 1870s. He estimated that there were about 400 'wild' Kubu in Palembang and that their numbers were slowly dropping. According to him, this was mainly caused by unremitting carnage brought about by the tigers, high child mortality and loss of tribe members who joined the ranks of the semi-sedentary agriculturalists. He does not mention any measures taken against tigers, such as elevated dwellings (as the Semang had), let alone expeditions to get rid of these animals. Their only 'adaptation' seems to have been that they slept lightly, being up and about with their spears as soon as a tiger was heard near the camp, as Verkerk Pistorius experienced himself when he stayed overnight. Here it seems that the 'devolved' Kubu were not as well adapted to sharing a habitat with the tiger as were the 'evolved' Semang.[19] It is nevertheless likely that the attitude of the Kubu regarding tigers has been similar to that of the Semang, namely that tigers could only be killed in self-defence, perhaps also in revenge for the taking of a human life, but never at random. Van Dongen's tiger trap – a contraption with a random effect – was found among the semi-sedentary Kubu but is not recorded in sources on the 'wild' Kubu.

Semi-sedentary groups in Malaysia

During the last three centuries, there were several peoples in Malaysia, Sumatra and even in densely populated Java whose main occupation was shifting cultivation (slash-and-burn agriculture). Most of these 'tribes' were small, but, due to their lifestyle, they occupied extensive territories. Taken together, they could, not so long ago, lay claim to the larger part of Sumatra and Malaysia. In terms of ethnicity, all groups are Austronesian-speaking 'Malays', with the exception of the Mon-Khmer-speaking Senoi of Malaysia. What they also have in common is that they are 'pagan' (animists); if they are Muslim, this takes a highly syncretistic form.

The Senoi

Starting with the Malaysian Senoi,[20] who have the Semang as neighbours to the north and the Aboriginal Malay to the southeast, we can say that, at least around 1925, they hunted less than the Semang, ate less meat and procured most of their meat from trapping instead of hunting. They spent more time on agriculture, particularly the cultivation of various roots and tubers.[21] Their knowledge of the forests was excellent, they felt absolutely at home and they had no objection to sleeping outside in the rain, either under a tree or 'up among its branches'. And yet, according to Martin, the Senoi, like the Semang, were terrified of tigers. The British hunter Theodore R. Hubback, who lived in the area around 1900, formulated their attitude towards large, dangerous animals as follows:

> Sakais [Senoi] never lose their terror of animals such as elephants and seladang [Indian bison]. Living as they do among wild beasts, armed only with weapons that are useless against the full-grown animals, they know full well their power and strength.
>
> (Hubback 1905: 184)

So here we find again, as with the Semang, a forest dwelling people very much at home in their habitat, but nevertheless living in fear of dangerous animals, including the tiger.

Turning now to the killing of tigers, we find a reference, ca. 1850, to blowpipes with poisoned darts being used against the rhino and the tiger. For the killing of these animals the Senoi had a very potent poison at their disposal. For a somewhat later period, Skeat and Blagden quoted 'a well-authenticated case of a leopard being killed near Gopeng in Perak by a Sakai [Senoi] with a blowpipe'. The same authors went on to say about the Perak Senoi: 'They are no less ready to attack the tiger and the rhinoceros'. But in the first place this sounds remarkably like the older source just quoted, and, secondly, they added a footnote citing another authority who doubted the veracity of this statement. They did not have the typical 'Malay' tiger trap.[22] That is all I could find about killing tigers (and leopards) among the Senoi. I did come across a long list of animals eaten by the Senoi, but although the tiger was among them there was a question mark against its name (Martin 1905: 733). So although they occasionally killed, perhaps in self-defence or in revenge, a leopard and probably a tiger as well, there is little evidence of systematic tiger-killing. Of course, tigers may have walked from time to time into their more effective all-purpose traps, such as the pitfall and the *belantek* (= belantik), but the Senoi stopped short of using specific tiger-traps.

If it is true that tiger-hunting was largely absent, this may have been related, as it was with the Semang and the Kubu, to the specific position of the tiger in the Senoi's belief system. This belief centred around the shaman (*hala, halak*). The shaman not only had a tiger spirit as 'familiar', guiding his own spirit when it went on a trance-induced journey, but could turn himself into a tiger at will. The shaman could also, after his death, become a tiger – or at least his soul would inhabit a tiger. This *hala*-tiger does not harm his people, and even helps them. It would therefore be worse than pointless to kill such a tiger.[23]

The Aboriginal Malay

We turn now to the southeastern neighbours of the Senoi, the Aboriginal Malay.[24] As with the Semang and the Senoi, they are a collection of smaller groups (tribes) with their own names, comprising, in turn, even smaller units, constituting camps. There is considerable variation between the groups, and some of the contradictions to be found in the literature can perhaps be traced to these real differences. One important contradiction, however, might be a question of development over time. The French missionary Favre, who visited a number of the Aboriginal Malay 'tribes' in 1846–47, stated explicitly that hunting was their most important activity. Schebesta, however, living among some of these groups in 1924, thought little of their volume of hunting. According to him, the Jakun hunted less than the Sakai [Senoi], who in turn were easily outdone by the Semang. However that may be, all authorities agree

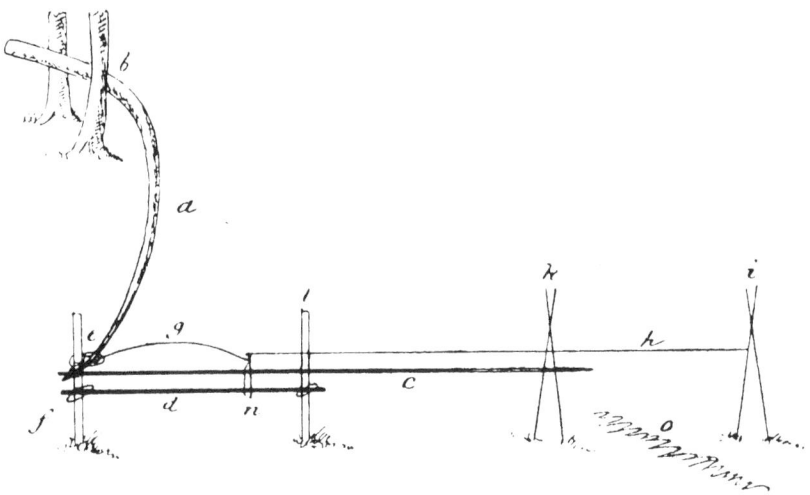

Figure 10.2 Trap photo (above). Trap diagram (below). Various traps of semi-sedentary Malaysian 'tribes'. These were not specific tiger traps but nonetheless they could and did kill tigers. (The photograph is from Martin (1905) and the diagram is from Skeat and Blagden (1906, Vol. I))

that the Aboriginal Malay were more involved in agriculture than the Senoi. Many of them planted fruit trees and cereals in addition to roots and tubers. They were also gatherers of commercial forest products, and around 1850 some of them were even (part-time) wage-labourers. Many of the better-off groups lived pretty much as the coastal Malay did, and Schebesta, somewhat overstating his case, said at one point that the Jakun's main difference from the Malay was that they were not Muslims![25]

An early source, dated 1642, gives the blowpipe with darts and the spear as their weapons. A Malay source, also dating from the seventeenth century, mentions blowpipes (*sumpitan*) and poisoned darts. The British lieutenant Thomas John Newbold, writing in the 1830s, also referred to poisoned darts. The 'Mallaye poison' was the strongest one they had. With three darts tipped with this poison in his body, a human being would, according to the Jakun, die in less than one hour, a tiger in less than three hours.[26] Around 1900, some groups had also started to use guns (Tower muskets), 'and this gun is then used in common by all members of the tribe'. The Aboriginal Malay were assisted during the hunt by two types of dogs, namely large ones for wild boar and deer and small ones for mouse deer. Finally, they used a 'formidable and effective snare', namely the spring-spear trap (*belantek*) (also mentioned when we were dealing with the other tribes).[27]

How far was this array of weapons brought to bear on the tiger? References to tigers being killed with blowpipe darts were made by Newbold (in the 1830s), Martin and Skeat and Blagden (in the 1900s). There is no explicit reference to spears being used for tigers, but one Jakun is on record as having killed a tiger with his *parang* when he was attacked. Tigers were also sometimes caught in a *belantek*. According to Favre (1840s), tigers were eaten by the Aboriginal Malay, and around 1900 Skeat and Blagden registered the killing and eating of tiger cubs. Tigers on the menu are not to be found in other sources, but neither is this explicitly denied, as was the case with the Semang. They may, indeed, have (had) fewer food taboos than, for example, the Semang, because they were also reported to eat elephant (although individual Jakun did not) and bear – both animals, according to most authors, taboo to the Semang.[28]

There is, alas, little reliable information on how great a problem tigers were to the Aboriginal Malay. The Portuguese author João de Barros, whose writings reflect the situation around 1510, reported that the thinly populated environs of Malacca were extremely vexed by tigers and that most people, therefore, slept in trees. This passage probably referred to the Benua. The same Benua, however, were said by other, somewhat later authors to be weretigers and 'charmers' of all sorts of dangerous animals such as tigers, who were thus forced to obey them. Wild animals did the Benua no harm. Although de Barros may have been right, it looks as if the other stories tell us more about what the coastal Malay told the Portuguese and the Dutch about the Aboriginal Malay than it does about the Aboriginal Malay themselves.[29]

As with the Semang and the Senoi, the Aboriginal Malay believe that tigers can be inhabited by the souls of dead or living people. As Newbold has it: 'The priests, incantators, or exorcists, are styled Poyangs [i.e. magicians]'. 'The soul of a Poyang after death is supposed to enter into the body of a tiger'. The spirit of a dead *batin* (chief) could also enter the body of a tiger. If, therefore, someone had been killed by a tiger, this person might have provoked the ire of a dead shaman or chief. In the guise of a tiger, the dead shaman normally would protect the people of his tribe. He was, moreover, able to change himself into a tiger when he was still alive. In this tiger-form he could harm people, presumably those against whom he nursed a grievance. One might have thought that these and other, related beliefs would have kept the Aboriginal Malay from killing tigers other than in self-defense, but this was perhaps not the case, at least according to Skeat and Blagden:

> The tigers are the slaves of the magician or Poyang. Although the Mantra [Aboriginal Malay] believe in this, as well as in the immortality of tigers, they nevertheless do not scruple to kill and eat the cubs whenever they find them.
> (Skeat and Blagden 1906, II: 324–325)

Around 1900, belief in weretigers was also supposed to exist among the Aboriginal Malay, reminding us of the stories told around 1600.[30] If this was indeed true, and not a somewhat botched rendering of the *poyang*-turned-tiger story, this belief might have been a cultural 'borrowing' from the sedentary Malay, amongst whom this is well documented. It might also explain why the Aboriginal Malay seem to have been less hesitant to kill – and eat – tigers, as killing a weretiger was a laudable action. A more mundane explanation for a higher incidence of tiger hunts than among the Semang and Senoi could be that the more open areas of the Aboriginal Malay were better suited to battue-style hunts, whereas the tigers themselves may very well have been more numerous than in the densely forested areas. Although the specific Malay tiger trap is nowhere mentioned, the killing and eating of cubs whenever they were found comes closer to a systematic attempt to 'cleanse' the neighbourhood of all tigers than anything found so far among the other tribes.

Semi-sedentary peoples in Sumatra

Turning now to Sumatra, I shall briefly deal with three groups, from west to east in Central Sumatra: the [*orang*] Lubu, Sakai and Mamak. Information on these groups is scant, sometimes restricted to one or two serious publications. This has the obvious advantage of producing fewer contradictions than we found among the better documented peoples, but it also makes it much harder to evaluate the information.

The Lubu

The Lubu lived among the Mandailing Batak in the more mountainous and forested areas. They were shifting cultivators, planting roots and tubers, rice and maize. They did not live in villages, but in simple, sometimes elevated huts near their *ladang* (temporary swidden field). Around 1880, part of their high-quality rice was sold at the nearest market. Some of their time was taken up with foraging, partly for subsistence, partly for the market. In addition, they were good hunters, with the bamboo blowpipe (*hina*, *ultop*) with poisoned darts as their weapon of preference. They also had spears and jungle knives, and around 1900 had started to buy old muzzle-loaders. Some game was trapped with nooses and nets, and *ranjau* were used to keep the pigs from the *ladang*. They also speared fish.[31]

The Lubu had few food taboos. They ate carrion, which they were even said to like more than fresh meat. A report, dated 1884, states explicitly that they ate tiger, and in a 1912 article the tiger is not listed among the animals forbidden to consume. Nevertheless, according to the same source, they did not hunt big game, such as elephant, rhino and tiger. However, as many Lubu had already around 1890 started to lead a more sedentary life, it is possible that they had hunted tigers earlier.[32] The specific Malay tiger trap is not recorded in the sources on the Lubu. The attitude of the Lubu towards tigers seems to have been relaxed: 'They often stay out in the forests for whole nights, and seem to be hardly afraid of tigers and other wild animals; thus, they often return in the evening or at night through the forest to their dwellings' (Dijk 1884: 157). So the Lubu were probably not all that afraid of tigers, and may have eaten them (perhaps as carrion, or when one had been killed in self-defense). But they did not use specific tiger traps or (at least around 1910) organize tiger hunts.

The Sakai

Our only source on the Sumatran Sakai is the German physician Max Moszkowski, who in 1907 visited the area around the river Siak (Moszkowski 1909). The Sakai were living upstream, in a typical tiger area, where one could hardly find a settlement that did not pay its annual tribute of human lives. Some groups had become more sedentary and lived in villages where, in addition to a number of annual crops, they planted coconut and areca palms, a sure sign of sedentary intentions. Most groups, however, were more or less nomadic, planted cassava, other roots and tubers and millet, but hardly any rice. The nomadic Sakai lived, as did the Lubu, near their *ladang*, in isolated, elevated dwellings. They gathered forest products for the market and hunted. Their Malay neighbours called them 'forest people', and they did, indeed, only feel themselves at home in the forest, if our one and only source is to be trusted. Their senses were at their sharpest in the forest and the Sakai could, for instance, smell a tiger at quite some distance.

In contrast to the Malaysian Sakai, they hardly knew the use of vegetable poisons and therefore had no poisoned darts, or blowpipes. Their only weapons were 'lances' (possibly spears) of various dimensions, which were thrust or thrown at the animal. One can assume that the Sakai, like the Kubu, gave up the use of the blowpipe when they had devolved'. In addition, they made use of spring-spear traps (*belantik*), nooses, nets and other snares. They also had dogs. Moszkowski does not mention guns, and although he gives a rather elaborate description of the specific tiger trap employed by the settled 'Malay' neighbours of the Sakai, he does not refer to it when dealing with the traps and snares of the latter. The Sakai themselves told the author that when two of them were armed with spears, they dared to attack a tiger, and Moszkowski was in possession of a 'lance' that had been used precisely for that purpose. However, he doubted the veracity of these stories because he had never heard of a tiger that would allow such an attack to take place. Among the Sakai, according to Moszkowski, tigers were usually caught in a spring-spear trap. So, with the Sakai too, we do not find much evidence of systematic attempts to free a tiger-ridden territory of tigers.

The Mamak

For the [*orang*] Mamak we have only two reporters, namely the Swiss naturalist Gustav Schneider and the Dutch civil servant V. Obdeyn. Schneider visited the heavily forested upstream areas of the Indragiri river, home to the Mamak, in 1898–99 and Obdeyn wrote his report in the 1920s.[33] Schneider describes the Mamak as typical forest dwellers, born hunters with an intimate knowledge of wild plants, roots and fruit. They ate practically everything, but the consumption of tiger meat is not mentioned. They also caught fish and other river creatures and collected commercial forest products, which they bartered for iron, cotton cloth, tobacco, gambier and salt. Their main food was (*ladang*) rice, that they had started to grow only thirty or forty years previously. The Mamak's main hunting weapon was the (iron-bladed) spear, of which they had two kinds, and various knives.[34] With their spears they killed large numbers of wild boar, but also occasionally elephants and rhinos. Other animals were caught in various, often rather specific, traps (including the *belantik*), snares, nooses, nets and *ranjau*, supposedly in pitfalls. In addition, the Mamak made use of hunting dogs. Schneider saw no evidence of blowpipes or guns, but Obdeyn mentioned that formerly they had had blowpipes, the use of which had become obsolete at the time of writing.

On 15 March 1899, Schneider visited the Mamak village of Sungai Raja, where a man had been killed by a tiger just the day before. He was told that tigers took many human and canine lives in the surrounding area, sometimes so often that people left their villages. The people of Sungai Raja had sent for a 'sorcerer' (i.e. a shaman), namely a *komantan harimau* or tiger-charmer, who arrived shortly after Schneider. The next morning, the sorcerer started preparations for the so-called *ancak* ceremony. The main feature of this ceremony was the offering of *makanan rimau*, tiger food, consisting of small portions of deer and chicken meat, eggs, uncooked rice and some water with small pieces of lemon. There were also offerings for a 'good spirit', namely rice, eggs, sirih (betel), lime, tobacco and fragrant wood. The latter is burned during the invocation of the 'good spirit', who is implored to bring to bear his influence on the 'evil tiger spirit', in order to make him accept the offerings of the poor villagers and, in return, leave them and their dogs alone in the future.

Schneider, who had often seen the specific tiger traps of the more sedentary 'Malay', was amazed that the Mamak made no use of such a contraption:

> It is not to be supposed that the Mamak, who otherwise are skillful snarers, would not be capable of constructing such tiger-traps, but I presume that their superstitious fear, that this would offend the evil tiger spirit and provoke even worse attacks, must be the reason that they are not used. From all the statements and stories about the tiger, with which they regaled me, it was apparent that they feared him greatly and that they felt defenceless when up against this dangerous predator, who had started to raid their villages, and that they tried to keep away the evil spirit, who, according to their beliefs, had caused this, by offerings and exorcism only.
>
> (Schneider 1958: 269)

The Mamak probably present the most extreme case of non-aggression against tigers of all the 'tribal' groups we have looked at. That they had no specific tiger traps was a feature they shared with all the other nomadic and semi-sedentary groups (with the exception of the sedentary Kubu). That they had no spring-spear traps, present among most of the tribes dealt with here (but not mentioned for the Semang and the Lubu), is already rather unusual. Although not a specific anti-tiger device, the spring-spear trap seems to have been fairly successful as such. One wonders whether the Mamak were so afraid of the tiger (spirit) that they wanted to avoid even the suspicion of having tried to harm it. They are the only group for which no tiger-killings whatsoever are recorded. Going after tigers in revenge for a human being having been killed was apparently out of the question, and it is highly unlikely indeed, given these perceptions, that they would ever consider tigers as food.

The attitude towards the tiger of the semi-sedentary tribes in Malaysia and Sumatra shows considerable variation. At one end of the spectrum we have the Aboriginal Malay, who ate tiger meat, seem to have no great fear of tigers and who, of the groups dealt with here, the nomadic Kubu and Semang included, came closest to a more or less random destruction of tigers. At the other end we find the Mamak, who feared the tiger (spirit) very much and did not even go after tigers when one of their own people had been killed by one. The attitude of the Lubu seems to have been close to that of the Aboriginal Malay, in that they also ate tiger meat and seemed unconcerned about the presence of tigers. The Sakai and the Senoi seem to take up an intermediate position between the Mamak on the one hand and the Aboriginal Malay and the Lubu on the other, but our information on both groups leaves much to be desired.

Conclusion

It is difficult to generalize about the data we have found for the seven groups dealt with here. A tentative conclusion could be that the more sedentary tribes – Aboriginal Malay, Lubu – were less hesitant to kill (and eat) tigers than were the more nomadic ones. This may have been related to a higher incidence of atrocities committed by tigers in the areas of these semi-sedentary peoples, where gaps had been created in the climax forest vegetation, thus attracting more game and, therefore, more tigers. These communities, still actively engaged in hunting even though a large share of their food requirement came from agriculture, may have captured so much game that tigers were forced to kill an occasional human, dog or goat, thus inviting retaliation.

One wonders whether the presence or absence of specific weapons may have had some influence on the attitude towards tigers. The two groups with the fewest inhibitions, namely the Aboriginal Malay and the Lubu, possessed both the spear and the blowpipe. The only other group with this combination were the Senoi, about whom we are not well enough informed to place them firmly in one category or another. According to our sources, the spear and the blowpipe could both be used to kill tigers, but it is conceivable that people armed with both – as opposed to just one of them – felt more confident that they were a match for the tiger. One is tempted to speculate that the same conditions that made for an increase in the quantity of game and, therefore, of tigers, namely the creation of ecotones, had also prompted the use of this combination of weapons. Another factor could have been that the nomadic groups ate carrion and may have seen the tiger as a provider of food. Most of the semi-sedentary groups are not on record as carrion eaters. As non-Muslims, the latter may have seen the tigers as competitors for wild boar, being therefore more inclined to go after them.

However that may be, most groups may have relied more on traps than on face-to-face contacts for catching tigers. This probably does not apply to the Semang, but the other tribes had either spring-spear traps or pitfalls with *ranjau*, or both. Although these are all-purpose traps, tigers could and did walk into them, and were thus killed or at least seriously injured. The typical Malay tiger trap, specifically designed for catching tigers, was not found among these groups, apart from the sedentary Kubu. Several explanations suggest themselves, of which the least likely is that they were not familiar with these traps or not able to make them. More plausible is the possibility that they were unwilling to spend the considerable time required to make them, or that they had insufficient dogs or goats (or were unwilling to part with them) to bait these traps. They may also have had 'ideological' objections. A tiger trapped in a spring-spear trap or a pitfall could be regarded as 'fate', a tiger caught in a tiger trap could not, and they may have feared that the use of such a trap would provoke the wrath of the spiritual entity embodied in the tiger. Finally, it is also possible that the tribal groups lacked the spiritual specialists who, according to most Sumatran peasants, were needed to make these traps work.

This brings us into the realm of the supernatural. Information on the Lubu and the Sumatran Sakai on this point is lacking, but all other tribes believed that tigers were inhabited by spirits, often those of shamans, dead or alive. One would be well advised not to kill these tigers, as they protected the camp, kept 'real' tigers away and even provided their people with game. Moreover, revenge would await those who killed such a tiger. However, tigers can also be inhabited by evil spirits, as witness the Mamak story. Among the Semang, such a spirit can be the 'soul' of a living, evil shaman. Apparently people do not dare to attack these tigers either. We have also encountered the notion that if a person is killed by a tiger, that person, surely, must have done something wrong, and the tiger killing him represents or embodies higher forces. We may suppose that in such cases revenge was not called for. The case of the Aboriginal Malay, however, seems to show that there were other ways around the prohibition to kill supernaturally endowed tigers, or that they were able to establish whether a specific tiger was real or supernatural before killing it. The fact that they did not have the specific tiger traps suggests that even the Aboriginal Malay stopped short of killing tigers at random.

Finally, I would like to compare the attitudes regarding tigers of the nomadic and semi-sedentary groups with those of the sedentary peasants. Summing up the behaviour of the latter, it can be said that systematic attempts to get rid of all tigers were rare in the Malay world. Only some individuals, such as the sultan of Johore and a Sumatran haji, seem to have been exceptions to this rule. The settled nineteenth-century Javanese peasants probably came close, although in some areas inhibitions were still quite strong around 1850. Sumatra's settled peasantry was much more hesitant to kill tigers, certainly if they had no human being on their conscience, but even then revenge was not an automatic response. Differences in natural environment and beliefs go a long way in explaining this.

The Sumatran peasants were not all that different from the semi-settled groups, but they were separated by the use of the specific tiger trap. This implies that even the Sumatran peasants, although more hesitant than the Javanese to confront such spiritually powerful beings as tigers, were less reluctant than their semi-sedentary brethren to become slayers of tigers. There are many indications that Javanese, Sumatran and Malaysian peasants feared and hated tigers. In this respect they were not much different from the European peasants with their hatred of bears and wolves. Fear was also felt by many of the groups that we have dealt with, although some (Aboriginal Malay, Lubu) seem to have had a more relaxed attitude than others (Semang, Mamak).

In the Malay world, fear and hatred of the tiger are not automatically translated into attempts to kill the animal. In that respect there are differences between some Malay regions and the European situation. However, some Malay regions are remarkably 'European' as regards the killing of tigers. This phenomenon may be partly explained by the fact that tigers, in the Malay world, are also spiritually powerful beings, or at least inhabited by such beings. Such feelings may have worked in favour of the tiger, but also against it. One would think twice before killing an 'ancestral' tiger, but killing a weretiger – a being comparable to the European werewolf – may have been regarded as meritorious (but also dangerous). Although some tigers were regarded as ancestors and many Malays avoided killing tigers, it is probably too romantic a notion to believe that tigers were not feared or hated. Calling the tiger 'grandfather' may have fooled the tiger, but it should not fool us.

Notes

1 In Indonesia, tigers were found in Sumatra, Java and Bali. By 2000, they were extinct in Bali and extinct or on the verge of extinction in Java. I deal with hunting kings and aristocrats, sedentary peasants and Europeans elsewhere (Boomgaard 2001). Here, I summarize my findings on the sedentary peasants.
2 Lee and DeVore (1968); Shrire 1984; Kent (1989); Boomgaard (1997).
3 Elsewhere (Boomgaard 2001) I have published a comprehensive account of the history of the tiger in the Malay world.
4 See, for example, MacKenzie (1987); Alter (1988); Hodges-Hill (1992).
5 See, for example, the title of Wessing's *The Soul of Ambiguity.*
6 It is rather amazing how much information actually exists on these tiny groups, although it is often impossible to ascertain its reliability.
7 Newbold (1839, II: 379); Schebesta (1928b: 180); Endicott (1979: 62).
8 Anderson (1824: App. Xxxix); Martin (1905: 796); Endicott (1979: 79, 119).
9 Anderson (1824: App. Xxxix); Martin (1905: 732); Schebesta (1928a: 237); Endicott (1979: 64).
10 Schebesta (1928a: 226–229, 237); Endicott (1970: 16–22); Endicott (1979: 44). *Hala'* are the souls of dead people or the original superhuman beings.
11 Schebesta (1928a: 197); Carey (1976: 72–73); Marks (1976); Yost and Kelley (1983: 191); Rambo (1985: 212–214).
12 Anderson (1824: App. Xxxix); Newbold (1839, II: 380); Skeat and Blagden (1906 I: 201); Schebesta (1928a: 115).
13 Schebesta (1928a: 197); Hubback (1905: vi, 63); Hames (1979: 226–228).
14 Martin (1905: 786, 794–796); Skeat and Blagden (1906, I: 278–279); Schebesta (1928a: 147); Evans (1937: 63); Rambo (1978: 210); Endicott (1979: 133).
15 Schebesta (1928a: 172, 229); Wiele (1930: 109).
16 Boers (1838: 290–291); Mohnike (1874: 198); Winter (1901: 230); Dongen (1906: 239); Hagen (1908: 108); Persoon (1994: 158).
17 Boers (1838: 295); Verkerk (1874: 151, 157–159); Dongen (1906: 243); Dongen (1910: 207–208, 217); Waterschoot (1915: 231); Schebesta (1928b: 232); Keereweer (1940: 366); Schneider (1958: 273, 276); Sandbukt (1982: 4); Yost and Kelley (1983: 212–213).
18 Boers (1838: 295); Dongen (1910: 208); Waterschoot (1915, 223).
19 Verkerk (1874: 156); Dongen (1910: 201).
20 Also called Semai and Sakai (not to be confused with the Sumatran Sakai). The two Sakai groups were, to my knowledge, not related. According to one author (Obdeyn 1929: 358) *sakai* is not really the name of a tribe, but just means 'nomadic tribe'.
21 Skeat and Blagden (1906, I: 206–207, 212–213); Schebesta (1928b: 36, 122–129).
22 Hubback (1905: 184); Martin (1905: 775, 796); Skeat and Blagden (1906, I: 206–207, 212).
23 Evans (1923: 210); Schebesta (1928b: 43, 70, 141); Endicott (1970: 17–22).
24 They were formerly often called Proto-Malay, also Jakun, and, earlier, Benua.
25 Favre (1865: 54–57); Martin (1905: 729–730); Schebesta (1928b: 180–182).

26 Menie (1862 [1642]: 132); Newbold (1839, II: 395–401).
27 Martin (1905: 794, 796); Skeat and Blagden (1906, I: 215, 220–223).
28 Favre (1865: 58, 63); Newbold (1839, II: 401); Martin (1905: 796); Skeat and Blagden (1906, I: 221–222, 262, 324–325).
29 A detailed analysis of these beliefs in Boomgaard (2001). Eredia (1930 [1613]: 41, 48); Menie (1862 [1642]: 132); Barros (1727: 119).
30 Newbold (1839, II: 387–388); Martin (1905: 953, 958, 969); Skeat and Blagden (1906, II: 305).
31 Müller (1855: 135, 156–157); Dijk (1884: 151, 157); Ophuijsen (1884: 99); Kreemer (1912: 308–310).
32 Dijk (1884: 156); Kerckhoff (1890: 576); Kreemer (1912: 307, 334).
33 Schneider (1905); Obdeyn (1929); Schneider (1958).
34 Schneider also mentions the lance as an important weapon, but the context suggests that it was used in combat, not for hunting.

References

Alter, S. (ed.) (1988) *Great Indian Hunting Stories*. Harmondsworth: Penguin.
Anderson, J. (1824) [1965] *Political and Commercial Considerations Relative to the Malayan Peninsula, and the British Settlements in the Straits of Malacca*. Prince of Wales Island: Cox.
Barros, J. De (1727) *Helddadige scheepstogt van Alfonso d'Albuquerque, 1506 en volgende*. In J. L. Gottfried (ed.) *De aanmerkenswaardigste en alomberoemde zee- en landreizen der Portugeezen, Spanjaarden, Engelsen (...) tot ontdekking van de Oost- en West-Indiën*. Vol. 1. The Hague and Leiden: Boucqueet *et al.*
Boers, J. W. (1838) 'De Koeboes'. *Tijdschrift Nederlandsch-Indië*, vol. 1–2, pp. 286–295.
Boomgaard, P. (1994) 'Death to the tiger! The development of tiger and leopard rituals in Java, 1605–1906'. *South East Asia Research*, vol. 2, pp. 141–75.
—— (1997) 'Hunting and trapping in the Indonesian Archipelago, 1500–1950'. In P. Boomgaard, F. Colombijn and D. Henley (eds) *Paper Landscapes: Explorations in the Environmental History of Indonesia*. Leiden: KITLV Press [Verhandelingen KITLV 178], pp. 185–213.
—— (2001) *Frontiers of Fear: Tigers and People in the Malay World, 1600–1950*. New Haven: Yale University Press.
Carey, I. (1976) *Orang Asli: The Aboriginal Tribes of Peninsular Malaysia*. Kuala Lumpur: Oxford University Press.
Dijk, P. A. L. E. van (1884) 'Rapport over de Loeboe-bevolking in de Onderafdeeling Groot-Mandheling en Batang Natal'. *Bijdragen tot de Taal-, Land- en Volkenkunde*, vol. 32, pp. 151–161.
Dongen, G. J. van (1906) 'Bijdrage tot de kennis van de Ridan-Koeboes'. *Tijdschrift Binnenlands Bestuur*, vol. 30, pp. 225–253.
—— (1910) 'De Koeboes in de onderafdeeling Koeboestreken der residentie Palembang'. *Bijdragen tot de Taal-, Land- en Volkenkunde*, vol. 63, pp. 181–336.
Endicott, K. M. (1970) *An Analysis of Malay Magic*. Oxford: Clarendon Press.
—— (1979) *Batek Negrito Religion: The World-View and Rituals of a Hunting and Gathering People of Peninsular Malaysia*. Oxford: Clarendon Press.
Eredia, G. E. de (1930) 'Description of Malacca, Meridional India and Cathay (1613) [transl. L. V. Mills]', *Journal of the Malayan Branch of the Royal Asiatic Society*, vol. 8, pp. 1–288.
Evans, I. H. N. (1937) *The Negritos of Malaya*. Cambridge: Cambridge University Press.

Favre (1865) *An Account of the Wild Tribes inhabiting the Malayan Peninsula, Sumatra, and a Few Neighbouring Islands*. Paris: Imperial Printing-Office.

Hagen, B. (1908) *Die Orang Kubu auf Sumatra*. Frankfurt: Baer.

Hames, R. B. (1979) 'A comparison of the efficiency of the shotgun and the bow in neotropical forest hunting'. *Human Ecology*, vol. 7, pp. 219–252.

Hodges-Hill, E. (1992) *Man-Eater. Tales of Lion and Tiger Encounters*. Heathfield: Cockbird Press.

Hubback, T. R. (1905) *Elephant and Seladang Hunting in the Federated Malay States*. London: Rowland Ward.

Keereweer, H. H. (1940) 'De Koeboes in de Onderafdeeling Moesi Ilir en Koeboestreken'. *Bijdragen tot de Taal-, Land- en Volkenkunde*, vol. 99, pp. 357–396.

Kent, S. (1989) 'Cross-cultural perceptions of farmers as hunters and the value of meat'. In S. Kent (ed.) *Farmers as Hunters: The Implications of Sedentism*. Cambridge: Cambridge University Press, pp. 1–18.

Kerckhoff, Ch. E. P. van (1890) 'Eenige opmerkingen betreffende de zoogenaamde "orang loeboe" op Sumatra's Westkust'. *Tijdschrift Nederlandsch Aardrijkskundig Genootschap*, 2nd series, vol. 7, no. 2, pp. 576–577.

Kreemer, J. (1912) 'De Loeboes in Mandailing'. *Bijdragen tot de Taal-, Land- en Volkenkunde*, vol. 66, pp. 303–335.

Lee, R. B. and I. DeVore (eds) (1968) *Man the Hunter*. Hawthorne: Aldine.

McDougal, C. (1987) 'The man-eating tiger in geographical and historical perspective'. In R. L. Tilson and U. S. Seal (eds) *Tigers of the World: The Biology, Biopolitics, Management, and Conservation of an Endangered Species*. Park Ridge, NJ: Noyes Publications, pp. 435–448.

MacKenzie, J. M. (1987) *Imperialism and the Hunting Ethos*. Manchester: Manchester University Press.

Marks, S. (1976) *Large Mammals and a Brave People*. Seattle: University of Washington Press.

Martin, R. (1905) *Die Inlandstämme der Malayischen Halbinsel: Wissenschaftliche Ergebnisse einer Reise durch die Vereinigten Malayischen Staaten*. Jena: Fischer.

Menie, J. J. (1862) 'De orang benoea's of wilden op Malakka in 1642'. *Bijdragen tot de Taal-, Land- en Volkenkunde*, vol. 8, pp. 127–33.

Mohnike, O. (1874) *Banka und Palembang nebst Mittheilungen über Sumatra im Allgemeinen*. Münster: Aschendorffschen Buchhandlung.

Moszkowski, M. (1909) *Auf neuen Wegen durch Sumatra: Forschungsreisen in Ost- und Zentral-Sumatra (1907)*. Berlin: Reimer.

Müller, S. (1855) *Reizen en Onderzoekingen in Sumatra (. . .) Tusschen de Jaren 1833 tot 1838 (. . .)*. The Hague: Fuhri.

Newbold, T. J. (1839), *Political and Statistical Account of the British Settlements in the Straits of Malacca (. . .) with a History of the Malayan States on the Peninsula of Malacca* [2 Vols.]. London: Murray.

Obdeyn, V. (1929) 'De lankah lama der orang Mamak van Indragiri'. *Tijdschrift Indische Taal-, Land en Volkenkunde van het Bataviaasch Genootschap*, vol. 69, pp. 353–425.

Ophuijsen, C. A. van (1884) 'De Loeboes'. *Tijdschrift Indische Taal-, Land en Volkenkunde van het Bataviaasch Genootschap*, vol. 29, pp. 88–100.

Persoon, G. A. (1994) *Vluchten of Veranderen: Processen van Verandering en Ontwikkeling bij Tribale Groepen in Indonesië*. Leiden: FSW.

Rambo, A. T. (1978) 'Bows, blowpipes and blunderbusses: ecological implications of weapons change among the Malaysian Negritos'. *Malayan Nature Journal*, vol. 32, pp. 209–16.

——— (1985) *Primitive Polluters: Semang Impact on the Malaysian Tropical Rainforest Ecosystem.* Anthropological Papers, Museum of Anthropology, University of Michigan. Ann Arbor: University of Michigan.

Sandbukt, Ø. (1982) 'Kubu conceptions of reality'. Paper presented at the Symposium on Southeast Asian Folklore, Denmark.

Schebesta, P. (1928a) *Among the Forest Dwarfs of Malaya.* London: Hutchinson.

——— (1928b) *Orang-Utan: Bei den Urwaldmenschen Malayas und Sumatras.* Leipzig: Brockhaus.

Schneider, G. (1905) *Ergebnisse zoologiscger Forschungsreisen in Sumatra. Erster Teil. Säugetiere (Mammalia).* Jena: Fischer [Abdruck aus den *Zoologischen Jahrbüchern, Abt. für Systematik, Geographie und Biologie der Tiere* 23/1].

——— (1958) 'Die Orang Mamma auf Sumatra'. *Vierteljahrsschrift der Naturforschenden Gesellschaft in Zürich*, vol. 103, no. 5, pp. 213–286.

Shrire, C. (ed.) (1984) *Past and Present in Hunter-Gatherer Studies.* Orlando: Academic Press.

Skeat, W. W. and C. O. Blagden (1906) *Pagan Races of the Malay Peninsula* [2 vols.]. London: Macmillan.

Verkerk Pistorius, A. W. P. (1874) 'Palembangsche schetsen: Een dag bij de wilden'. *Tijdschrift Nederlandsch-Indië* N. S., vol. 3, no. 1, pp. 150–160.

Waterschoot van der Gracht, W. A. J. M. van (1915) 'Eenige bijzonderheden omtrent de oorspronkelijke orang Koeboe in de omgeving van het Doewabelas-gebergte van Djambi'. *Tijdschrift Koninklijk Nederlandsch Aardrijkskundig Genootschap*, vol. 32, pp. 219–225.

Wessing, R. (1986) *The Soul of Ambiguity: The Tiger in Southeast Asia.* Monograph Series on Southeast Asia: special report, 24. DeKalb: Center for Southeast Asian Studies, Northern Illinois University.

Whitney, C. (1905) *Jungle Trails and Jungle People: Travel, Adventure and Observation in the Far East.* New York: Scribner's.

Wiele, H. (1930) *Indische Jagdabenteuer.* Leipzig: Verlag Deutsche Buchwerkstätten [1st ed. 1925].

Winter (1901) 'Ook onderdanen onzer Koningin (Een bezoek aan de tamme Koeboes)'. *Indische Gids*, vol. 23, no. 1, pp. 208–247.

Yost, J. and P. M. Kelley (1983) 'Shotguns, blowguns and spears: the analysis of technological efficiency'. In R. B. Hames and W. T. Vickers (eds) *The Adaptive Responses of Native Amazonians.* New York: Academic Press, pp. 189–224.

11

THE RAJ AND THE NATURAL WORLD

The war against 'dangerous beasts' in Colonial India

Mahesh Rangarajan

Introduction

South Asia has been a major arena for conflicts between people and predators but, unlike the examples of England and North America, their history has hardly been told (Worster 1977; Thomas 1983). The focus has mainly been on the changing attitudes and practices of the imperial rulers (MacKenzie 1988; Rangarajan 1996: 138–198). This is inevitable given the extent of literature available on shikar, or hunting. Useful as this may be in understanding the culture of empire, it refers to only a small part of the picture. But the issue of the decline of wildlife raises broader questions about the nature and impact of colonial rule.

Hunting for sport was integral not only to the lifestyle of officials but also to their self-image as men who believed in fair play. The Raj was seen as powerful enough to contain danger (MacKenzie 1988). Yet the self-regulatory character of colonial power is then used to blame the increase of population and the breakdown of norms after 1947 for general ecological decline (Gen. J. G. Elliot 1973: 4; Vernede 1995: 142–143). In a sense, this is a double blind, for British officials in the past often assumed that they alone were brave enough to face large and hostile beasts. Fair play meant that strength was tempered by mercy. Such attitudes are not confined to the past (Davidar 1986: 67–71). This view of imperial stewardship needs to be tested against the evidence.

The question can be posed at another level. The disappearance of free-ranging wildlife could simply have been a by-product of the expansion of agriculture. Greater mobility and better weapons for the hunter may have been incidental to the decline of the Bengal tiger (*Panthera tigris tigris*). In that case, the fate of wildlife then becomes simply a part of a larger drama, and specific drives against particular species are incidental to the general impoverishment of a region's ecology. The killing of tigers or leopards (*Panthera pardus*) for bounties depended on Indian cooperation, but their reasons for doing so – or not doing so – have hardly been looked into. The schemes to wipe out 'dangerous beasts' offer insights into the interaction of the government with rural land users.

Indian legacies

It is essential to have a perspective on the relationship between large mammals and people in the pre-colonial era. This may elide and tend to flatten out the significant shifts over time, such as changes in technologies of warfare or production. But there is sufficient reason to assert that the dynamics of people–nature relations in the pre-colonial period were very different from what was to follow the consolidation of British power, especially after 1857. The agrarian frontier had long been the site of intense conflict, with people working out various ways to minimize contact with or even to retaliate against carnivores. Further, the tiger and the Asiatic lion (*Panthera leo persica*) were more than just the embodiment of ferocity; they also had religious or even magical attributes.

The relationship with large predators was multi-faceted. Animals like the lion and tiger were often metaphors for power even as they were a source of danger.[1] The importance of the lion in the Asokan pillar (third century BC) is well known. The Valmiki Ramayana, possibly dating to the fourth to sixth centuries AD, refers to Dasratha as a 'lion among kings' but also describes Rama and Lakshmana as 'tigers among men' (Pollock 1984: 18; 1991: 184).[2] Even at a much later date, the killing of large mammals was a sign of a warrior's prowess. Farid's slaying of a tiger earned him the name Sher Khan (1540–45) and the fact that it was killed with a spear on foot was a mark of bravery (Abbas Khan Sherwani, in Elliot and Dowson 1979: 316). Akbar (1556–1605) was on a pilgrimage to the Chishti shrine at Ajmer in July 1572 when he hunted down a man-eating tiger (Abu al Fazl 1972: 539). The religious dimensions of perceptions of the natural world were often significant. Mughal chroniclers cited how Akbar stared at a tiger which 'cowered down from that divine glance' (Abu al Fazl 1972: 294). An early European account from 1670 mentions the mangrove swamps of eastern Bengal, an area notorious for man-eaters, and refers to a place where, 'every Thursday night a tiger comes out and salaams a fuckeer's [fakir] tomb there' (Eaton 1994: 209).[3] Islam had clearly co-opted and assimilated older traditions of tiger worship into a new framework.

The religious dimension should not detract from more prosaic ways of dealing with threats. Buchanan found villages in Kanara that had been 'formerly much infested with tigers' now devoid of them due to forest clearance (Habib and Raychaudhuri 1982: 48). Animal attacks could limit the use of forests. The collection of lac and sandalwood had been given up in some woodlands in Kanara due to loss of lives to tigers. Cattle were carefully penned in at night to protect them from the big cats (Buchanan 1807: 75). Soldiering and hunting went hand in hand. The Bedas recruited into the army by Tipu Sultan of Mysore (d. 1799) were skilled marksmen who also pursued tigers (Buchanan 1807b:

178–179). Nor did Indians necessarily lack a notion of agrarian 'improvement'. Bishop Reginald Heber met a zamindar who exulted that tigers had given way to 'better things like corn fields, villages and people'. Nature retreated before the advance of the plough (Heber 1993: 14, 148).

There was no innate contradiction between such notions and a religious awe of wild animals. The Raja of Maihar near Panna, central India told William Sleeman that there were two kinds of tigers. Ordinary ones that turned man-eaters could be tracked and shot. But the ones along the Jabalpur-Mirzapur road were 'tail-less tigers' with magical properties. The only recourse was to pay a Gond tribal 10 to 20 rupees to conduct a sacrifice and placate the beasts (Sleeman 1980: 126–127). The long exposure to carnivores made it inevitable that various cultures and rulers in the sub-continent would perceive the animals in a host of ways. There were mutually contradictory strands, with the same animal being revered and feared, hunted and worshipped even at the same time. It would be anachronistic to see the past in terms of an easy coexistence of people and predators, but the dynamics of the interaction were about to undergo a very significant shift.

A comparison with European attitudes

In contrast with South Asia, Britain had a very different history of relations with large wild mammals. In common with much of Europe, there had been concerted campaigns against specific animals (Braudel 1981: 67). In the 1620s, tenants in parts of Scotland were still paying a tribute in iron each year to their lord to forge weapons against wolves (Smout 1969: 131). The extermination of the species in the British Isles preceded its elimination in large parts of continental Europe (Zimen 1981: 296). The animal vanished in areas of intensive human settlement on the North Sea shores of the continent.

Other dimensions of extermination were significant. As game became an object of leisure, landed groups sought to exclude the lower classes, while training their guns – literally as well as metaphorically – on the wild animals that lived on game birds and fish. Otters and wild cats were explicitly excluded from protection under the Game Act of 1671. They were vermin, to be killed whenever possible (Ritvo 1990: 13). In the eighteenth century, the King of England had an official rat-catcher, whose embroidered uniform showed figures of mice devouring wheatsheaves (Thomas 1983: 274). Some carnivores were seen as lawless beasts. But the tiger was especially ferocious even among other flesh eaters due to its fondness for people as prey (Ritvo 1990: 28). The broadly negative attitude to carnivores was to acquire a new significance in the sub-continental context.

South Asia must have seemed like a menagerie of fierce free-ranging wild animals (Courtenay 1980: 47). The death of the young army officer Munro, eaten by a tiger on Saugor island, was assimilated into British folklore. Staffordshire potteries even had a chimney ornament showing a tiger with the head of the hapless officer (ibid.: 52). As British power expanded into the sub-continent, the tiger loomed even larger in the consciousness of the conquerors. When Tipu Sultan was eventually defeated in 1799, the victory was commemorated with a special medallion which showed the imperial lion of Britannia overpowering a tiger (Bayly 1991: 156, 159). Another predator that had an extensive range in the sub-continent, but unlike the tiger was directly known within Europe, was the wolf. In Kanpur cantonment, wolves were such a menace that they 'frequently' carried off children and attacked sentries (Forbes 1813: 81). Buchanan found them common in Shahabad, but they were elusive (Buchanan 1986: 229). Such deep fears were not limited to large beasts of prey. The civet cat, for instance, is a small mustelid the size of a domestic cat and often reared in captivity for its musk and to control mice. But to Williamson, it was an insensate marauder, 'killing as it were, merely for sport' (Williamson 1807–8: 109).

In the Company Raj, different remedies were tried out in areas where the depredations of carnivores were seen as a threat to human life or as a barrier to cultivation (Heber 1993: 15). Wild creatures that harmed humans violated what the conquerors held to be the order of God's creation. Their affinity to human rebels was all too obvious. There were two options. One was to pay rewards for killing the animals; in Madras, bounties were paid out for a variety of beasts from 1815 onwards by the Board of Revenue, including a larger bounty for the female of the species.[4] In Indore, the bounty on a tiger's head was the equivalent of four months' pay of a court officer.[5] The other option was to use military force to curb errant animals. Such instances were less common and the offer of rewards and incentives was the main instrument of policy (Campbell 1880: 30). Such efforts were mainly at a local or provincial level, and it was only after 1857–58 that such schemes became more systematic.

A war against vermin

Provincial level schemes were extended after a debate on the ravages of wild beasts. The wider context of the debate was the disarming of Indians after the Rebellion of 1857–58. There were fears that the denial of modern firearms to people in rural areas was indirectly contributing to 'serious depredations' of cattle and crops. In the Madras Presidency there were frequent complaints of increased damage by tigers, wolves and other animals.[6] In Bombay the authorities went a step further and even distributed arms in certain localities.[7] But such efforts were sporadic at best. An image of Indians as people incapable of self-defence meshed well with the kind of distrust that became the norm in the aftermath of 1857.

The discussion itself was in response to a bizarre proposal from a former Army officer, B. Rogers, who wanted the entire campaign against carnivores placed on a war footing. In each district, the local shikaris were to be organized into a corps under the command of a civil or preferably a military officer. Their sole task would be to eliminate large carnivores. He first won some support but the proposal eventually came to nought.[8] Most officials agreed that something should be done, but they disagreed on what this line of action ought to be.

Dr Joseph Fayrer explicitly saw the Thugee Department of Colonel William Sleeman as the model. The affinity of human and beastly outlaws was stretched to a point when even the remedies proposed against the latter began to mirror those used against the former (Fayrer 1878).

Parallels were also drawn between vermin control in the home country and the tasks at hand in India. Major Tweedie wanted to employ tiger-killers 'just like' the mole-catchers and rabbit-killers in England.[9] But senior officials frowned upon ambitious programmes that could involve more government spending. Alfred Lyall opposed the creation of an 'asphyxiating department'.[10] Rogers's proposal was aimed at his 'own nourishment', not the destruction of beasts.[11] In Madras, a police officer, Colonel Caulfield, was paid an extra stipend to kill tigers in the Coimbatore district.[12] He killed seven tigers using traps made by Toda tribals and was then dispatched to the Vizag region in the Eastern Ghats, an area known for tigers.[13] But it all cost too much.[14] The experiment was not a success. In contrast, the use of poisons to kill carnivores in Coimbatore won encomiums from officials.[15] The poisoning of carcasses was in any case fraught with dangers and was 'perilous even in England'.[16]

The criticisms of a special force focused on the comparative advantages of drawing on local hunters who would function more autonomously. Mobile corps of hunters would move from one tract on to another. The village shikari was the best bet against 'both evils' – crop-raiding deer and man-eating tigers. In general, the existing system of bounties was seen as effective.[17] The idea of the temporary deployment of 'men from outside' remained as an option, though it was not to be imposed on the various provinces.[18] In Dinajpur in north Bengal, zamindars shared the cost of hiring specialist tiger-killers from outside the district. This significantly increased the numbers of tigers slain in the dry season.[19] In the North West Provinces and Awadh, District Officers employed Kanjers to kill snakes, paying a regular salary as well as two annas a snake in excess of twenty specimens. It cost much less than a special corps and allowed flexibility at a local level.[20]

Others contended that the expansion of agriculture would automatically resolve the issue by destroying the habitat of wild beasts. But the protection of new settlements often entailed vigorous measures to exterminate wildlife. In the plains region of the Bombay Presidency, 'dangerous animals' had became locally rare in areas where they had been common a few decades earlier.[21] Officials in Assam and Berar were confident that clearance of the jungle would ensure that animals would 'gradually but surely disappear'.

Princely states

Much of India was not directly administered by the British but was under princely rule. There were also several regions, such as Bengal, Bihar and Orissa, where the Permanent Settlement made the zamindar, the landed intermediary, a key player in the countryside. Hunting was a way of life for the landed gentry and princes, but their reactions to the programme to wipe out vermin were diverse. The responses in princely India offer an insight into contrasting attitudes to the programme.

Most rulers in Rajputana in western India either denied that carnivores were a major problem or claimed that sportive hunting was adequate as a means of control. The ecology and the forms of land-use in these areas were often crucial in defusing the intensity of conflict between people and predators. Alwar included excellent tiger habitats in the thorn forests of the Aravalis, but rewards were only given in case of man-eaters.[22] In Bikaner it was reported that wolves 'do not hunt people'.[23] Again, the large herds of blackbuck (*Antelopa cervicapra*) and gazelle (*Gazella gazella*) and livestock probably provided an adequate prey base. Elsewhere, carnivores had been eliminated with the extension of agriculture and some forests were already largely bereft of game.[24] In three years, not a single tiger had been shot in Sirohi despite the efforts of sportsmen.[25] Sportive hunting was often seen as an adequate defence against wild predators.[26]

Captain Walker, the British Agent in Marwar, was confident that the Rajputs were 'more or less true sportsmen' and could keep animals in check.[27] This was also an expression of a confidence in what was then considered a so-called 'martial race'. Alwar's ruler, Sheodan Singh, claimed that was he willing to wipe out tigers from the state if only government would 'have the goodness to cause masonry "odeys" to be built at the shooting places'. These odeys were small fortresses, with a place where the hunter could wait in safety several feet above a buffalo tied up as bait.[28] This was clearly a bid to use the programme to satisfy the raja's hunger for trophies.

Some princes had already initiated or were willing to follow through with a scheme of rewards for killing vermin. It is possible that the predators were a threat to livestock in certain areas.[29] Elsewhere in India, Mysore and Hyderabad were prominent in their extensive system of bounties for wild animals and snakes.[30] Religious beliefs may have been crucial in resisting such measures in Marwar where the raja refused to help in 'any organised attempt' to get rid of venomous reptiles.[31] In Jhallawar, the Raj Rana was already giving out rewards for tigers and Bundi was ready to follow suit.[32] Kotah had a comprehensive system of bounties, with 10 rupees for a tiger and five for bears, wolves and leopards. The state was unusual as it appears to be the only princely house that still rewarded the killing of lions, which were already very rare. The bounty was as high as 25 rupees.[33] What is important is that such instances were still exceptional – until the 1870s when the schemes were promoted by the British.

The tiger: ally or enemy?

While there was broad accord on the need to control carnivores, even at its zenith the project did not have universal support. G. P. Sanderson was emphatic in his condemnation of the policy of eliminating large predators. Tigers lifted cattle but the advocates of 'a war of extermination' ignored a simple fact. The tiger was entitled to 'present his little account for services rendered in keeping down wild animals which destroy crops' (Sanderson 1983: 307, 312). There was no easy answer to the question raised by such objectors: was the tiger a friend or foe? Tolerance for the tiger was qualified and within the logic of agrarian expansion and consolidation. The Commissioner of Rajshahi explained that a tiger that killed game was fine but, 'when he travels outside the forests and seeks his prey in the neighbourhood of villages, he becomes destructive'.[34]

Sportsmen became more critical of extermination as a policy with the decline in numbers of trophy-worthy specimens. The zamindar of Punganur in North Arcot was concerned that increased movement of traffic on roads had driven tigers away from their age-old haunts.[35] Similarly, the veteran big-game hunter Captain A. E. Wardrop favoured abolition of rewards for killing tigers.[36] As the range and availability of large carnivores declined, there was also a move to classify the tiger as a game animal instead of as vermin. Lord Curzon was concerned that 'foreign sportsmen not resident in India were bagging too many trophies',[37] but he warned against making forest reserves 'into a sort of sanctuary' where animals enjoyed protection. This would be 'indefensible and immoral'. There was no doubt that the war against the tiger was far from over.[38]

The anxieties about the availability of tigers did not prevail over the deeper concern about the need to control a predator with an extensive range over various parts of the sub-continent. The idea that they were evil because of 'the fear they spread [more] than the damage they do' was not going to die down so quickly.[39] The worth of a species as a trophy became crucial in enabling or thwarting the change of status from vermin to game. Advocates of conservation of carnivores referred to the plight of the Asiatic lion.[40] But it was doubtful if tigers would be seen in the same light.[41] Three decades earlier, when the lion was already vanishing across much of its range in northern and central India, the Foreign Department had approved of rewards in Kotah.[42] The debate was shifting from efforts at examination to the applicability of protection. Protection was motivated by an ethic of use (Fleetwood Wilson 1921: 137). Protection of the cheetah was said to be justified due to its rarity.[43] The idea that a carnivore could die out forever was now seen, at least by some, as a matter of regret. But they remained a powerless, if vocal, minority.[44]

Hunting for sport and fishing for leisure even pitted Forest Officers in particular against smaller carnivores. The forest was to be made an ordered landscape that could yield trophies and game meat to sportsmen. Civets were dubbed as 'active game-destroyers'. Owls and eagles were 'egg-thieves' and 'chick-destroyers' (Stebbing 1906: 220–223). In the Konkan, sportsmen were worried about the large increase in the numbers of birds of prey (Anon. 1891: 119–123; Hodgson 1893: 530–533). Anglers feared their sport was in danger due to otters (*Lutra lutra*) (OC 1893: 34–35). Three peons were set to work to kill fresh water crocodiles (*Crocodilus parustrius*) and destroy their eggs in the Tulsi and Virar lakes near Bombay. H. M. Phipson of the Bombay Natural History Society blamed the 'loathsome reptiles' for destroying fish and wild fowl (Phipson 16/1/1892; Barrow

1895: 144). Scientific expertise, aesthetic concerns and sporting interests broadly concurred on this issue.

Critics of extermination often shared the values of its proponents. They saw carnivores as allies in the drive to transform jungle into farmland. But game-destroyers were still marginal to official efforts. It was animals like the tiger and wolf that were a threat to domestic stock and human life.

Tiger versus people

Tigers were in a very different category from mere game-destroyers. It was unlikely in any case that a case for their protection would win enough support to bring about a change of policy in general. The schemes for elimination continued without modification in most parts of British India. There is little doubt that there were problems in the coexistence of people with the striped cat in many areas. Given the diversity of habitats and of land-use systems, the intensity of conflicts varied in different areas.

In general, the species vanished from the plains areas: extermination and the extension of agriculture went together and were complementary. But vermin slaughter gave the process an added impetus. For example, the Bombay Presidency witnessed a serious diminution in the range of the big cat. Tigers were still found in the Sahyadaris in the first half of 1870s. About 20 tigers were killed each year not far from Bombay. By 1900, it was claimed, though inaccurately, that the tiger was confined to the Dangs, the Satpuras and the Hatti hills. They were still found in Trombay in 1907 (Campbell 1984: 44–45); tigers swam across from the mainland (Prater 1929: 973–974). The Amir of Bahawalpur shot 13 specimens in Sind, and the tiger became extinct in these galley forests (Roberts 1976: 144–145). A similar picture emerges in the Punjab, where the species was 'almost extinct', only rarely found even in the hilly parts of Ambala district. H. Maude recalled how in 1866–69 no less than 84 tigers had been killed, of which as many as 50 were in Pindi and 22 in Ambala.[45]

The 20 per cent of the land area of British India that was government forest by 1900 could have helped the tiger. However, the existence of Reserved Forests was in itself no guarantee for survival. Foresters oversaw special efforts to exterminate or at least to limit the numbers of carnivores. This had a pincer-like effect: if the prey base and tree cover outside government forests was reduced by agrarian extension, the latter offered limited shelter due to the killing of vermin. Smaller predators could still survive in the patches of dhak (*Butea frondosa*) and scrub forest, but not the tiger.[46] The Forest Department rules allowed access into fire-protected forests for killing carnivora even when such tracts were normally closed.[47] In Naini Tal Division, tigers could be shot at water holes and salt licks.[48] Nor were such rules in any way specific or unique to the province.

Elsewhere, the level of predation on human beings was showing clear signs of decline. Among those places where its incidence declined sharply was the Madras Presidency.[49] Even in the mid-1920s, Madras topped the country in lives lost to tigers, with two out of every three attacks being fatal. But predation was highly localized.[50] In general, there was a decline in the tiger's range in the south, but it is possible that, in the short run, increased human intrusion into forests exposed more people to the possibility of confrontation. In Vizag, for instance, the construction of the East Coast Railway opened up the country and increased demand for wood for sleepers.[51] Despite the scepticism about rewards, Vizag and Ganjam were the districts with maximum tigers killed.[52] By 1945, an experienced sportsman could argue that there were very few instances of man-eaters south of a line drawn across the peninsula at Bombay (O'Brien 1945: 231–232). The pieces in the story point to a pattern. Increased market integration of forest areas led to a higher level of conflict in the short-term, but these were resolved with the extinction of tigers outside very small parts of their former range. In turn, a region once known for its man-eating tigers now became renowned for the absence of attacks on people (Sankhala 1978).

The reverse phenomenon was taking place in the United Provinces, where man-eating was relatively rare till the 1920s, after which it increased sharply. But the species as a whole fared better than in the west and the south, possibly due to the nature of the habitat. The combination of sal forests and open grasslands with large ungulate populations supported many big cats. As late as 1904, John Hewett noted that two-thirds of all money spent on tiger bounties was expended in the hill areas of Kumaon. Yet only three persons were annually killed there by tigers.[53] Despite this, in the 1920s and 1930s, the area became synonymous with man-eaters. Corbett himself wrote of 'the stress of circumstances', including disability and wounds, as being a major factor for the spurt in man-eating. The other factor was possibly that leopards began living on human flesh in the aftermath of epidemics of cholera and influenza, the latter in 1918 (Corbett 1995: x–xiii). Destruction of individual tigers in one area simply opened up opportunities for others to move in from adjacent forest (Prater 1940: 881–888). This was a resilient species, capable of making a comeback and even of modifying its behaviour as it came under intense pressure. Tigers seem to have become more nocturnal and secretive in their habits as hunting pressures increased (McDougall 1978: 155–159). These populations could survive as long as cover and prey were adequately available in an area (Singh 1993: 81–82).

The scarcity of wild prey could compel carnivores to turn on cattle and even human beings.[54] The dry and moist deciduous forests that comprise much of tiger habitat in India have an extensive prey base, including large animals like the sambhar (*Cervus unicolor*) and gaur (*Bos gaurus*). The wet grasslands of the Indo-Gangetic plains had an even higher concentration of prey species, including the wild buffalo *Bubalis* and great one-horned rhino (*Rhinoceros unicornis*). Their steady retreat to small portions of their erstwhile range by 1900 may have led to increased tiger predation on cattle or people. Interestingly, in Java, with its long history of habitual man-eating by tigers, the most common prey species was the much smaller wild boar (Boomgaard n.d.).

Many observers in India linked tiger predation on cattle to human induced scarcity of wild prey. In the Santhal inhabited areas of Chhotangapur, officials believed the tribal hunts had left vast stretches of sal forests with no deer or wild boar.[55] This factor should not be seen in isolation.[56] The decline of prey species may also have been due to more sportive hunting with modern weapons. Cattle may have been killed for opportunistic reasons as they were easily available in forest tracts at certain times of the year (Forsyth 1889: 95). Tigers in many areas had probably had a long history of preying on cattle which were slow moving in comparison with most wild prey species. But any decrease in deer or an increase of cattle numbers in the home range of a tiger could lead to more predation (Shahi 1978: 13–14). Many man-eaters shot by Corbett were probably sub-adult or mature individuals edged out by their rivals into marginal habitats (McDougall 1987: 435–448). The territorial behaviour of the species often led to the dispersal of some tigers, sparking off fresh conflicts in areas with a greater presence of people.

Bengal was the only province with a long history of predation on humans by tigers but even here there were specific regions where conflicts were very sharp. In 1901–5, over one-third of all deaths due to tigers in British India were in Bengal.[57] It was not so much a question of attacks in the villages or in cultivated fields as in the mangroves. The mangrove forests of the Sundarbans had a distinctive ecology. Unlike elsewhere in tiger country, chance encounters were more common, and the prey base was quite narrow. Attacks were likely to continue in Khulna as the Sundarbans were 'more and more opened up'.[58] Similarly, in the 24 Paraganas district, the large numbers of poor people who collected fuel and thatching reeds were especially vulnerable.[59] These two districts accounted for the major increase in the levels of tiger predation on people.

Such 'tiger-plagues' were highly specific in timing and occurrence even in the one region notorious for man-eaters.[60] The time for gathering forest produce coincided with the time of year when tigresses had cubs. Wood-cutters in the Sundarbans ventured out into the mangroves in large parties accompanied by a fakir, who propitiated wild animals with offerings to the presiding deity. The system was not foolproof. The fakir decamped if any of the party was carried off by a tiger (Hunter 1973: 36–37, 312). In 24 Paraganas, the wood-cutters worked only on specific sites sanctified by a fakir whose prayers were to scare away tigers. 'Gazi Sahib and his brother Kalu' were guardian deities 'venerated' by both Muslims and Hindus (O'Malley 1914: 74–75). These were not expressions of fatalism as much as responses to a large predator in a harsh environment.[61] Going out in large parties was a way of scaring away all but the most determined of tigers (O'Malley 1908: 20, 121). The paucity of wild prey species in comparison to other habitats and the vulnerability of gatherers of forest produce and fisherfolk may explain the frequency of fatal attacks by tigers in the Sundarbans (Rishi 1988: 9–14).

The Sundarbans were still exceptional. The reality was that conflicts of tigers with people were common in some areas but virtually unknown in others. It is even possible that changes in the colonial period contributed to increased man-eating in the short run. Tigers wounded by sportive hunters could turn to human prey as they were unable to chase wild animals. Prater estimated that as many as one in five tigers shot at in the late 1930s escaped with just a wound (Prater 1940). Given the sheer scale of hunting, such wounding of the quarry may have contributed to man-eating on a major scale. One estimate of the number of tigers shot as trophies between 1860 and 1960 is 20,000, and a fifth of that would mean 4,000 wounded tigers (Thapar 1994)! Also, the reduction of the prey base was partly related to sportive hunting. It was easy to blame the cultivator and the native shikari but many officials felt that 'the common ground' in the decline of game throughout India was 'the depredations of European sportsmen'.[62] Even a slight decline of the prey base would affect a large carnivore like the tiger.

The ecology and behaviour of the tiger gave it both strengths and weaknesses. Due to its diversity of habitats, the tiger could weather the onslaught much more successfully than the lion or the cheetah, which, at least in India, were primarily animals of open plains and scrub country. Nevertheless, the tiger was a large predator, weighing 200–230kg, and a stable population required adequate numbers of large prey. The bounty system had funded the killing of 16,573 tigers in the period 1879–88 alone. By computing the numbers killed between 1875 and 1925, the total comes to over 57,000 tigers. Yet this excludes 13 years for which the files are missing and does not account for animals killed for which no reward was paid. Even at a conservative estimate, the total computed by averaging the number killed in years before and after comes to over 80,000.[63] The numbers were notoriously unreliable but observations on man-eating tigers in Kumaon suggest (though not conclusively) that males were more prone to be aggressive.[64] Still, it is revealing that the bounty system in general was weighted against tigresses. Further, even if the number killed for rewards was assumed to be less than reported, there were several cases where kills were not reported or rewards refused for lack of sufficient evidence. The conclusion is inescapable: bounty killing was far more important in the decline of the feline than was realized by most observers.

Contrary to the picture suggested for Java, there was no inherent conflict between people and tigers across the South Asian sub-continent. Tigers may have survived partly because attitudes to them could range from unremitting hostility to tolerance. Some urban Indians already saw the elimination of wild animals as both necessary and desirable. In April 1880, the *Maharashtra Mitra*, a newspaper of Satara, asserted that tigers were 'creating havoc' in the district. The government was criticized for not allowing people to obtain modern firearms for self-defence. The newspaper went so far as to allege that it would be better if the British government were to kill its native subjects itself instead of allowing them to be devoured by animals.[65] The issue of marauding predators also provided a means to condemn the inequities of the Arms Act. The Indian language press expressed resentment at the disparity between the conquerors and the conquered. The rural poor lacked weapons to defend themselves from 'dacoits' or wild beasts'.[66] The Bengal landholder Govind Charan Das blamed the absence of guns in the village. This enabled big cats to hunt without any great danger or trouble to themselves.[67] The critics agreed with the objective of curbing tigers but wanted people to be equipped to accomplish the task.

But differences often went further, especially in the forest regions where tigers were seen by inhabitants as part of the landscape. Some groups had religious qualms about killing certain animals. The Baghel Rajputs claimed descent 'from a royal tiger' and protected them 'whenever' they could. They even refused to provide bait for white hunters (Forsyth 1889: 28). But religious beliefs could change over time. The Khonds of Ganjam dragged their feet when it came to setting traps for carnivores 'for fear of destroying their ancestors'. But they overcame their reticence when officials increased the amount of the bounty.[68] The Mikirs of Nowgong, Assam made no attempt to kill or net tigers, for fear of offending their deity.[69] There were differences too in the extent of opposition. After all, tigers and people often lived in close proximity without threatening each other. Sanderson knew of a male tiger which lived close to a village and only lifted cattle, never mauling a cow herd. This tiger, known as 'Don', knocked down and killed a man when a group of people surrounded him with nets, but he still 'lost nothing in public esteem'. His effigy was in the precincts of the Koombappa temple. When the Don was finally killed, a tracker recalled how he had 'never hurt any of us'. Even those who lived in the haunts of man-eaters refuted the idea that all tigers were dangerous (Sanderson 1878: 307, 312).

An affinity with the animal may also have stemmed from a realization of its role as a controller of deer and wild boars (Forsyth 1889: 259). The Nawab of Rampur in the North West Provinces objected to the elimination of tigers because they 'do no harm to crops and keep down the numerous animals that do'.[70] Pioneer cultivators in the forest 'trust to the tigers' to hold down the deer (Sanderson 1878: 237–238). True, the bulk of killing for rewards was done by Indians, but the impulse to exterminate the striped cat was not shared by all. Many hesitated. Others were indifferent and for some, strategies of avoidance seemed a better bet. In the net, these strategies may well have made the difference between extinction and survival. The absence of a uniformly negative image in cultures of the sub-continent stood the tiger in good stead. The war against the tiger had its moments of peace.

Wolf – the unseen predator

This was not so with a different animal with a wider range: the wolf. Only an excessive reliance on shikar literature and our own cultural disposition towards the big cats has obscured the significance of the wolf as a co-predator and victim of human intervention. Until the decline under human pressures over the last century, the grey wolf was the most widespread large carnivore in the world. The Indian sub-species of the wolf (*Canis lupus pallipes*) is much smaller than its North American counterpart and weighs about 20–22kg (Zoological Survey of India

1994: 45–49). Wolves preyed on smaller domestic animals than the tiger and they were pre-eminently not inhabitants of the forest. Their habitat is described vividly in hunting memoirs that provide details on the wolf. In the Bombay Presidency the maximum numbers of grey wolves were killed in Sind, Pune and Ahmadnagar. On the other hand, tigers were mainly reported from Kanara, Kolaba and Thana, all in regions with higher rainfall and dense tree cover.[71] This simple fact points to a very significant feature of wolf ecology in the sub-continent: it was 'extremely common' in areas with gazelle and blackbuck but 'very seldom' seen in forested patches (Forsyth 1889: 68–70; see also Pythian Adams 1949: 646–655).

Major Ray found wolves in large acacia (*babool*) plantations that were several hundred acres large. The country was 'absolutely flat' with only a few nullahs (dry stream beds) and a few stones here and there, crops down and 'not a bush or anything' to obstruct the view (Major Ray 1893: 145–148). Waddington rode down and speared wolves in flat, open country near the seashore (Waddington 1893: 554–555). A pair of wolves was found harrying railway labourers in the 'low jungles' outside Jabalpur. In Damoh, low-caste villagers helped a forester track down and shoot a female and a sub-adult male which were lifting children (Forsyth 1889: 68–70). These two kinds of terrain, undulating open lands and hill-dotted rocky jungle, were the main habitat of the wolf on the peninsula and in the west (Krishnan 1984).

The regional spread of the conflicts between people and wolves requires careful attention. Figures can convey only a partial view of the scale of conflict, but in 1875 more people were killed by wolves than by tigers (Fayrer 1878: 14). The North West Provinces (NWP) and Bihar were among the worst affected areas by wolf attacks. In 1876, in the former province, as many as 721 human lives were lost to wolves and in turn 2,825 wolves were slain for bounties.[72] The loss of lives to wolves was also heavy in Bihar. In 1876, all but a dozen of the 185 deaths were from Patna and Bhgalpur Divisions. In turn they accounted for over half of the wolves killed.[73] In northwestern, central and southern India, the main conflict with the species had not so much to do with attacks on children but depredations on livestock, especially sheep and goats.[74] Similarly, the animal was common in many parts of the Deccan, but here human casualties were rare or unknown.[75]

The extermination of wolves remained a priority in the NWP and Awadh (later the United Provinces) well into the 1920s. The Lt Governor singled out the elimination of the wolf as 'the single most important part' of measures for destroying wild animals.[76] Wolves in the NWP accounted for 'twenty to fifty times' the human lives lost to other beasts of prey. Their persistent efforts to carry off children were 'almost incredible ... If cubs are perseveringly destroyed, which they would be for good rewards, the race would be perceptibly diminished in a few years'. The bounty for a female wolf cub was as high as 12 annas as compared to only 8 annas for a male. The cubs could be located in their dens in the breeding season and smoked to death. The Commissioner of Jaunpur favoured even higher rewards of 5 rupees for an adult animal and 1 rupee for each cub. Higher bounties, with larger amounts for females and cubs, were one way to encourage the destruction of the carnivore. In Gorakhpur where fatalities were highest in summer 'when people are in the habit of sleeping in the open air', the reward for a wolf was therefore fixed at 4 rupees in contrast to only 3 rupees for the former.[77] In 1883, the Allababad district paid as much as 8 rupees for a female cub, twice that for an adult male.[78]

The question of whether to give more incentives to local shikaris or deploy special forces often arose with respect to wolf-killing measures. The cash prizes were considered insufficient in areas where attacks on people and stock were common. Nats, Kanjers and Musahirs, low-caste hunters who caught small mammals and reptiles, were paid a retainer in Fatehpur in the NWP to kill wolves. The scales could be tilted against the animal if it was possible to support anyone who made a business of wolf-catching'.[79] But such steps were often unsuccessful.[80] Groups like the Pardhis were still better equipped against a wary carnivore than was a soldier, as they had a deep knowledge of its habits and the local terrain (Hurst 1931: 51). They often hunted wild boars and wolves and had an added incentive due to the rewards (Hobbart 1932: 22–28).

The fact that wolves could take shelter in fields is itself testimony to their adaptability and versatility. Even a large male was a only little over 2 feet, 2 inches tall. This would explain its continued survival in the Indo-Gangetic plains well into the twentieth century (Prater 1965: 121, 125–126; Burton 1915: 16). When the NWP government sent Gorkhas to kill the packs, they were asked to hunt 'when the crops are off the ground' in order to obtain good results.[81] Until the harvest, wolves could take cover in the standing crop. In Badaun, Gorkha troops were unsuccessful, as they went out after the animals in the first three months of the year. Local shikaris did not bother to try shooting wolves until the end of the harvest and were more successful.[82] The breeding cycle of the species coincided with the time when the winter crop was still uncut; the wolf mates after the end of the monsoon and cubs are born around December (Prater 1965: 121, 125–126). Recent research suggests that single-crop cultivation of millets such as *bajra* or *jowar* provides ideal cover for cubbing (Jhala and Giles 1991: 476–483). Cubs could also be born in scrub jungle that adjoined cultivation.

There were doubts over the need to kill off wolves, but the arguments were utilitarian rather than aesthetic or ethical. The District Magistrate of Hamirpur was sceptical of the claim that wolves were killing people on a large scale. In all, seventeen lives had been lost in less than three years but he felt that several deaths were due to hyenas (*Hyena hyena*). Over 450 wolves had been killed for a sum of over 700 rupees disbursed as rewards. But this idea was rejected by his superiors who shared the commonly held view that hyenas were eaters of carrion and would 'never' attack human beings or children. There was also a realization that particular individual wolves may have 'tasted human flesh' and become dangerous to people. In Rae Bareli, special parties of hunters were deployed to eliminate the marauding pack.[83] Attempts to blame hyenas or stigmatize particular individual wolves made little headway.

Programmes to wipe out the species had to contend with unforseen obstacles. Local shikars in Berar obtained licenses on condition that they would shoot the carnivores, but in fact they rarely used the guns for this purpose.[84] In several provinces, the carcasses of wolves that had been paid for were dug up and turned in again for a fresh bounty. In Gonda, the wolf-killers discovered where the ears were buried and sewed them back neatly. They managed to get a second reward for the same beast.[85] Often, the golden jackal (*Canis aureus*) was killed and produced to claim the bounty on the grey wolf.[86] Officials poorly versed in natural history were hard put to tell the difference. The young Phillip Mason in Saharanpur in Uttar Pradesh certified that thirteen animals produced before him by Kanjers were wolves and had the tail and ears cut off and burnt to prevent fraud. Only later did he realize that the Kanjers had sun-dried the carcasses of wolves and inserted freshly killed jackals inside the rib cages, 'as the chef of a Victorian duke would stuff a quail inside an ortola – and sewn on ears and tails manufactured from hessian and smeared fresh blood' (Manson 1978: 72). Such evasion frustrated official campaigns.

It is not easy to assess the impact of the slaughter on the wolf as a species. Over 100,000 were killed for rewards in British India between 1871 and 1916. The records for as many as 19 years (1877–79, 1889–95 and 1905–15) are incomplete and this may be a gross underestimate.[87] As several of the so-called wolves killed were actually jackals, the figures are less useful as an index of status than they could have been. But there is little doubt of a decline in wolf numbers in some areas.[88] In areas around Delhi, where as many as 300 wolves were killed for rewards in the period 1878–83, only three bounties were claimed in the year 1913.[89]

There were factors working in favour of the wolf that were absent in the case of the tiger. Except in the case of the Tibetan sub-species, their pelts were not valuable.[90] The smaller body size of the Indian wolf also enabled it to survive in areas with a high degree of human activity, often in low jungle and scrub on the outskirts of villages. This did not confer on them any kind of immunity. A shift to double cropping with irrigation could probably reduce breeding sites. Above all, wolves retained a negative image in the eyes of most herders, who would not hesitate to kill them. Evidence of private vendetta is limited, but Le Mesurier referred to the sheep owners of Chunar paying a rupee for a she-wolf. This was in addition to the reward of 6 rupees given by the authorities (Le Mesurier 1865: 185).

Despite this, there were substantial areas where the species could survive. In much of western India, the blackbuck still survived in large numbers well into the 1930s (Jhala and Giles 1991). The clearance of forests for cultivation has often been cited as a major threat to the survival of the tiger, but the wolf weighing a bare 20kg could eke out an existence where the 230kg big cat could not possibly survive. The former was 'a difficult animal to exterminate'.[91] It could become increasingly nocturnal in order to avoid detection (Dunbar Brander 1931: 27) and its absence from timber forests was possibly a blessing in disguise, for the

Forest Department's writ in such areas included intensive programmes to kill off carnivores.

But the pressure on the wolf could also have been more intense on account of cultural biases and prejudices against it. The 'vigorous campaign' against the wolf was stepped up at precisely the same time as the tiger began to win a reprieve.[92] S. H. Butler, who favoured abolition of a general reward on the tiger, lobbied for a higher bounty for a grey wolf.[93] If shepherds were not positively disposed to the species, officials and princes were no great fans either. In the Punjab hills, princes gave a bounty of 3 rupees for a wolf, using not only villagers but also armed Gorkhas to hunt them down. The objective was to build up stocks of game birds for sportive hunting (Ellison 1928: 124). The value of a wolf skin may have been low, but conversely, there was no great sporting merit in maintaining its population. The wolf was seen not only as a killer of goats and sheep but often as a marauder of children. Its survival was a tribute to its own capacity to be an unseen predator.

Dissenters and defenders

The revulsion against the slaughter of predators was more marked by the 1920s and 1930s when the tiger and leopard found articulate defenders. The early revulsion was often to do with the killing of cubs and pregnant females. Keeping wild animals as pets did not make their owners admirers of them in the wild, but this practice did represent a stark contrast with the general policy of slaying cubs. In the 1880s, a military officer kept a pet leopard in the Bundelkhand cantonment. He would turn it loose in his garden where he enjoyed watching it 'stalking ponies in the long grass' (Birdwood 1941: 51). There were many other expressions of concern for slain predators. Lady Curzon recounted how a pregnant tigress 'was skinned and cut up. They found three little ones inside her nearly ready to be born. Isn't it sad?' (Bradley 1985: 76). Lord Reading shot a maned lion in Gir Forest, Kathiawar but was moved by the sight of a lioness and her six cubs playing around the base of a tree, all the more so because '[t]he cubs played like kittens' (Butler 1969: 151). Big cats were seen in a more favourable light than reptiles. 'Such things', wrote Vernede, 'are repulsive to man. The cat tribe is not. Between us there is not the same cold-blooded element' (Vernede 1911: 85).

The idea that watching animals in the wild was better than shooting them on sight was gathering ground. Jim Corbett, who would later win renown for his accounts of man-eaters in Kumaon, expressed his admiration for the tiger. He had sat up 'with the most modern rifle across my knee' as he watched 'a tigress and two fully grown cubs eat up the sambhar stag they killed and counted myself no poorer for having taken no trophy' (Corbett 1931: 61–62). The first attempts to rein in tiger shooting had admittedly been 'unpopular' but there was really no alternative (Champion 1927: 45). Part of the new tolerance was also a response to the advances in technology that made hunting down the tiger much easier than in the past. The keen sportive hunter was now assisted by 'good motor roads and even telephones in places' (Champion 1933: 45). The revised game rules began to take note of the scarcity of tigers, though this was only true at a regional level. In the United Provinces, a favourite hunting ground of the official elite and a major battleground in the struggle for the survival of the species, limits on the bags came into force in 1921 (Stracey 1963: 103). In Mysore, the 'fear of total extinction' led to controls on shooting (Hayavadana Rao 1984: 78). There were glimmers of a more nuanced understanding of the relationship between people and predators. Champion blamed the elimination of tigers for the rapid growth of populations of wild boar and deer in the forests and on the borders of cultivated lands. He rebutted the idea that tigers were cruel beasts and instead upheld them as 'merciful' in comparison to 'man, the avowed hater of cruelty' (Champion 1933: 36).

The transformation of the deadly hunter of humans into an animal citizen endowed with the quality of mercy was not going to be easy. Corbett's own writings were to help reinforce the image of marauding man-eaters that could only be curbed by a hunter with superhuman qualities (Sankhala 1978: 138). It is difficult to see how readers would remain unmoved by the descriptions of the cunning of errant and devious predators on the rampage in the villages of India (Mukherjee 1993: 212). There was, of course, no shortage of opponents of empire, and Mason explicitly compared some of them to tigers. Officials in the 1920s and 1930s hoped that they would not have to do to Pathan tribes of the frontier what they did to disobedient tigers. 'They (the tigers) could kill what deer they liked in a park', wrote Mason, 'but they risked a bullet if they came outside and took village cattle' (Mason 1974: 292). A tiger that knew its place was all right. This was a shift from seeing all tigers as potential man-killers. The view also encapsulates in kernel a major change that had come about in the previous 50 years. Instead of being presumed guilty, it was now possible for tigers to be on parole. A victor could afford to be magnanimous. The domain of mercy was not very far removed from the world of blood and gore.

Conclusion

The programmes to control vermin provide insights into the wider process of imperial intervention in rural India. The Indian legacy of dealing with carnivorous animals was rich and diverse. In contrast to Britain and Western Europe, there was no long history of state-sponsored projects to eliminate carnivores. Rulers did kill man-eaters in order to help in the process of agrarian expansion, but a degree of tolerance was perhaps easier in the context of relatively dispersed rural settlements with a large proportion of the land area under forest cover. There is evidence of recourse to religious and magical methods to ward off tigers. But there were a variety of other means of self-defence, ranging from avoidance to the simple elimination of marauding animals.

British attitudes were not static. Initially, the extermination of these creatures was left to local authorities. Curbing the tiger seemed the logical corollary to the campaign against human outlaws and to be essential for agricultural extension. But it was more than just that. The 'improvement' of India under the Raj could only proceed if such obstacles were removed. In the aftermath of the Rebellion of 1857, Indians were not to be trusted with modern firearms. Some officials felt that only a white man would be resourceful enough to oversee the hunting of predators. 'Martial races' like the Rajputs and Gorkhas were seen as capable of standing up to dangerous beasts, just as they were seen as capable in war. Village headmen and cultivators were sometimes even issued guns for self-protection, but the main thrust of policy was to get villagers to eliminate carnivores without using modern firearms, and the general policy of denying them such weapons was retained intact, except in special cases.

The bounties on female tigers and cubs aimed at more than control: the 'elimination' of species was seen as an index of success. Princely states varied in their responses to the programme, but many followed suit. Cattle-lifting was cited as a brake on the expansion of agriculture, as cattle were the prime source of draught power. Tigers had to be sacrificed in the interest of land revenue. However, there was no unanimity on how much damage predators actually caused to the economy. A few officials argued that the control of herbivores by leopards and tigers actually assisted cultivators. It was more than a simple case of identifying barriers to the expansion of the plough. The grey wolf rarely threatened cattle, but often preyed on sheep and goats. In this case, the conflict was with animal rearing rather than cultivation. Sportsmen saw carnivores as competitors for game. Elite hunters carried the war against vermin into the heart of the forest.

There was a strong cultural bias against an animal like the tiger with its alleged liking for human flesh. But there is little evidence that tigers in all areas habitually consumed people. The Sundarbans in Bengal were exceptional in having a very high degree of man-eating by tigers throughout our period and beyond. Elsewhere, man-eating was situation-specific and not a general phenomenon. The factors that led to man-eating included a paucity of natural prey, opportunities for attacks on cattle, wounds inflicted by sport hunters and the dispersal of younger tigers into areas with greater human use. There is little evidence of an irreconcilable conflict of existence between people and tigers. There were also clear inter-regional disparities with the wolf. After all, tigers and wolves had shared their habitats with people for centuries.

Bounty hunting contributed to the decline of predators, but its impact was uneven, depending on the region and the species. Its impact on the tiger cannot be seen in isolation from hunting for sport. Sportsmen often saw elimination as their great 'achievement'. Far from simply being an adjunct to expanding cultivation, vermin-killing carried the battle well beyond the borders of arable land. Any tiger anywhere was better dead than alive. For a time, no place in British India was a sanctuary for such vermin. Preliminary figures indicate that far more tigers were killed for reward than were shot for sport, but this does not minimize the latter as the two were inseparable for a long time. The extension of agriculture, as in the Indus basin, also exerted a powerful influence on the fate of the species. Given its large body size and its need for forest cover, even an animal as resilient as the tiger found it difficult to survive. Unlike the tiger, the wolf could survive in scrub jungle and switch to living off smaller prey. The carnivore managed to survive in many areas by remaining elusive.

The strategies of avoidance and self-defence at the local level meant that people often stopped short of a desire to wipe out the tiger. Religious and cultural objections put a brake on vermin-killing but were perhaps specific to certain regions and cultures. What was more widespread was a willingness to coexist with the animal. It would be misleading to romanticize such situations especially when many such people turned up to claim bounties. The extent to which specific kinds of land-use enabled or hindered coexistence is worth detailed study at the micro-level. Tolerance and 'good will' were not universal. Middle-class Indians opposed to the Arms Act cited the need for self-defence against wild animals. Often the terms of discord in village and forest hamlet were over how much should be paid out as reward. Specialists in wolf-killing and snake-catching took advantage of the opportunity to earn extra cash, and this brought them under government control to a greater extent than in the past. The growth of market linkages and the quickening pulse of economic interaction brought new pressures. Despite such extensive changes, there is enough evidence to suggest that there were alternatives to extermination. Certain forms of land-use could coexist with wild animals. The animosity of the Raj to the feline was not shared by all. Many zamindars and some princes did much less than the British to wipe out dangerous beasts. Clearly, a romanticized view, however attractive, would be as misleading as a picture of all-out conflict.

The shift in attitudes by the 1920s was in part a consequence of the success of vermin extermination operations. Though the number of lives lost and beasts killed for rewards was still high in the mid-1920s, a decision was taken to stop collating any information at the all-India level.[94] Provinces were free to pursue their own line of action. If the tiger found breathing space, the wolf did not. The former won admirers who saw it as an embodiment of gentlemanly virtue. Others acknowledged the positive role of predators in controlling herbivores in forests and on the fringes of the cultivated arable. But such views were still held by a minority and programmes to wipe out vermin continued for much longer (Pythian Adams 1949; Stracey 1963: 168, 181). Still, the era of branding a species as vermin was drawing to a close. It was the tragedy of the tiger that it was winning small battles but losing the war.

Acknowledgements

This chapter is part of a larger research project undertaken by the author at the Nehru Memorial Museum and Library, New Delhi. The writer is also grateful to the editors of the journal *Studies in History* for publishing a longer and more extended version of this text (Vol. 14, 1998). Dr John Knight, Prof. Peter Boomgaard, Prof. R. Kumar, Dr Ravi Chellam, Prof. M. Gadgil and Dr Ram Guha helped enormously with their suggestions. The usual disclaimers apply.

Notes

1 The' wolf is less prominent in lore than either the lion or the tiger. See Gandhi (1992: 504–505, 389–390, 506–507).
2 The lion motif was not an Islamic import as suggested by Bayly (1991: 156).
3 For a later period, see Bhattacharya (1947).
4 National Archives of India, New Delhi, India (hereafter NAI), Home (Public) (hereafter H (P), Dec. 1890. A, nos. 360–407, 'Results of the measures adopted for exterminating wild animals and poisonous snakes in British India during the year 1889', no. 363, pp. 37–39: J Grose, Board of Revenue, Madras to Sec., Home Dep, GOI, 8 May 1890 (hereafter Vermin 1890).
5 NAI, Foreign (Political), 19 Sep. 1845, no. 72: 'Reward for killing tigers in Indore', Sec., govt. NWP to Assistant to Gov., Nimar, 29 July 1845; Sec. Govt. Bombay to Foreign Dep., 19 Sep. 1845.
6 NAI, H(P), 16 May 1864, B, nos. 86–87: 'Orders regarding arms to be retained by villages suffering from the ravages of wild beasts', C. Wood, India Office, to Gov. Gen. (G. G.), 26 Mar. 1864.
7 Ibid., Home Office, GOI to NWPs and Presidency govts., 16 May 1864.
8 NAI, H(P), July 1875, nos, 151–152, 'Proposal of Major B Rogers regarding the organization of a system of spring guns for the destruction of wild animals in India', Aug. 1869; also H(P) Sep. 1871, A. no. p. 6: Capn. Rogers, Aug. 1869.
9 NAI, H(P), Jan. 1875, A nos. 286–311, no. 297, no pagination, 'Destruction of noxious beasts in India', Major Tweedie, First Assistant Resident, Berar to Sec., GOI, Home, 5 June 1874 (hereafter Vermin 1875).
10 NAI, H(P), June 1876, A, 127, 'Use of aconite for destroying tigers', A. C. Lyall, Note, 26 Oct. 1876.
11 NAI, H(P), Oct. 1873, nos. 287–289, pp. 2457–2458, 'Measures taken for the destruction of wild animals', A. C. Lyall, Home Dep. Resolution, 25 Aug. 1873.
12 NAI, H(P), Vermin 1875, no. 307, R. A. Dalyell, Offg. Chief Sec., Madras, 4 Aug. 1873.
13 Ibid., Lt. Col. W. S. Drewer, Acting Inspector General of Police to Chief Sec., Madras Govt. 17 Nov. 1873.
14 NAI, Vermin 1890, no. 365, p. 46: C. A. Galton, Sec. Board of Rev., 12 June 1890.
15 Vermin 1875, no. 307, W. S. Whiteside, Mag., North Arcot, to D.F. Carmichael, Sec., Bd. of Rev., 17 July 1874.
16 Ibid., note by A. C. Lyall, pp. 1–2, 26 Oct. 1874.
17 NAI, H(P), Dec. 1882, A, nos. 32–70, 'Results of the measures adopted for exterminating dangerous beasts and poisonous snakes in British India in 1881', no. 38, p. 13: C. A. Galton, Board of Rev., 4 Mar. 1882, Extract from Procs., Board of Rev., (hereafter Vermin 1882).
18 NAI, H(P), Dec. 1882, A, nos. 32–70, no. 68, p. 82: G.G. to A. MacKenzie, 8 Nov. 1882.
19 Ibid., no. 38, p. 4, note, JM, 19 Oct. 1882.
20 Ibid., no. 54, p. 47: Resn., NWP and Oudh, 2 June, 1882.
21 Ibid., no. 292, F. C. Chapman, Chief Sec., Bombay Pres. to Sec., GOI, 2 Oct. 1873, no pagination.

22 NAI, Vermin 1875, no. 301,Cap. T. Cadell, Pol. Agent, Alwar to Agent to G. G., Rajputna, 30 Dec. 1873.
23 Ibid., Col. C. W. Burton, Asst. Agent, Soojangarh to Agent, Gov. Gen., Raj., 26 Nov. 1873.
24 Ibid., Cap. W. A. Roberts, Pol. Agent, Eastern States to Agent, G. G., Raj., 26 No. 1873.
25 Ibid., Lt. Col. W. Carnell, Supt., Sirohi to Agent, G. G., Raj., 27 Nov. 1873.
26 Ibid., JC Berkeley, Pol. Agent, Harrowtee and Tonk to Agent, G. G., 30 Dec. 1873, quoting the Nawab's letter to the former dated 1 Nov. 1873.
27 Ibid., KM, Walker, Offg. Pol. Agent Marwar and Jaisalmer to Agent, G. G., Raj., 20 Oct. 1873.
28 Ibid., no. 301, Cap. T. Cadell, Pol. Agent, Alwar to Agent to G. G., Rajputana, 30 Dec. 1873. Quoting a letter from Raja Sheodan Singh to the former, 4 Oct. 1873. Emphasis added.
29 NAI, Vermin 1877, no. 289: A. C. Lyall, Offg. Chief comm., Ajmer-Mhairwara to Sec., Foreign Dep., 2 May 1877.
30 NAI, Vermin, 1875, no. 303: Cap. G. H. Trevor, Offg. Ist Asst. Resident, Hyd. to Sec, Foreign Dep., 10 Mar. 1874; H(P), Dec. 1877, A, nos. 269–292, no. 284, p. 1829, Mjr. T. G. Clarke, Sec., Chief Comm., Mysore to Sec., Home Dep, GOI, 16 Nov. 1877 (Hereafter Vermin 1877).
31 NAI, Vermin 1877, no. 301: L. Pelly, Pol. Agent, G. G., Raj. To C. I. Aitchison, Sec., Foreign Dep., 21 Feb. 1874.
32 Ibid., no. 301, Cap. J. C. Berkeley, Pol. Agent, Harrowtee and Tonk to Agent, G. G., 30 Dec. 1873. Quoted letter of the Raj Rana, Jhallawar, 10 Dec. 1873; Berkeley, letter again, 17 Jan. 1874.
33 Ibid., Berkeley, 10 Apr. 1874. The scheme was endorsed by the Foreign Dep. on 11 May 1874.
34 NAI, H(P), Mar, 1908, A, 27–53, no. 29, p. 228, 'Game law for India,' Annexure II, Bengal, CR Maridin, Comm., Rajshahi Div. to Sec., Bengal, Rev., 9 Jan 1905 (hereafter Game 1908).
35 NAI, Vermin 1875, no. 307, W. S. Whiteside, Magistrate, North Arcot to D. F. Carmichael, 17 July 1874.
36 NAI, Game 1908, no. 30, p. 374: Captain A. E. Wardrop, Commanding Officer, European Battery, Royal Horse Artillery, Meerut, 23 July 1904.
37 NAI, H (P), Aug. 1904, A, 266–78, 'Preservation and protection of game and fish', p. 45, Note by Curzon, Keep-With section, no date (hereafter Game 1904).
38 NAI, Game 1904, no. 266, Notes, Letters from local Govts., Curzon, 1 Aug. 1902.
39 NAI, Vermin 1890: C. J. Lyall, Note, 19 Oct. 1890, p. 27.
40 NAI, Game 1908, no. 30, p. 329: G. Bower, Collr., Saharanpur to Chief Sec., UP, 21 July 1904.
41 Ibid., no. 30, p. 155: Conservator, Central Div., Bombay Pres., to Comm., 27 July 1904.
42 NAI, H(P), Jan. 1875, nos. 286–311, no. 304, no pagination: Capt. J. C. Berkeley, Pol. Agent, Harrawtee and Tonk to Gov. Gen, 10 Apr. 1874.
43 NAI, Game 1908, no. 30, p. 329: G. Bower, Collr., Saharanpur to Chief Sec., UP, 21 July 1904.
44 NAI, Game 1911, no. 197, no. 106: Lt. Gen. W. Osborn to Asst. Comm., Kulu 20 Oct 1908.
45 NAI, Vermin 1890, no. 379, p. 81: H. Maude, Offg. Jr. Sec., Punjab to Sec., Home Dep., GOI, 28 Apr. 1890.
46 NAI, Game, 1908, 30: W. H. L. Impey to Sec., Home Dep., 2 Feb. 1905.
47 NAI, Game, 1904, no. 269, J. O. Miller, Chief Sec., NWP and Oudh, 28 Mar. 1895. Misc. Forest Dep., 15 Mar. 1895.

48 NAI, Dep. of Revenue and Agriculture (Forests) Dec. 1909. A, nos. 1–2, 'Proposed Rules to regulate hunting, shooting and fishing for Naini Tal and municipal forests which it has decided to make Protected Forests'.

49 NAI, Vermin 1890, p. 13: Home Dep. Res., Dec. 1890.

50 NAI, H(P), 1926, File 126, no. 16, Board of Rev., Madras, 25 Mar. 1926.

51 NAI, Vermin 1890, no. 363, p. 39: J. Grose, Board of Rev., 8 May 1890.

52 NAI, H(P), Dec. 1885, A nos. 69–101, no. 69. Board of Rev. Procs., Madras. 19 June 1885

53 NAI, Game 1904, no. 266. 'Letters from Local Governments', John Hewett, Notes, 19 Aug. 1902.

54 NAI, Game 1904, no. 277, p. 281: H. H. Risley, Sec., Home Dept. to Local Govts., 23 May 1904.

55 Game 1904, no. 29. P. 257: J. Taylor, Sett. Officer, Burdwan to Chief Sec., Bengal, 20 Sep. 1904

56 NAI, Game 1904, no. 269, Notifications by the Bengal Govt., 'District Officers' power in the Protected Forests of Santhal Parganas, Chotangpur, Khurda and Rohtas', 24 Sep. and 16 Dec. 1895.

57 NAI, Vermin 1905, no. 269: Home Dep. Resn., 23 Sep. 1905 (hereafter Vermin 1905).

58 NAI, Vermin 1905, no. 67: H. W. C. Carduff, Offg. Sec., Bengal to Sec., Home Dep., GOI, 3 May 1904.

59 NAI, H(P), Sep. 1902, A, nos, 281–88, 'Measures adopted with a view to destroying wild animals and poisonous snakes in India during the year 1901', no. 285, p. 23: J. A. Bourdilon, Chief Sec., Bengal to Sec., Home Dep, 21 Apr. 1902.

60 The phrase is from Boomgaard (2001).

61 This view is very different from Boomgard (2001).

62 NAI, Game 1908, no. 30, p. 22: W. S. Morris to Chief Sec., UP, 2 Aug. 1905.

63 Vermin 1890, 'Keep-with Notes': JP Hutchins, 30 Sep. 1890. The totals are computed on the basis of the Home (Public) documents for succeeding years 1873–1926. The missing years are 1891–5, 1905–11 and 1917–18.

64 Of the 11 man-eaters taken by Corbett, six were male (Sukumar 1991).

65 NAI, H (P), May 1880, B, Nos. 67–68. P. 4, 'Allegation in Maharashtra Mitra about the inability of residents of Valva taluka, Satara district to defend themselves from attacks by tigers for lack of the arms,' G. W. Kulkarni, Reporter, Native Press, 20 Apr. 1880.

66 NAI, H(P), June 1881, A, nos. 10–13, 'Amendment of Arms Act (IX of 1878)', Note by the Sec., Viceroy, 14 Jan. 1881.

67 NAI, Game 1908, no. 29, p. 288: Govind Chandra Das, Sec., Eastern Bengal Land-Holders Association, to Sec., Rev., Bengal, 24 Sep. 1904.

68 NAI, H(P), Dec. 1885, A, nos., 69–101, 'Results of measures adopted for exterminating wild animals and poisonous snakes in British India during the year 1884', no. 69, Procs., Board of Rev., Madras, 19 June 1885 (hereafter Vermin 1885).

69 NAI, H(P), Dec. 1884, A, nos., 109–140, 'Results of measures adopted for exterminating wild animals and poisonous snakes in British India during the year 1883', no. 130, p. 61: Sec., Chief Comm., Assam to Sec., GOI, Home Dep., 16 May 1884.

70 NAI, Vermin 1875, no. 302, no pagination, C. A. Ellison, Sec., NWP to A. C. Lyall, 20 Mar. 1874.

71 NAI, H(P), Sep. 1915, nos. 100–116, 'Results of measures adopted for exterminating wild animals and poisonous snakes in British India during the year 1914', no. p. 18: J. L. Rieu, Bombay Govt., 14 June 1915.

72 NAI, H(P), May 1877, A, nos. 60–85, 'Results of measures adopted for exterminating wild animals poisonous snakes in British India during the year 1876', no. 66: B. W. Colvin, Offg. Sec., NWP, 29 May 1876 (hereafter, Vermin 1877).

73 NAI, Vermin 1877, no. 274, S. C. Bayley, Sec., Bengal, 6 Aug. 1877.

74 Vermin 1877, no. 69, L. Griffin, Offg. Sec., Govt. of Punjab and its dependencies to A. P. Howell, Sec., GOI, Home, 12 May 1876. 4 people and 431 head of stock were killed by wolves that year.

75 NAI, Vermin 1877, no. 274, S. C. Bayley, Sec., Bengal, 6 Aug. 1877.

76 NAI, H(P), Dec. 1882, A, 32–70, 'Results of measures adopted for exterminating wild animals and poisonous snakes in British India during the year 1881', no. 54, p. 46: Resn., Gen. Dep, NWP and Oudh, 2 June 1882 (hereafter Vermin 1882).

77 NAI, Vermin 1875, no. 302, C. A. Ellison, Sec., NWP to A. C. Lyall, 20 Mar. 1874.

78 NAI, H(P), Dec. 1883, A, 29–64, 'Results of measures adopted for exterminating wild animals and poisonous snakes in British India during the year 1882', no. 56, p. 52: Resn., Gen. Dep., NWP, 1 Aug. 1883 (hereafter Vermin 1883).

79 NAI, Vermin 1882, 54. p. 47: Resn., Gen. Dep., NWP and Oudh, 2 June 1882.

80 NAI, Vermin 1883, no. 56, p. 52: Resn., Gen. Dep., NWP, 1 Aug. 1883.

81 NAI, H(P), Sep, 1902, A, nos. 281–92, 'Results of measures adopted for exterminating wild animals and poisonous snakes in British India during the year 1901', no. 287, pp. 32–33: WHL Impey, Chief Sec., UP, Resn., Gen Admin., 20 May 1901.

82 NAI, Vermin 1903, no. 241, p. 29: Resn., Gen. Admin., UP, 20 May 1902.

83 NAI, H(P), Dec. 1899, nos. 272–85, 'Results of measures adopted for exterminating wild animals and poisonous snakes in British India during the year 1898', no. 276, p. 37: Chief Sec., NWP to Sec., GOI, Home, 19 May 1899.

84 NAI, H(P), Nov. 1893, A, nos. 1–43, 'Results of measures adopted for exterminating wild animals and poisonous snakes in British India during the year 1892', no. 11, p. 26: K. J. L. Mackenzie, Comm., Hyderabad Assigned Districts to Sec., Berar, 11 Mar. 1893.

85 NAI, Vermin 1885, nos. 176–77, J. W. Reid, Sec., Govt., NWPs 18 Aug. 1885.

86 NAI, H(P), Oct. 1891, nos. 316–53, 'Results of measures adopted for exterminating wild animals and poisonous snakes in British India during the year 1890', no. 350, p. 93: Chief Sec., Bengal to Sec., GOI, Home, 29 July, 1891.

87 These are from the totals given in the Home (Public) records.

88 Vermin 1914, no. 7, pp. 51–52, 57: Procs., Lt. Gov., Punjab, 4 May 1914.

89 Ibid: W. M. Hailey, Chief Comm., Delhi, 18 Mar. 1914.

90 E. D. (1893).

91 NAI, H(P), Nov. 1914, A, nos. 1–19 no. 10, p. 75: D. C. Jabalpur, quoted in the CPs Misc. Dep. Resolution, 14 May 1914 (hereafter Vermin 1914).

92 NAI, H(P), Sep. 1905, A, nos. 255–70, 'Results of measures adopted for exterminating wild animals and poisonous sankes in British India during the year 1904', no. 241, pp. 29: EPL Winter, Chief Sec., Gen., Admin., Resn., UP, 19 May 1905.

93 NAI, H(P), Mar. 1907, A nos. 122–131, 'Results of measures adopted for exterminating wild animals and poisonous snakes in British India during the year 1906', no. 126, p. 21: S. H. Butler, Sec., UP, 1 April 1905.

94 NAI, H(P), 1926, File 126.

References

Abu al Fazl (1972) *The Akbar Namah*, Vol. II. H. Blochman (tr. and ed.). Calcutta: Asiatic Soc. Bengal (reprint, Delhi.).

Anon. (1891) 'The proposed introduction of game into the neighbourhood of Bombay'. *JBNHS*, vol. 6, pp. 119–123.

Barrow, H. W. (1895) 'Crocodiles in artificial reservoirs'. *JBNHS*, vol. 10.

Bayly, C. A. (1991) *An Illustrated History of Modern India, 1600–1947*. Bombay: OUP.

Bhattacharya, A. (1947) 'The tiger cult and its literature in Lower Bengal', *Man in India*, vol. 27, pp. 44–56.

Birdwood, Field Marshall (1941) *Khaki and Gown, An Autobiography*. London: Ward, Lock and Co.

Boomgaard, P. (n.d.) 'Man-eating: the deadly encounters between people and tigers/ leopards in Indonesia, 1650–1950'. Paper presented at the Conference on the Environment in South East Asia, School of Oriental and African Studies, London, 28–30 March 1994.

—— (2001) *Frontiers of Fear: Tigers and People in the Malay World, 1600–1950*. New Haven: Yale University Press.

Bradley, J. ed. (1985) *Lady Curzon's India, Letters of a Vicerine*. New York: Beaufort Books.

Braudel, F. (1981) *Civilization and Capitalism, 15th to 18th Centuries: The Structures of Everyday Life*, vol. I. London: Collins, Fontana.

Buchanan, F. (1986) *An Account of the District of Shahabad in 1812–13*. Reprint, Delhi: Usha Jain.

Buchanan, H. (1807) *A Journey from Madras through the Countries of Canara, Mysore and Malabar*. London: W. Bulmer and Co.

Butler, I. (1969) *The Viceroy's Wife, Letters of Alice, Countess of Reading, from India, 1921–25*. London: Hodder and Stoughton.

Burton, R. W. (1915) 'Weights and measurements of game animals'. *JBNHS*, vol. 24.

Campbell, J. (1880) *Gazetteer of the Bombay Presidency, vol. XII, Khandesh*. Bombay: Govt. Central Press.

—— (1984 [1882]) *Gazetteer of the Bombay Presidency, vol. XII, Part ii*. Bombay: Thana (reprint, Pune, Government Press).

Champion, F. W. (1927) *With a Camera in Tigerland*. London: Chatto and Windus.

—— (1933) *The Jungle in Sunlight and Shadow*. London: Chatto and Windus.

Corbett, J. (1931) 'The Pipal Pani tiger'. *Hog Hunters Annual*, no. 4.

—— (1995 [1944]) *The Man-eaters of Kumaon*. Delhi: OUP, reprint.

Courtenay, N. (1980) *Tiger, Symbol of Courage*. London: Quartet Books.

Davidar, E. R. C. (1976) 'Wildlife conservation in Tamil Nadu'. *Journal of the Bombay Natural History Society*, vol. 83, pp. 67–71.

Dunbar Brander, A. A. (1931 [1923]) *The Wild Animals of Central India*. London: Edward Arnold.

E.D. (pseudonym) (1893) 'Tiger skins'. *Indian Forester*, vol. 19, p. 69.

Eaton, R. (1994) *The Rise of Islam and the Bengal Frontier, 1204–1760*. Delhi: OUP.

Elliot, Gen. J. G. (1973) *Field Sports in India, 1800–1947*. London: Gentry Books.

Elliot, H. M. and J. Dowson (1979) *The History of India as Told by Its Own Historians*. Allahabad: Kitab Mahal.

Ellison, B. C. (1928) 'Game preservation and game experiments in India'. *JBNHS*, vol. 33.

Fayrer, J. (1878) *Destruction of Life by Wild Animals and Venomous Snakes in India*. London: Royal Society of Arts.

Fleetwood Wilson, Sir G. (1921) *Letters to Nobody, 1908–12*. London: John Murray.

Forbes, J. (1813) *Oriental Memoirs*. London: William, Constance and Co.

Forsyth, J. (1889) *The Highlands of Central India: Notes on their Forests, and Wild Tribes, Natural History and Sports*. London: Chapman and Hall.

Gandhi, M. (1992) *The Penguin Dictionary of Hindu Names*. Delhi:Viking Penguin.

Habib, I. and T. Raychaudhuri (eds) (1982) *The Cambridge Economic History of India, Vol. I. c. 1200–1750*. Delhi: Cambridge University Press.

Hayavadana Rao, C. (1984 [1927]) *Mysore Gazetteer, Vol. I, Descriptive*. Mysore, B. R. Pub. Co., reprint.

Heber, R. (1993 [1827]) *Narrative of a Journey through the Upper Provinces of India from Calcutta to Bombay, 1824–25*. Delhi: Low Price Publications, reprint.

Hobbart, C. (1932) 'Pig-rearing Pasis and kindred clans'. *Hog Hunters Annual*, no. 5.

Hodgson, H. (1893) 'Preservation of harmless wild animals in Malcompeth (Mahabaleshwar)'. *JBNHS*, vol. 7, pp. 530–53.

Hunter, W. W. (1973 [1875]) *A Statistical Account of Bengal, Vol. I, 24: Paragans and Sundarbans*. Delhi: DK Publishers.

Hurst, G. (1931) 'Pardhis', *Hog Hunters Annual*, no. 4.

Jhala, Y. V. and R. Giles (1991) 'The status and conservation of the wolf in Gujarat and Rajasthan, India'. *Conservation Biology*, vol. 5, pp. 476–483.

Krishnan, M. (1984) 'The plains wolf'. *The Statesman*, Calcutta, 15 July.

Le Mesurier, H. P. (1865) 'Wolves'. Murwala, 21 Jan., *The Field*, 18 Mar.

MacKenzie, J. M. (1988) *The Empire of Nature, Hunting, Parks and British Imerialism*. Manchester: Manchester University Press.

Manson, P. (1978) *A Shaft of Sunlight: Memories of a Varied Life*. Delhi: Vikas.

Mason, P. (1974 [1954]) *The Men Who Ruled India: The Guardians*. London: Jonathan Cape reprint.

McDougall, C. (1978) *The Face of the Tiger*. London: Rivington Books.

McDougall, C. (1987) 'The man-eating tiger in geographical and historical perspective'. In R. L. Tilson and U.S. Seal (eds) *Tigers of the World: The Biology, Biopolitics, Management, and Conservation of an Endangered Species*. Park Ridge, NJ: Noyes Publications, pp. 435–448.

Mukherjee, S. (1993) *Forster and Further, The Tradition of Anglo-Indian Fiction*. Hyderabad: Orient Longman.

O'Brien, Lt. Col. (1945) 'Where man-eating tigers occur'. JBNHS, vol. 45, pp. 231–232.

O. C. (1893) 'Notes on the otter'. *Indian Forester* vol. 19, pp. 34–35.

O'Malley, L. S. S. (1908) *Bengal District Gazetteers, Khulna*. Calcutta: Bengal Sect.

—— (1914) *Bengal District Gazetteers, 24 Paraganas*. Calcutta: Bengal Sect.

Phipson, H. W. (1892) 'The crocodiles in our reservoirs'. Letter to the Editor. *The Times of India*, Bombay, 16 Jan.

Pollock, S. E. (ed.) (1984) *The Ramanaya of Valmiki, An Epic of Ancient India, Vol. 1, The Balakanda*. Princeton: Princeton University Press.

Prater, S. H. (1929) 'On the occurrence of tigers on the island of Bombay and Salsettee'. *JBNHS*, vol. 33, pp. 973–974.

—— (1940) 'The number of tigers shot in the Reserved Forests of India and Burma during the year 1937–38'. *JBNHS*, vol. 41, pp. 881–888.

—— (1965 [1948]). *The Book of Indian Animals*. Bombay: OUP.

Pythian Adams, Lt Col E. G. (1949) 'Jungle memories'. *JBNHS*, vol. 48, pp. 646–655.

Rangarajan, M. (1996) *Fencing the Forest: Conservation and Ecological change in India's Central Provinces, 1860–1914*. Delhi: OUP.

—— (ed.) (1999) *The Oxford Anthology of Indian Wildlife, Vol. I: Hunting and Shooting* and *Volume II: Watching and Conserving*. Delhi: OUP.

—— (2001) *India's Wildlife History: An Introduction*. Delhi: Permanent Black.

Ray, Major J. W. (1893) 'Wolf-hunting in southern Maratha country'. *JBNHS*, vol. 8.

Rishi, V. (1988). 'Man, mask and man-eater'. *Tiger Paper*, vol. 15, pp. 9–14.

Ritvo, H. (1990) *The Animal Estate: The English and Other Creatures in the Victorian Age*. London: Penguin.

Roberts, T. J. (1976) *The Mammals of Pakistan*. London: Ernst Benn.

Sanderson, G. P. (1983 [1874]) *Thirteen Years Among the Wild Beasts of India*. New Delhi: Mittal Publications, reprint.

Sankhala, K. S. (1978) *Tiger! The Story of the Indian Tiger*. London: Collins.

Shahi, S. P. (1978) *Backs to the Wall, The Saga of Wildlife in Bihar*. Delhi: East–West Affiliated Press.

Singh, A. (1993) *The Legend of the Man-eater*. Delhi: Ravi Dayal.

Sleeman, W. (1980) *Rambles and Recollections of an Indian Official (1844)*. Karachi: OUP.

Smout, T. C. (1969) *A History of the Scottish People, 1560–1830*. London: Collins.

Stebbing, E. P. (1906) 'Review of W. S. Burke, "The Indian Shikar Field Book"'. *Indian Forester*, vol. 32, pp. 220–223.

Stracey, P. D. (1963) *Wildlife in India: Its Conservation and Control*. Delhi: Manager of Publications.

Sukumar, R. (1991) 'The management of large mammals in relation to male strategies and conflicts with people'. *Biological Conservation*, vol. 55, pp. 93–102.

Thapar, V. (1994) *The Tiger's Destiny*. London: Elm Tree Books.

Thomas, K. V. (1983) *Man and The Natural World: Changing Attitudes in England, 1500–1800*. Harmondsworth: Penguin.

Vernede, R. E. (1911) *An Ignorant in India*. London: William Blackwood.

Vernède, R. V. (1995) *British Life in India*. Delhi: OUP.

Waddington, R. C. (1893) 'Wolf hunting'. *JBNHS*, vol. 7, pp. 554–555

Williamson, T. (1807–8) *Oriental Field Sports*, vol. II. London: Edward Orme.

Worster, D. (1977) *Nature's Economy, the Roots of Ecology*. New York: Sierra Club Books.

Zimen, E. (1981) *The Wolf, His Place in the Natural World*. London: Souvenir Press.

Zoological Survey of India (1990) *The Red Data Book of Indian Animals, Part I: Vertebrata*, Calcutta: Government of India.

12

WOLF REINTRODUCTION IN JAPAN?

John Knight

Introduction

Wolf conservation initiatives have often generated conflicts with local populations. In Minnesota in the 1970s, 'people choked Eastern timber wolves to death in snares to show their contempt for the animal's designation as an endangered species' (Lopez 1995: 139), and cattle ranchers in this same state reportedly 'shoot, shovel, and shut up' when they encounter protected wolves (DiSilvestro 1991: 105). Proponents of wolf reintroduction in Yellowstone National Park have declared wolf conservation to be a test case of public commitment to conservation as such. Although wolf reintroduction in America appears to enjoy considerable public support, it has also attracted opposition among the local people most directly affected, especially ranchers who see the reintroduced wolves as a threat to their livestock herds as well as an unwarranted national interference in local affairs (Paystrup 1993). Wolf conservation in Sweden has led to conflict with Saami reindeer herders whose herds are threatened by the protected wolves, and to high-profile public protests by the herders in the national capital (Lindquist 2000: 170). In Norway wolf reintroduction is condemned as an illegitimate attempt by the central state to dominate sheep farmers: 'They know that if they can get farmers to accept and adapt to the wolf, they can get the farmers to accept *anything!*' (Brox 2000: 391, emphasis original). It is increasingly recognized that the success or failure of such conservation initiatives hinges on local reactions to them.

This chapter examines the issue of wolf reintroduction in Japan. Wolves became extinct in Japan at the beginning of the twentieth century. In 1993 an organization called the Japan Wolf Association (*Nihon ōkami kyōkai*, hereafter JWA) launched a campaign to reintroduce wolves to Japan. Influenced by wolf reintroduction debates abroad (and the Yellowstone wolf reintroduction campaign in particular), the JWA has proposed that colonies of wolves from continental Asia be established in a number of upland areas across Japan. In the examples above, wolf conservation or reintroduction appears as a threat to local livelihoods, but a central argument advanced by the JWA is that in Japan the reintroduced wolf would protect local livelihoods, while restoring Japan's natural heritage. In what follows I examine the JWA reintroduction proposal and the arguments supporting it, along with the various responses (including objections) to the plan, particularly among mountain villagers on the Kii Peninsula in western Japan, one of the areas designated as a possible site for reintroduction. This chapter draws on ethnographic fieldwork on the Kii Peninsula (specifically the municipality of Hongū-chō) in the late 1980s and the 1990s.

Wolves in Japanese culture

Two species of wolves inhabited the Japanese archipelago: the Hondo wolf (*Canis lupus hodophylax*) on the main islands of Honshu, Shikoku and Kyushu, and the Ezo wolf (*Canis lupus hattai*) on the northern island of Hokkaido. Both animals are now extinct. The Hondo wolf disappeared at the beginning of the twentieth century, against the background of a rabies epidemic. The last Hondo wolf was killed in 1905 in the Yoshino area of the Kii Peninsula, and the carcass is now held by the Natural History Museum in London. The extinction of the Ezo wolf occurred in the late nineteenth century, with the establishment of livestock ranches in Hokkaido. As a threat to the growing numbers of horses and cattle, wolves were systematically eradicated through the use of strychnine-poisoned bait and had disappeared from Hokkaido by 1889 (Chiba 1995: 166–172). Not everyone in Japan accepts that wolves are extinct. Over the years there have been many reports of wolf sightings, wolf tracks, wolf howls and even wolf remains, along with claims to have discovered wolf-dog hybrids. Amateur naturalists and wolf enthusiasts in different parts of the country devote much time and energy to trying to prove that wolves still exist by searching the remote mountains for signs and evidence of wolves. None of these claims has ever been verified. In the absence of a *verified* sighting for nearly one hundred years, such anecdotal challenges to the wolf extinction orthodoxy have little credibility (especially when extinction is understood in terms of whether or not a reproductive *population* of animals exists). There is a general, albeit not universal, acceptance in present-day Japan that wolves are extinct.

The wolf appears as a good animal in much Japanese folklore. The folklorist Chiba Tokuji argues that up until the second half of the seventeenth century the wolf was considered an *ekijū* or 'benign beast' in Japan (Chiba 1995: 183). The goodness of the wolf is in fact directly suggested by the Chinese character used to write it, 狼, which consists of two parts: the wild animal radical 犭, and the character for 'good' (*ryō*) 良, together giving the meaning of 'good animal'. Sometimes when hunters are asked about the wolf (its extinction, the prospects for its reintroduction etc.), they respond by referring to or writing down the character 狼 to show that the wolf was traditionally considered to be a 'good animal' (*ii dōbutsu*). The wolf is also associated with *kami* spirits. One illustration of this is the existence of wolf shrines in Japan, Shinto shrines where the wolf serves as the *otsukai*, or messenger, of the *kami*. There are estimated to have been more

Figure 12.1 A stamp commemorating the extinction of the Japanese wolf

than twenty important wolf shrines on the main Japanese island of Honshu, including the nationally famous shrines of Mitsumine and Yamazumi (Maruyama *et al.* 1996: 199; Hiraiwa 1992: 90). *Ōkami*, the Japanese word for wolf, is homonymic with the word for 'great spirit' (even though they are written with different characters), and it has been suggested that the wolf was imagined by earlier generations of Japanese as a spirit or *kami* (Saitō 1983: 22). The wolf has often been specifically identified with the *yama no kami* or 'mountain spirit' in rural Japan.[1] Nakamura Teiri suggests that in ancient Japan the wolf was viewed as 'the dog belonging to the mountain spirit' (*yama no kami ni shitagau inu*) (Nakamura 1987: 66). In Japanese folk religion the *yama no kami* is often a protective deity, guarding the *murazakai* (village boundary) from demons and other evil or harmful influences (Yagi 1988: 137–138), and the guardian role of the wolf would accord with this logic.

The principal functional basis of the wolf's spiritual identity in Japan lies in the tradition of upland farming. 'In particular, the wolf, by catching the wild boar, a harmful animal that destroys crops, becomes a benign animal for human beings. It is this on which the mountain dog worship found in the different regions of Japan is based' (Nomoto 1990: 66). Japanese villagers have long protected their fields through a variety of practical measures, such as field-guarding, patrolling, scarecrows, noise, light and sound repellents and so on. But they also employed ritual means, including the use of charms to invoke the power of the wolf to protect their farms from forest herbivores. Cultivators would use wolf charms (*ofuda*) obtained from wolf shrines and place these in the local shrine or bury them in the fields to be protected. The farmers' preoccupation with safeguarding these fields, as their means of subsistence, formed the basis for the representation of the wolf as a guardian spirit. On the Kii Peninsula, the villagers of Shinohara carried out an annual festival before the village shrine in which they petitioned village *kami* (*ujigami*) to protect fields from damage by wild boars and other animals (MS 9/4/1989). The wolf also offered protection from other

大滝温泉 三峰神の湯

三峯神社興雲閣

Figure 12.2 A depiction of Mitsumine Shrine

kinds of threat, including theft and robbery, house fires and illness, in urban areas as well as rural ones.[2] Wolf skulls and other wolf charms were used in folk religion to expel animal spirits (for example, those of dogs, foxes and snakes) and other spirits possessing human beings.[3] On the Kii Peninsula, people suffering from fox possession were brought to the Tamaki Shrine, where the wolf spirit, with its superior power, would drive out the fox from the possessed person.

In line with the view of the wolf as a protector of farmland, Japanese farmers welcomed any increase in wolf numbers. This is most strikingly illustrated by the custom of offering a gift of food to the wolf when wolf cubs are born – the 'wolf's birth gift' or *inu no ubumimai* (Ōfuji 1968: 368; Matsuyama 1977: 187–189). Usually, *sekihan* (rice with azuki beans) is offered; *sekihan* is a ceremonial food traditionally served to celebrate a human birth and on other felicitous occasions such as New Year or festivals for the village deity. In some regions the *ubumimai* practice included the belief that the wolf would likewise, in return, make a congratulatory offering on the occasion of a human birth in the village (Ishizaki 1991b: 236). This Japanese celebration of wolf increase contrasts starkly with institutionalized wolf eradication in Europe and North America. The *ubumimai* custom illustrates the importance of the principle of reciprocity in human-wolf relations in Japan. The theme of reciprocity between people and wolves is widely found in wolf folklore in Japan. In myths and folktales the wolf is characterized as a 'dutiful' (*girigatai*) animal, which reciprocates any human kindness it receives (Maruyama 1994: 139). The wolf that has been helped by human beings appears extremely grateful – in some tales the wolf bows repeatedly to its human saviours before disappearing into the forest (Yanai 1993: 163–164). In many tales the wolf offers some kind of animal prey to the village in return for the earlier help it received. One story from the Hongū village of Heichigawa tells of a wolf caught in a pitfall used for catching wild boar. Villagers take pity on the wolf and help it out of the pit, allowing it to return to the forest, and a few days later hear a wolf-howl from the vicinity of the pit, in which they discover a large deer – the wolf has made its return gift (*ongaeshi*) (Hongū-chō 1969: 12–13). Similar examples of the wolf's sense of reciprocity, involving offerings of other animals or their parts (a wild boar, a pheasant, a bear's paw, etc.), can be found elsewhere in the region and beyond.[4]

Reinforcing this positive image of the wolf are references in Japanese folklore to wolves nurturing people, recalling the Romulus and Remus legend in which the founders of Rome are suckled and raised by a she-wolf. Thus, an infant (of the court noble Fujiwara Hidehara), abandoned in the mountains of the Kii Peninsula, is brought up and protected by wolves.[5] In the Kantō area, the milk of captured pregnant wolves was fed to children: 'if a child is fed with wolf's milk, it grows up strong' (Yanai 1993: 137). It should be noted that this motif is not widely found in Japanese folklore; indeed, in his book, *Yama no jinsei*, the great Japanese folklorist Yanagita Kunio, while mentioning reports of 'Romulus' children raised by wolves in India, specifically denies their existence in Japan (Yanagita 1961: 167). However, this motif of the nurturant wolf finds a more recent expression in Japanese popular culture. One example is the 1964 animated television series *Ōkami shōnen Ken*, 'Ken the Wolf Boy', recalling the story of Mowgli in *The Jungle Book*.[6] The wolf could also protect the spirits of dead foetuses or infants. In one part of the Hongū forest there exists an *ōkami jizō*, the 'Wolf Jizō'. In Japan, Jizō is the compassionate bodhisattva associated with children; appearing childlike itself in its statue form, Jizō has long been viewed as a protector of children, and is ubiquitous in present-day Japan in the many temples dedicated to *mizuko kuyō* or memorials for aborted foetuses. In poor upland areas like Hongū, abortion and infanticide have long been resorted to as a necessary means of limiting family size, and one interpretation of the Wolf Jizō in Hongū might therefore be that it was petitioned by mothers to care for the

Figure 12.3 An *ōkami kuyō* ('wolf memorial') ritual carried out in 1987 in the village of Takada on the Kii Peninsula (Courtesy of Naruji Toshiaki)

spirits of miscarried or aborted foetuses or dead infants buried in the area – i.e. to protect them from the attentions of other forest animals (cf. Chiba 1977: 143).

There is, however, another side to wolf imagery in Japan. By the nineteenth century, a strongly negative view of the wolf is apparent in written works. From the late seventeenth century, there are reports of wolf attacks on livestock, especially horses (for example, Chiba 1995: 175; Maruyama *et al.* 1996: 199). Wolves are reported to have attacked, killed and even eaten village dogs on the Kii Peninsula and elsewhere (Suzuki 1986: 239; Hiraiwa 1992: 130). Reports also emerge of wolf attacks on people in different parts of Japan. In the early nineteenth century, Suzuki Bokushi, the author of *Hokuetsu seppu* (*Snow Country Tales*), graphically described a horrific wolf attack which took place in Echigo (Niigata), in which three wolves attacked a remote household and killed and devoured three family members (Suzuki 1986: 239–241). There also appear to have been occasional incidents in which wolves snatched children. In the Shinshu area, while adults gathered grass near the forest, the children accompanying them were made to climb trees and to wait there, as protection against wolf attacks (*Mainichi Shinbun* 27/5/1990). One category of people particularly vulnerable to wolf attacks were field guards. Remote swidden fields adjacent to the forest were guarded overnight to protect them from crop-raiding wild boar and deer, but the crop-raiders were not the only forest animals drawn to the night-time fields. As the field crops attracted hungry wild boar and deer, these same animals in turn attracted hungry wolves to feed on them. Battles between field guards and wolves are reported in the eighteenth century (*Mainichi Shinbun* 21/1/1990). Wolves even posed a threat to the dead. Reports of grave robbing by wolves, between the late seventeenth century and the late nineteenth century, tell of corpses exhumed and eaten by wolves.[7]

Various measures were taken by upland dwellers to protect themselves from the threat of the wolf. The physical guarding of livestock was one anti-wolf measure. In what is now Nagano Prefecture, 'wolf watches' (*yamainuban*) were undertaken to guard horses put out to pasture (Matsuyama 1977: 178). One response to the perils of night-time field guarding was to construct specially elevated field huts to secure the field-guard from wolf attacks (Matsuyama 1977: 163; Miyamoto *et al.* 1995: 394). Wolf trapping, involving the use of poisoned-bait traps, pitfalls and baited ambushes, also took place. Bounty systems were established: from the early eighteenth century 'rewards' (*hōbi*) were offered by the feudal *han* authorities for dead wolves (Hiraiwa 1992: 132–133). More informally, wolf catchers known as *ōkamitori* were rewarded with *sake* by grateful villagers (ibid.). Battue-style wolf hunts were another response (for example, Chiba 1993; Matsuyama 1994: 137–140). In Akita in 1866, for example, large numbers of villagers, beating drums and launching fireworks, systematically drove wolves out of the forest into a clearing where they surrounded and destroyed the animals (Ōta 1997: 206–207). More often, a wolf attack would lead to specialized wolf hunters being called in (Hiraiwa 1992: 129–130).

There are many well-known Japanese word compounds containing the Chinese character for wolf 狼 which have highly negative meanings, including 'wolf's heart' or *sairō no kokoro* (to be cruel), 'wolf's voice' or *rōsei* (a frightening voice), 'wolf sickness' or *rōshitsu* (a terrible sickness) and 'a mountain wolf' or *nakayama ōkami* (a savage) (Iwase Momoki, Suzuki 1986: 244). Iwase Momoki concludes emphatically that '[a]ll of this shows without a doubt that wolves are the most hated of animals' (in Suzuki 1986: 244). This negative view of the wolf also existed on the Kii Peninsula: up until the end of the nineteenth century, parents would deal with the unruly behaviour of their children by warning them that 'the wolf will carry you off if you do that' (Nakamori 1940: 33–34). A great many rural tales and legends suggest that wolves were an object of considerable fear. This fear of wolves appears to have intensified in the modern period. With the spread of rabies in late seventeenth-century Japan, the image of the wolf as a benign spirit started to change to that of a mortal threat. The first report of rabid wolves (in Kyushu and Shikoku) occurred in 1732 and the disease then spread eastwards (Maruyama *et al.* 1996: 200). For Tokugawa Japanese, the word *yamainu* – 'mountain dog', a long-established popular term for the wolf – became synonymous with the rabid dog that attacked people (Chiba 1995: 51). It was this ubiquity of rabies in eighteenth-century Japan that 'badly affected the Japanese people's attitude toward wolves; they started to fear wolves, and began to control wolves as nuisance animals' (Maruyama *et al.* 1996: 200). Other factors included deforestation and changes in Japanese

Figure 12.4 A pamphlet produced by a rural municipality on the Kii Peninsula, which claims to be the last place in which the wolf was sighted. The pamphlet shows a local statue that commemorates the wolf (Courtesy of Higashi Yoshino)

agriculture that deprived wolves of the predation opportunities of the upland swidden fields.

For most of the twentieth century, Japanese mountain villagers have lived without wolves. There have been claimed sightings of wolves, and even claims to have discovered wolf carcasses, but none of these has ever been verified and there is a general acceptance that Japan's indigenous wolves have become extinct. But in recent years the possibility has been raised that in the twenty-first century wolves might once again inhabit the Japanese mountains.

The wolf reintroduction proposal

The Japan Wolf Association (JWA), is an organization dedicated to the goal of wolf reintroduction in Japan. In the first issue of the Association's newsletter, the objective of wolf reintroduction was outlined as follows:

Shouldn't We Call the Wolf Back to the Japanese Forest?
The wolf, along with the monkey, the hare and other wildlife is an indispensable part of the forest ecosystem. But it will be nearly one century since the wolf disappeared in Japan. A forest where the wolf cannot be seen is not real nature. . . . The extinction of the wolf came about because, with the growth of the human population and the development of industry, humanity invaded the territory of nature and did not consider the possibility of co-existing with the wild environment. Now, we human beings, as a result of the environmental destruction we have caused ourselves, fear for our own existence. We are now starting to realize that, because humans and the Earth are part of the same ecosystem, human prosperity can only come about by living our lives as a part of nature. And so, in recent years, there has flourished a movement to protect the nature that remains and to restore that which has been lost. Shouldn't we call the wolf back to the Japanese forest?

Maruyama Naoki and the other founding members of the JWA were inspired by the international debates on predator reintroduction, and actual reintroduction programmes in North America. The JWA has proposed that wolves from Chinese Inner Mongolia be used to establish grey wolf colonies (of up to thirty animals) in a number of upland areas across Japan. Ten (national park) areas have been proposed as possible candidate sites for wolf reintroduction (Takahashi and Maruyama n.d.: 15–16).[8] The JWA's time frame for wolf reintroduction would appear to be in terms of decades rather than years. The strategy of the JWA is to build up public support for wolf reintroduction in the short and medium-term, and to lobby the government to establish an effective wildlife administration. The JWA has a largely middle-class membership of around 550 people (ca. August 2000), drawn largely from the professional classes, including university professors, schoolteachers, journalists, doctors and students. The JWA leader, the zoologist Maruyama Naoki, is an articulate and forceful spokesman who has succeeded in helping the issue achieve a high public profile through extensive media coverage. The association has a regular newsletter that carries specialized articles as well as contributions from ordinary members. There is a dedicated e-mail message board on the JWA home page where members communicate with each other on wolf-related matters. They also exchange information on other wolf sites, wolf literature and wolf memorabilia, deer damage to forests, environmental and wildlife issues in general, opposition to wolf-culling overseas, experiences of wolf-watching opportunities around the world and so on.

A number of arguments are advanced by the JWA for wolf reintroduction. The first reason offered is that it would restore Japan's 'forest ecosystem' (*shinrin seitaikei*). Based on a key notion of modern ecology, the JWA defines the 'forest ecosystem' in terms of the existence of top-line wild carnivores such as wolves which, through their predation, keep in check the numbers of wild herbivores. Ecological reasoning thus has a prominent place in the justification of the wolf reintroduction proposal, as the following excerpt from the JWA newsletter makes clear.

> As medium and large herbivores increase, they are destroying the forest ecosystem. These herbivores, by ruining their own habitat, will in the end bring about their own destruction. This kind of argument is well known, but in Japan there are now dangerous signs of this happening. A Japanese forest without the wolf is one where monkeys, serow and deer lead dull, tedious lives and lose their state of alertness, and where they start to ruin the land. For the forest ecosystem and for the prey animals within it, the wolf, as predator, is an indispensable presence.
>
> (JWA 1994: 2)

The extinction of the Japanese wolf, it is argued, has resulted in a fractured ecosystem in which forest herbivores can no longer be kept in check by a top-line predator. In the absence of their natural predator, deer, serow and other categories of forest wildlife have proliferated in number. For a long time, the effects of the extinction of the Japanese wolf were concealed by the actions of human hunters, who in effect carried out the wolf's role of regulating herbivore numbers. But with the decline in hunting in recent decades the full effects of wolf extinction in terms of the proliferation of numbers of deer, serow and monkeys have become apparent. The reintroduction of wolves would restore the functioning forest ecosystem of the past and ensure that the numbers of these animals are contained.

Second, wolf reintroduction promises to alleviate the problem of wildlife pestilence. Wildlife depredations are a serious problem in upland Japan; intrusive wildlife is a cause of economic loss and a major nuisance. Wildlife problems have even led, in some cases, to the abandonment of farming and have contributed to the depopulation of the remote, forest-edge villages that are worst affected. In response, upland farmers and foresters expend a great deal of time and energy to defend their fields and plantations from wildlife pests. The JWA claims that upland farmers would benefit from wolf reintroduction because wolves, by preying on wild crop-raiders such as the wild boar, the monkey and the deer, would relieve this increased wildlife pressure on mountain villages.[9] Similarly, the JWA claims that foresters, as well as farmers, would benefit from wolf reintroduction. As Maruyama Naoki recently put it, reintroduced wolves would be the 'saviour' (kyūseishu) of the foresters who presently suffer from deer pestilence (Chūnichi Shinbun 7/3/2000). This is because the reintroduced wolves would prey on deer and serow, the two main forestry pests. In a forestry magazine, Maruyama points out to foresters that forestry pestilence is not solved by fencing and other measures, which merely redirect the problem elsewhere and fail to address its cause – the large numbers of deer and serow (Maruyama 1994: 4). A restored wolf presence would reduce the numbers of these herbivores. This same point was graphically made in a television documentary (shown on NHK, Japan's public broadcasting channel on 7 November, 1994) entitled 'The Deer is Eating Up Nature' (Shika ga shizen o kuitsukushimasu) which looked at the reintroduction proposal. Focusing on the 'explosive growth' of deer numbers across Japan, the scale of economic damage (to timber plantations and to farms) caused and the failure of the various attempts made to tackle the problem, the documentary concluded with footage of wolves killing deer in Polish forests.

Another claim made by proponents is that wolf reintroduction would protect the natural (primary and secondary) forest, as well as commercial plantations, from herbivore damage. In Japan deer are responsible for damage to trees and other vegetation of the natural forest, a cause of some concern both to environmentalists and to the municipal authorities who see the remaining natural forest as a valuable tourist resource.[10] The JWA argument is that, by containing deer numbers, the wolf would help to protect the forest flora threatened by the deer. A further environmental benefit claimed for prospective wolf reintroduction relates to bear conservation. Japan's wild bear population has sharply declined in recent decades, and some regional bear populations (such as on the island of Shikoku) have become extinct, while others (such as on the Kii Peninsula) are in danger of becoming so. According to the JWA, wolf reintroduction would benefit the black bear because this opportunistic omnivore would be able to scavenge from wolf-kills and so increase the meat content of its diet (Maruyama 1995: 6).[11]

Wolf reintroduction is justified as an act of cultural, as well as natural, restoration. The existence of wolf shrines, the popular recognition of the wolf's role in farm protection, and the benign image of wolves in Japanese folklore lead reintroductionists to argue that the wolf is an important constituent element in traditional Japanese culture. Since the 1970s there has been a growing interest in the forest, or mori, as a crucible of ancient Japanese culture (in contrast to the conventional emphasis that historians and others have hitherto placed on rice cultivation). The wolf serves as animal symbol of the mori, as is clear from the JWA literature (in which the mori is mentioned almost as much as the wolf itself). The reintroduction of the wolf would serve as a tangible expression of the (ecological) revival of the forest. But, because of the *mori*'s associations with Japanese cultural 'roots', the restored wolf would also herald a kind of cultural revival.

Responses to the proposal

What of the actual views about reintroduction among the people who live in upland Japan? In the mid-1990s an article hostile to wolf reintroduction appeared in a leading local newspaper on the Kii Peninsula. It cited a declaration of opposition by the Nara Prefecture Wildlife Protection Committee to the JWA wolf reintroduction plan. The journalist summarizes the objections made to the JWA proposal in the declaration.

> The Japanese wolf is different from the foreign breed. The irresponsible plan of introducing an animal that was not originally in Japan is an abdication of the scholar's work and amounts to conduct that is difficult to forgive. There is absolutely no understanding of the Japanese wolf. To simply carry out *introduction* without any scientific research survey whatsoever will make Japanese zoology, and even the Japanese as a whole, a laughing stock around the world.... There is a hidden background to this [plan] that starts to become apparent: the belief that money can solve everything and the Japanese tendency to give priority to economic efficiency over everything else.
>
> (*Kii Minpō* 12/3/1996, p. 11, emphasis added)

A couple of weeks later the following reader's letter appeared in the same newspaper.

> If foreign wolves are released into the Kii Mountains, I think that, rather than the hoped for reduction of deer numbers, what is more likely is that the animal and plant ecology would be destroyed and that a sharp increase in wolf numbers would lead to damage, to danger, and to disease-causing germs. Also, terrible things could happen that at present we cannot even imagine.... They want to restore the Japanese wolf, don't they? But then why are they bringing in a foreign wolf in place of the Japanese wolf? I know that the increase in deer numbers is having an effect on nature and forestry. But do the people who live on the Kii Peninsula and who until now have protected nature on the Kii Peninsula – do they want foreign wolves to be set loose? At present, the Nara Prefecture Wildlife Committee is opposed. We the [Wakayama] prefectural people, to whom the deep mountains of the west of the peninsula belong, should also firmly oppose it and continue to protect the nature and the way of life of the Kii Mountains.
>
> (*Kii Minpō* 30/3/1996, p. 7)

In both the original article and the letter above, the wolf reintroduction plan is represented not as restoring what was lost but as importing something new and alien. The journalist pointedly seems to avoid use of the word *saidōnyū* or 'reintroduction', and instead refers to *dōnyū* or 'introduction', thus challenging the way in which the JWA defines the issue. The effect of this rhetorical switch of emphasis is to represent the JWA's initiative as the bringing in of a new, alien animal rather than the restoration of a traditional one. The letter similarly depicts the plan as having nothing to do with bringing back the *Japanese* wolf, but is instead about setting loose a foreign animal on national territory. While the JWA claims that its plan would restore nature (the 'natural ecosystem'), the journalist refers in passing to the complaints among 'people living in mountain areas' that the plan 'will lead to the destruction of ecology' (*seitai no hakai ni tsunagaru*). As noted above, some distinguished Japanese mammalogists have stressed the uniqueness of the Japanese wolf. This uniqueness argument has been used to criticize any proposed reintroduction on the grounds that the indigenous Japanese ecosystem would be undermined by the presence of an alien species (see Hatano 1996: 47). The JWA counters this argument by claiming that the introduced wolves would be genetically close to Japanese wolves. This would mean that wolves from China (i.e. Inner Mongolia), rather than, say, Canada, would be reintroduced to the Honshu mainland (Takahashi and Maruyama n.d.: 6). If the continental wolves to be introduced belong to the same species as the Japanese wolf, the above objection to reintroduction would be overcome.

A common criticism of reintroduction programmes is that of misadaptation: insofar as they 'involve translocating genotypes across geographic ranges', they can 'introduce "incorrect genotypes" where they do not belong' (Bowles and Whelan 1994: 3). Similarly, on the Kii Peninsula, the foreignness of the reintroduced wolf fuels local doubts about its ability to live in the Japanese mountains. One local journalist (the author of the critical newspaper article above) believes that Japanese mountains are quite different from mountain ranges found on the Asian continent. In Japan, mountains 'have many folds' – he likened them to the ribbed underside of a *shiitake* mushroom cap (with its recurrent undulations). To introduce large foreign wolves accustomed to the plains of the continent into such a landscape would be 'cruel' (*zangyaku*). Another wolf conservation group, the small and more recently formed Ōkami no Nakamatachi (Friends of the Wolf), also cites cruelty to the wolves, which would be forcibly displaced and relocated to alien surroundings, as one of its reasons for opposing reintroduction.[12] Doubts about habitat extend to the issue of environmental degradation. Ōkami no Nakamatachi has argued forcefully that the combined impact of forestry and other forms of development (golf courses, ski slope and resorts) on upland Japan precludes the possibility of successful reintroduction. Instead of planning to bring in foreign wolves, the priority should rather be to control or even halt this development and to restore the earlier natural habitat (OGHK n.d.: 109). Only then could wolf reintroduction be sensibly considered. From this critical perspective, given all the media attention it has received, the issue of wolf reintroduction is a damaging distraction from the all-important fact of development-related loss of wildlife habitat in Japan. Similarly, by suggesting that Japan's deer problem solely has to do with the absence of the wolf, the culpability of commercial forestry in deer proliferation is obscured (Okami no Nakamatachi n.d.; OGHK n.d.: 109).

Human presence is the other key consideration in relation to the issue of suitable wolf habitat. A common objection to the idea of wolf reintroduction in Japan is that there is no longer any space left for wolves (see Kanzaki *et al.* 1995: 1). The argument is that the scale of the human population in present-day Japan precludes the introduction of a wolf population. The JWA response is to point to future population trends, and in particular to the projected decline of the Japanese population in the twenty-first century – to less than 80 million by 2050. This change would make the restoration of a wolf population much more feasible (Maruyama n.d.: 5). The argument is that just as the original extinction of wolves in Japan at the beginning of the twentieth century was related to a rapid increase in the human population, so conversely a necessary condition of wolf reintroduction would be fewer human beings. In practice, this point about human numbers has less to do with national than with regional (more particularly, upland) space. In this connection, the JWA argues that the depopulation of mountain villages in the postwar decades is something that improves the prospects for wolf reintroduction. Typically, between 1955 and 1995, the upland municipalities of the Kii Peninsula lost over half their population through urban out-migration, and within these municipalities the remoter settlements have been disproportionately affected, leading in many cases to total abandonment. As these mountainous areas are vacated by humans, so they become available for a restored wolf population (ibid.).

Although appearing to skilfully rebut the arguments of its critics, in fact the JWA comes very close to adopting their reasoning. For it attempts to overcome the objection to a wolf presence in Japan based on the scale of the contemporary human population by referring to projected trends towards a lower population, both nationally and regionally. The problem with this line of argument is twofold. First, it relies on an implicit zero-sum logic, according to which fewer people becomes the condition of wolf restoration, something which seems to be potentially at odds with the notion, otherwise stressed by reintroductionists, of human–wolf coexistence. Second, to invoke rural depopulation as an enabling condition of wolf reintroduction risks offending local sensibilities. In many upland areas efforts are being made (by local government as well as local groups) to stop and even reverse depopulation. Certainly, many upland settlements have been totally abandoned and are unlikely ever to be resettled, and to this extent rural depopulation does tend to appear irreversible; it may well be that in the long run upland areas in Japan are being vacated by their erstwhile human residents and are, therefore, available for occupation by a different set of inhabitants. But as this is a prospect greeted with little enthusiasm locally, arguments for reintroduction based on it are unlikely to commend themselves to mountain villagers. In fact, it resonates with a long established fear among villagers that ongoing depopulation will erase their communities altogether. In another part of the Kii Peninsula this sentiment was captured in the grim prophesy of an informant of the Japanese cultural anthropologist Yoneyama Toshinao about the future of his village: '[S]oon this place will become land only for the snakes' (Yoneyama 1969: 9). Arguably, the presence of wolves in the mountains would symbolize a similar fate.

Despite the JWA's claim of a Japanese cultural affinity with the wolf, the prospect of wolf reintroduction, even at this early stage, is greeted locally with a certain amount of trepidation. Some villagers say that wolf reintroduction would not find favour with forest workers. Forest workers aside, villagers (including many women) frequently go gathering in the forest for mushrooms, herbs, grasses and orchids; the presence of wolves, it is feared, might well deter them from going. Another fear is that old people might become victims of the wolf. A boar-hunter explained to me that because wolves prey on weak or old animals, were they to be released into the peninsular mountains they might well turn their attention to the elderly inhabitants of the remoter villages. The point could be extended to children. In the newspaper article above, the journalist, recounting a local anxiety, asks: '[I]sn't there a worry that when the foreign wolf's stomach is empty it will attack children?'

Tourism is another consideration. At a time when there is a strong local self-consciousness that much of the mountain forest has lost its natural character due to the spread of timber plantations, wolf reintroduction might conceivably appeal as a means of re-naturalizing the area. Wildlife tourism on the Kii Peninsula, in the form of serow parks, monkey parks, safari parks and so on, is becoming increasingly important. Some town hall staff believe that the presence of wolves in the *yama* would contribute to the appeal of the area. The national parks, already popular with visitors, would arguably become an enhanced tourist attraction. However, there is also the worry that a wolf presence might add a new, and ultimately unacceptable, level of danger to the parks. Many tourists hike through the mountains or go herb and mushroom picking. A single, serious, wolf attack might stop people visiting the area altogether! The fear is that, given the number of tourist guesthouses, any reintroduced wolves, failing to find enough to eat in the mountains, might descend to feed on the refuse of tourist villages and in the process frighten and endanger the visitors.[13]

Another cause for concern is the history of past wildlife introductions in Japan. There have been many examples of strategic animal introductions aimed at pest control that have proved to be problematic. After weasels were introduced to control populations of small rodents on the island of Miyake, the weasels preyed instead on birds and lizards, which declined in number as a result (Brazil 1992: 334–335). The introduction of sable on the island of Sado to control hares also went awry as the numbers of sable proliferated (KSY 13/3/2000). Other attempts at interfering with wildlife populations have backfired, such as the pest eradication measures applied to hares and voles, where poison was scattered in large quantities. In addition to its intended victims, the measure killed weasels and foxes; as they prey on young wild boar, the elimination of these predators led to an increase in wild boar numbers and an increase in agricultural damage (Ichikawa and Saitō 1985: 173–174)! On the southern Kii Peninsula there are vivid local memories of the effects of mistaken animal introductions in recent decades. In the early 1960s Wakayama Prefecture decided that foxes should be released in the forest in an effort to reduce the number of hares that, at the time, were deemed responsible for much of the damage to timber plantations. But the omnivorous foxes, instead of staying put in the forest to chase the hares, opted to come to the village where they caused great damage to farm crops (Ue 1980: 84–5)! Given these earlier failures, the prospect of the introduction of what many people believe to be the most dangerous wild animal of all arouses considerable trepidation. Moreover, some people point out that it was such human interference in nature that led (in Hokkaido) to wolf extinction in the first place!

An indication of possible negative reactions to wolf sightings is the furore that arises in Japan whenever a bear is seen. Bear sightings and encounters typically lead to panics and tend to result in the dispatch of the animal in question (see Knight 2000). Another indication of the near-hysteria that wild predators can arouse was provided by an incident in the Kantō area in 1979, in which two pet tigers escaped from captivity: the response was public panic and an enormous mobilization of policemen and hunters to track the animals down and destroy them (Mitani 1979). More recently, a furore over feral dogs broke out in Aomori Prefecture over sightings of packs of dogs killing serows, incidents which reportedly disgusted the tourists and spoilt the image of the area (*Kahoku Shinpō* 21/5/1996). The likely response to the appearance of feral dog packs is an intensification of efforts to eradicate them. These bear, tiger and feral dog examples point to a deep-seated intolerance of wild predators in Japan, at least among the public authorities which attach overriding importance to public safety. Such extreme intolerance of wild predators raises serious questions about the prospects for any wolves reintroduced to the Japanese forest.

Conclusion

The JWA's wolf reintroduction proposal contains a twofold discursive claim: that it would be both a *natural* and a *cultural* restoration. But, as we have seen, it is open to question on both counts. In the JWA reintroduction proposal, the wolf appears as the missing link in the Japanese forest – and, by extension, Japanese nature – and the return of the wolf appears as the means by which the forest can be made whole and complete once more. One objection to this kind of claim would be on the grounds of the mechanistic assumptions about 'nature' on which it seems to rest – that the forest is rendered fundamentally incomplete and ecologically dysfunctional by the absence of a single species. This style of reasoning has become the object of growing criticism in recent years; ecologists appear to be increasingly skeptical about the ecosystem concept and its mechanistic vocabulary (Budiansky 1995: 182–183). More specifically, the earlier orthodoxy of predator regulation of prey numbers, on which the JWA reintroduction proposal is clearly based, has also been widely called into question by zoologists and ecologists. There is a basic disagreement among specialists over just what the predatory impact of wolves on wild herbivores is (whether they simply prey on weak or old animals or additionally target healthy animals) (Talbot 1978: 309–310; Steinhart 1995: 61–77).

Another objection to the JWA proposal would be to the way it tends to downplay, if not neglect, the anthropogenic character of the forest environment, which comes to be equated with 'nature'. In particular, the impact of forestry has tended to be downplayed by proponents of wolf reintroduction. Of course, this environmental degradation of the mountains can be seen as making the restoration of the forest ecosystem, typified by the wolf, all the more urgent. But it also calls into question whether there still is suitable habitat to which the wolf can be restored. Proponents of wolf reintroduction in Japan have yet to convincingly address this issue. Their invocation of the 'forest ecosystem' (to be restored via the wolf) tends rather to obscure the degree to which forests (even those in the designated national park areas) have been affected by forestry. To the extent that wolf reintroduction is promoted without proper reference to this wider context of current land-uses and habitat alteration, it becomes vulnerable to the charge that it would amount to no more than a kind of cosmetic restoration. In other words, the return of the new wolves would not *restore* nature or wilderness to its earlier completion as claimed, but offer a kind of symbolic compensation for its loss.

The idea that wolf reintroduction would be a cultural restoration is also open to challenge. The JWA proposal holds out the promise that the return of the wolf would make the upland environment manageable once more for those who live in it. One of the benign Japanese images of the wolf from earlier times is that of the *banken* or 'watchdog' in the forest. The picture of the wolf presented by the JWA, as a sort of pest control officer in the forest that stops harmful animals from spilling over into the villages below, recalls this earlier image. But this claim of the supposed cultural compatibility of the wolf in premodern Japan must be set against the evidence that there was also fear and, latterly, persecution of the wolf. Moreover, the implicit claim of the JWA that wolf reintroduction elicits a cultural endorsement in Japan is, at the least, questionable, notwithstanding the existence of wolf shrines and wolf-related folklore. Many upland dwellers hold negative views of wolves, which the fledgling (and largely metropolitan-based) JWA campaign has yet to come to terms with. The second half of the twentieth century saw large-scale out-migration from the remote areas of Japan. If negative popular perceptions of the wolf – rural as well as urban – are not addressed, wolf reintroduction could well reinforce this trend towards the abandonment of the Japanese mountains as a space of human settlement. Rather than restore the upland culture of Japan, wolf reintroduction might end up hastening its demise.

Acknowledgements

This is a revised and updated version of a paper published in 1998 in *Japan Forum*, vol. 10, no. 1: 47–65. An extended analysis of the issue of wolf reintroduction and its implications for Japanese mountain villagers is to be found in my book, *Waiting for Wolves in Japan: An Anthropological Study of People-Wildlife Relations* (Oxford University Press, 2003). I would like to thank Maruyama Naoki, Kanzaki Nobuo and Takahashi Masao of the JWA and Suzuki Atsuko of *Ōkami no Nakamatachi* for sharing with me their views and expertise on wildlife matters and for sending me various written materials on the issue of wolf reintroduction. I owe a special debt of gratitude to Tanagami Kazusada of the newspaper *Kii Minpō* for his assistance in field research and for the many stimulating conversations about wildlife issues we have had. Japanese names appear in Japanese order, with family name first.

Notes

1 Kaneko *et al.* (1992: 22); Satō (1990: 153–155); Naumann (1994: 34–35).
2 MS (14/5/1989), MS (28/5/1989); Hiraiwa (1992: 89).
3 Kitajo (1994: 4); Komatsu (1988: 37); Matsutani (1994: 161, 163).
4 For regional examples, see Nakamori (1941: 30–31), Wada (1978: 265), WKMK (1987: 62) and HYMKI (1992: 247–249). See also Morita (1994: 141–142) and Sakai (1986) on the motif more generally.
5 Nakamura (1987: 67); Tabuchi (1992: 84–85); KHI (1980: 63); cf Tanigawa (1980: 32).
6 A second example is the popular 1997 animation film *Mononoke Hime* ('Princess Mononoke'), an eco-fable directed by Miyazaki Hayao, in which the leading character, 'San', is a wild girl abandoned as an infant in the mountains and raised by wolves. San devotes her life to waging war on the human society which discarded her. In this film wolves (along with other forest animals) appear as the protectors of the forest against a human society which threatens it through its destructive modern weapons.
7 Hiraiwa (1992: 135–136); *Mainichi Shinbun* (18/11/1990); *Mainichi Shinbun* (22/12/1990); Ōta (1997: 205).
8 These are Shiretoko National Park, Akan National Park, Hidaka Mountains Quasi-National Park, Mount Daisetsu National Park, Nikkō National Park, Chichibu Tama Quasi-National Park, Southern Alps National Park, Yoshino-Kumano National Park, Kōya-Ryūjin Quasi-National Park, and Kyūshū Central Mountains Quasi-National Park.
9 Maruyama *et al.* (1995: 24); Maruyama (1995: 4–5); Watanabe (1995: 50); Mizuno (1995: 15–16).
10 This is a concern shared by local authorities on the Kii Peninsula, worried about the effect of deer numbers on the well-known and much visited Ōdaigahara forest, but reluctant to take action for fear of the controversy that would likely arise.
11 This is a response to the fears voiced that the black bear – the numbers of which have already declined sharply on the Kii Peninsula – would be further imperilled by the presence of wolves.
12 See Anon. (1995), Ōkami no Nakamatachi (n.d.) and Togashi (1994). A similar ethical criticism, on the grounds that relocation is at odds with the interests of the relocated animals themselves, has been made in relation to wolf reintroduction initiatives in North America (Finsen 1995: 29).

13 Wolves elsewhere have been reported as feeding on human garbage (in parts of southern Europe in particular), although this seems to be related to low numbers of wild ungulates, the wolf's preferred food source (Meriggi and Lovari 1996). In Japan, of course, as there is a relative abundance of wild ungulates, this may be much less likely.

References

Anon. (1995) 'Konna hatsuyume o michatta' [I had this New Year's dream]. *Oikos File*, no. 36, pp. 6–7.

Bowles, M. L. and C. J. Whelan (1994) 'Conceptual issues in restoration ecology'. In M. L. Bowles and C. J. Whelan (eds) *Restoration of Endangered Species: Conceptual Issues, Planning and Implementation*. Cambridge: Cambridge University Press, pp. 1–7.

Brazil, M. (1992) 'The wildlife of Japan: a twentieth-century naturalist's view'. *Japan Quarterly*, vol. 39, no. 3, pp. 328–338.

Brox, O. (2000) 'Schismogenesis in the wilderness: the reintroduction of predators in Norwegian forests'. *Ethnos*, vol. 65, no. 3, pp. 387–404.

Budiansky, S. (1995) *Nature's Keepers: The New Science of Nature Management*. London: Phoenix.

Chiba, T. (1993) 'Kinsei Kanazawa-Hirano chiiki no yajūgai: toku ni nihonōkami ni tsuite' [Wild animal damage in the Kanazawa-Hirano area in the modern era]. *Rekishi Chirigaku*, vol. 16, pp. 38–47.

—— (1995) *Ōkami wa naze kieta ka* [Why did the wolf disappear?]. Tokyo: Shinjinbutsu Ōraisha.

DiSilvestro, R. L. (1991) *The Endangered Kingdom: The Struggle to Save America's Wildlife*. New York: John Wiley.

Finsen, S. (1995) 'Here we go again'. *The Animals' Agenda*, vol. 15, no. 4, pp. 29.

HYMKI (Higashi Yoshino-mura Kyōiku Iinkai) (ed.) (1992) *Higashi Yoshino-mura no minwa* [Folktales of Higashi Yoshino Mura]. Higashi Yoshino: Education Committee.

Hatano, I. (1996) 'Motto deta o, motto giron o' [More data, more discussion]. *Forest Call*, no. 3, pp. 46–53.

Hiraiwa, Y. (1992) *Ōkami – sono seitai to rekishi* [The wolf: its ecology and history]. Tokyo: Tsukuba Shokan.

Hongū-chō (1969) *Yamabiko* [Echo]. Hongū: Town Hall.

Ichikawa, T. and I. Saitō (1985) *Saikō – nihon no shinrin bunkashi* [The history of Japanese forest culture reconsidered]. Tokyo: NHK Books.

Kaneko, H., Konishi M., Sasaki K., and Chiba T. (1992) *Nihonshi no naka no dōbutsu jiten* [A dictionary of animals in Japanese history]. Tokyo: Tōkyōdō Shuppan.

Kanzaki, N., N. Maruyama and T. Inoue (1995) 'Ankēto chōsa: ōkami ni taisuru nihonjin no ishiki' [Questionnaire survey: Japanese consciousness of wolves]. *Forest Call*, no. 2, pp. 1–3.

KHI (Kumanoji Hensan Iinkai) (1980) *Kumano nakahechi densetsu* [Legends of Kumano Nakahechi]. Tanabe: Kumano Nakahechi Kankōkai.

Kitajo, S. (1994) 'Minkan shinkō no naka ni ikiru ōkami' [The wolf alive in folk religion]. *Inadani Shizen Tomo no Kaihō*, no. 52, p. 4.

KMG (Kinki Minzoku Gakkai) (ed.) (1985) *Kumano no minzoku – Wakayama-ken Hongū-chō* [The folk customs of Kumano: Hongū Town, Wakayama Prefecture). Osaka: Kinki Minzoku Gakkai.

Knight, J. (2000) 'Culling demons: the problem of bears in upland Japan'. In J. Knight (ed.) *Natural Enemies: People–Wildlife Conflicts in Anthropological Perspective*. London: Routledge, pp. 145–169.

Komatsu, K. (1988) *Nihon no noroi* [Curses in Japan]. Tokyo: Kōbunsha.

Lindquist, G. (2000) 'The wolf, the Saami and the urban shaman: predator symbolism in Sweden'. In J. Knight (ed.) *Natural Enemies: People–Wildlife Conflicts in Anthropological Perspective*. London and New York: Routledge, pp. 170–188.

Lopez, B. H. (1995 [1978]) *Of Wolves and Men*. New York: Touchstone.

Maruyama, N. (n.d.) 'Extermination and recovery of wolves in Japan', unpublished manuscript in author's possession.

—— (1994) 'Dōbutsu ni yoru shinrin higai wa naze okiru no ka?' [Why does animal damage to forests occur?]. *Ringyō Gijutsu*, no. 633, pp. 2–6.

—— (1995) '"Nihon no honyūruigaku no mōten o tsuku – ōkami fuzai no ekoroji" o kaisai' [Opening address to the conference on 'Challenging the blindspot of Japanese mammalogy: ecology without wolves']. *Forest Call*, no. 2, pp. 4–7.

Maruyama, N., K. Kaji and N. Kanzaki (1996) 'Review of the extirpation of wolves in Japan'. *Journal of Wildlife Research*, vol. 1, no. 2, pp. 199–201.

Maruyama, N., K. Wada and N. Kanzaki (1995) 'Dai 38 kai shinpojiumu "ōkami fuzai no ekorojii" ni tsuite no shusaisha sokatsu' [Review of the 38th symposium of the Mammalogical Society of Japan, 'ecology without wolves']. *Honyūrui Kagaku*, vol. 35, no. 1, pp. 21–27.

Matsutani, M. (1994). *Ōkami, yamainu, neko* [Wolves, mountain dogs and cats]. Tokyo: Rippū Shobō.

Matsuyama, Y. (1977) *Kari no kataribe – Ina no yamagai yori* [Hunting storytellers: from the Ina Gorge]. Tokyo: Hōsei Daigaku Shuppankyoku.

—— (1994) *Inadani no dōbutsutachi* [The animals of Inadani]. Tokyo: Dōjidaisha.

Meriggi, A. and S. Lovari (1996) 'A review of wolf predation in southern Europe: does the wolf prefer wild prey to livestock?' *Journal of Applied Ecology*, vol. 33, no. 6, pp. 1561–1571.

Mitani, K. (1979) 'Tora sōdō tokuhō daiichidan' [The tiger furore – the first special report]. *Shuryōkai*, vol. 23, no. 10, pp. 175–182.

Miyamoto, T. *et al.* (eds) (1995) *Nihon zankoku monogatari 2 – wasurerareta tochi* [Tales of cruelty from Japan, volume II: forgotten land]. Tokyo: Heibonsha.

Mizuno, A. (1995) 'Hakusan chiiki no engai to inu' [Dogs and monkey damage in the Hakusan region]. *Wildlife Forum*, vol. 1, no. 1, pp. 11–17.

Morita, M. (1994) 'Ōkami hōon' (The wolf's obligation to return). In Inada K. *et al. Nihon mukashibanashi jiten* [Dictionary of old Japanese tales]. Tokyo: Nihon Hōsō Shuppan Kyōkai, pp. 141–142.

Nakamori, S. (1940) 'Yoshino no ōkami no hanashi' [Tales of wolves in Yoshino]. *Dōbutsu Bungaku*, no. 68, pp. 26–36.

—— (1941) 'Yoshino no ōkami no hanashi (II)' [Tales of wolves in Yoshino II]. Edited by Hiraiwa Yonekichi. *Dōbutsu Bungaku*, no. 76, pp. 30–38.

Nakamura, T. (1987) *Nihon dōbutsu minzokushi* [Japanese animal folklore]. Tokyo: Kaimeisha.

Naumann, N. (1994) *Yama no kami* [The mountain spirit]. (Translated from the German by S. Nomura and Y. Hieda) Tokyo: Gensōsha.

Nomoto, K. (1990) *Kumano sankai minzokukō* [A treatise on the mountain and coastal folk customs of Kumano]. Kyoto: Jinbun Shoin.

OGHK (Okami Gyakusatsu ni Hantaisuru Kai) (n.d.) 'Ōkami dōnyū yori kaihatsu tomeru no ga saki' [Stopping development comes before wolf introduction], pp. 109–111.

Ōkami no Nakamatachi (n.d.) *Opposing the Reintroduction of Wolves to Japan*. Leaflet in author's possession.

Ōkami no Nakamatachi (1996) 'Watashitachi wa nihon ni ōkami ni saidōnyūsuru keikaku ni hantai shimasu' [We oppose the plan to reintroduce the wolf to Japan]. *Wolf Tsūshin*, December.

Ōta, Y. (1997) *Matagi – kieyuku yamabito no kiroku* [Matagi: a record of a disappearing mountain people]. Tokyo: Keiyūsha.

Paystrup, P. (1993) *The Wolf at Yellowstone's Door: Extending and Applying the Cultural Approach to Risk Communication to an Endangered Species Recovery Plan Controversy.* Unpublished PhD Thesis, Purdue University.

Saitō, H. (1983) 'Aiken monogatari' [Tales of favourite dogs]. In *Zenshū nihon dōbutsushi 12 (Record of Japanese Animals Vol. 12)*. Tokyo: Kōdansha, pp. 5–102.

Sakai, S. (1986) 'Ōkami hōon' [The return made by the wolf]. In *Nihon denki densetsu daijiten (Dictionary of Japanese Romantic Legends)*. Tokyo: Kadokawa Shoten, pp. 175–176.

Satō, S. (1990) *Yama no kami no minzoku to shinkō* [Mountain spirit folklore and religion]. Tokyo: Chūōkōron Jigyō Shuppan.

Steinhart, P. (1995) *The Company of Wolves*. New York: Vintage Books.

Suzuki, B. (1986) *Snow Country Tales: Life in the Other Japan.* (Translated by J. Hunter with R. Lester). New York and Tokyo: Weatherhill.

Tabuchi, J. (1992) 'Chichishiro mizu' [The teat's white water]. In Kenichi Tanigawa (ed.) *Dōshokubutsu no fōkuroa 1* [Animal and plant folklore, Vol. 1]. Tokyo: Sanjūichi Shobō, pp. 9–94.

Tada, T. (1996) 'Baransu no toreta shinrin kankyō o' [Towards a balanced forest environment]. *Forest Call*, no. 3, pp. 44.

Takahashi, N. and N. Maruyama (n.d.) 'Is there room left for reintroduced wolves in Japan?' Paper presented to the conference *Coexistence of Large Carnivores with Man*, Saitama, Japan.

Talbot, L. M. (1978) 'The role of predators in ecosystem management'. In M. W. Holdgate and M. J. Woodman (eds) *The Breakdown and Restoration of Ecosystems*. New York and London: Plenum Press, pp. 307–319.

Tanigawa, K. (1980) 'Karigotoba ni michita sekai' [The world of hunting language]. In *Tanigawa Kenichi chosakushū 1* [Collected works of Tanigawa Kenichi, vol. 1]. Tokyo: Sanichi Shobō, pp. 284–295.

Togashi, S. (1994) 'Ōkami no ekorojii' [The ecology of the wolf]. *Namae no nai Shinbun*, no. 59, 1 December.

Ue, T. (1980) *Yamabito no ki: ki no kuni, hatenashi sanmyaku* [Diary of a mountain person. The Hatenashi Mountain Range, Tree Country]. Tokyo: Chūkō Shinsho.

WKMK (ed.) (1987) *Kīshū-Ryūjin no minwa* [Folktales of Kishū-Ryūjin]. Gobō: Wakayama-ken Minwa no Kai.

Wada, H. (1978) 'Wakayama no minwa, densetsu' [Folktales and legends of Wakayama]. In S. Andō (ed.) *Wakayama no kenkyū 5* [Research on Wakayama, vol. 5]. Ōsaka: Seibundō Shuppan.

Watanabe, K. (1995) 'Chiiki ni okeru yasei nihonzaru hogo kanri no mondaiten to kongo no kadai' [The problems of protecting and managing wild Japanese monkeys in the regions and tasks for the future]. *Reichōrui Kenkyū*, vol. 11, pp. 47–58.

Yagi, Y. (1988) 'Mura-zakai: the Japanese village boundary and its symbolic interpretation'. *Asian Folklore Studies*, vol. 47, no. 2, pp. 137–151.

Yanagita, K. (1961) 'Yama no jinsei' [Mountain lives]. In *Sekai kyōyō zenshū 21* [Collected works of world learning, volume 21]. Tokyo: Heibonsha, pp. 97–209.

Yanai, K. (1993) *Maboroshi no nihonōkami* [The phantom Japanese wolf]. Urawa, Saitama: Sakitama Shuppankai.

Yoneyama, T. (1969) *Kaso shakai* [Depopulated Society]. Tokyo: NHK Books.

Newspapers

Chūnichi Shinbun (7/3/2000) 'Shokugai no shika – ōkamitsukai taiji' [The deer pest – overcoming it with the wolf]. Evening edition.

Kahoku Shinpō (21/5/1996) 'Murenashi kamoshika osou, no noinu shutsubotsu' [Feral dog outbreak in Yakita Aomori – pack attacks serow].

Kahoku Shinpō Yūkan (13/3/2000) 'Shika shokugai taisaku ni okami yobō' [Let's call on the wolf as a countermeasure against deer damage].

Kii Minpō (12/3/1996) 'Nara yasei i ga hantai mōshiire' [Nara Wildlife Committee Declares Opposition], p. 11.

Kii Minpō (30/3/1996) 'Koe – dokusha no ran' [Voice: readers' column], p. 7.

Mainichi Shinbun (12/3/1989) 'Yamagoya' [Mountain hut].

Mainichi Shinbun (21/1/1990) 'Shūgeki' [Attacks].

Mainichi Shinbun (27/5/1990) 'Kusakariba' [Grassfields].

Mainichi Shinbun (18/11/1990) 'Hōrōsha' [Wanderers].

Mainichi Shinbun (22/12/1990) 'Yamai' [Illness].

INDEX

The page numbers in bold refer to illustrations

Abashiri 75
abhayadana 40–1
Aboriginal Malays 194–7, 200, 201, 202
Achenese 187
adat 149–50, 158
Agta 172–5
ahimsa (non-violence) 5–6, 116–117, 139–140
Akita 239
Anderson, E. N. 5, 90, 91, 92, 97
Anderson, John 189, 190
animals, Chinese definition of 16–21
antelope 36
Arikawa 75–6, 83
asiatic lions 208, 211, 213, 217
Assam 211
Ayukawa 75–6, 80, **81**, **82**, 84, 85n8

Bakels, Jet 8–9, 147–64
Bali 4, 187
Bangladesh 1
Barros, João de 196
Batak 187
bears 5, 10, 23, 92, 137, 138, 139, 185, 186, 196, 202, 243, 248, 250n11
Bengal 208, 211, 216, 218, 224
Bhutan 2, 8, 131–45
Bihar 211, 219
Bird-David, Nurit 147, 161
blackbucks 212
blowpipes 190, 191, 196, 198, 199
Boers, J. W. 192
Bombay 210, 213–14, 215
Boomgaard, Peter 147, 157, 185–206, Bonai 170
Borneo 170
bow and arrow 190

Buddhism 3, 4, 5–6, 7, 9–10, 97, 104; in Bhutan 131, 133–4, 139, 142; in Japan 67–9, 77; in Thailand 116, 117; in Tibet 37–44, 45, 46–51
Burma 4

Chapple, Christopher 3, 5–6, 117
Chiba, Tokuji 68, 234, 238, 239
China 15–31, 36, 46, 49, 79, 88, 89, 92, 93, 95, 98, 104, 241
Chinese animal zodiac 4
Chinese cuisine 90–1
Chinese diaspora 91
Chinese pharmacopoeia 91–2
civet cats 213
coconut palms 113–14, 118
Coimbatore 211
Confucianism 20
Confucius 17, 28, 30
Corbett, Jim 215, 216, 222, 223
crocodiles 25–6, 160, 162n16, 213–14
Curzon, Lady 222
Curzon, Lord 213

Dalai Lama 44, 46, 47, 48, 49, 49–50
deer 36, 60, 97, 137, 139, 140, 151, 152, 161; excess numbers of 250n10; as pests 243
Doctrine of Signatures 96, 106–7n9
Dolphin and Whale Action Network 80
dolphins 2, 7, 79, 80, 81, 82, 83–4; compassion for 83, 84; dolphinaria 79
domestication 178, 179–80
Dongen, G. J. van 192–3, 203n16–19
Donovan, Deanna 7–8, 88–111
dukun 152–4, 159

eagles 213
ecosystem 241, 242
ecotourism 142
elephants 4, 148, 170, 186, 189, 191, 193, 196, 198, 199
Ellen, Roy 148, 156,
Else Nature Conservancy 80
Endicott, Kirk Michael 189–90, 203n7–8, n14 and n23
Fishery Protection Act (US) 80
Friends of the Wolf 245–6
Fujiwara, Hidehara 237

gaur 215
gazelle 212
Greenpeace 80, 85n6
Guangxi 92

Han Wudi 21
Han Ying 17–18
Han Yu 25–6
Hanuman 4
Harris, Marvin 169, 180
Hayakawa, Kōtarō 65, 70n1
Headland, Thomas 173
Hinduism 3, 4, 5–6
Hmong 99, 103
Ho Chi Minh City 88
Hokkaido 234, 248
Hong Kong 92, 95, 96, 101, 102
Huaulu 156
Hubback, Theodore R. 193, 203n13 and n22
Huber, Toni 6, 36–55
hunters; curse on 67–9; samurai spirit of 63–4
hunting; in southeast Asia 98–9, 102; in Japan 56–70; laws on 38–50; *matagi* tradition of 57; as pest control 64–5, 65–6; *ranjau* method 151–5, **154**, 198; and rhetoric 56; and ritual 68–9; royal hunting 21–2; as sport 56, 59–60, 69, 151–2, **152**, 212, 213, 217, 222–3, 224–5; as threat to public safety 66–7; as war 60, 65, 70; Westernization of 57
Hyderabad 212
hyenas 220

India 2, 4, 6, 10–11, 134, 186, 207–25
Indonesia 2, 4, 8–9, 10, 147–61, 165–72, 175–81, 185–203
Inner Mongolia 241, 245

International Fund for Animal Welfare (IFAW) 80
International Whaling Commission (IWC) 76, 79, 80
Iongh, Hans de 9, 165–84
Islam 9, 115, 116, 124n6, 151, 152, 169, 170–1, 172, 178–9, 188, 208, 214
Iwase, Momoki 239

jackals 221
Jainism 3, 5–6, 116–17
Jambi 191
Japan 2, 3, 4, 5, 6, 11, 56–70, 73–85, 233–49
Japan Whale Conservation Network 80
Japan Wolf Association 233–4, 241–9
Japanese culture; opposition to nature 83–4
Java 171, 186, 187, 188, 202, 215, 217
Jigme Dorji National Park 131–45
Jungle Book, The 237

Kagoshima 67
Kalland, Arne 7, 73–87
Kampuchea 4
Katsumoto 83–4
Kerinci 147–61
Kerinci Seblat National Park 148, 157
Kii Peninsula 57–8, 61, 63, 234, 235, 236, 238, 239, **240**, 243, 244–8
Knight, John 1–12, 56–72, 162n14, 233–54
Korea 1, 79
Kubu 154, 156, 170, 188, 190, 191–3, 198, 200, 201

Laos 1, **90**, **94**, 99, 103
leopards 23, 49, 138, 189, 207, 215
Lhasa 47, 48, 49, 50
Lu Qiaoru 24
Lubu 197–8, 200, 201, 202
Luzon 167, 169, 172–5

Madras 210, 211, 214–15
Malaysia 1, 10, 98, 185–203
Mamak 199–200, 202
Martin, Rudolf 189, 193, 194, **195**, 196, 203n8–9, n14, n22 and n25, 204n27–28 and n30
Maruyama, Naoki 239, 241–3, 245, 250n9
Mencius 17, 18–19, 19–20, 29
Mentawai 6, 9, 154, 156, 162n11, 167, 169, 175–9

Minangkabau 172, 175, 178–9
Mindanao 167
Miyake 247
Miyazaki, Hayao 68, 250n6
Mohnike, Otto192
monkeys 2, 4, 8, 137, 150, 172; and
    Buddhism 117; conservation of 122–3;
    as crop-raiders 112; handlers of
    118–19; as Hanuman 115, 124n5;
    Islamic view of 116; as meat 122; and
    tourism 122; training of 119; trade in
    122; as a 'weed species' 112, 120–21,
    123
Moszkowsi, Max 198–9
Mysore 212, 223

Nagano 65, 239
Nakamura Teiri 235, 250n5
Nara Prefecture Wildlife Protection
    Committee 244
Natadecha-Sponsel, Poranee 8, 112–28
national parks 3, 131–45, 148, 157, 233
nature conservationism 131–2
Natural History Museum (London) 234
Nepal 134
New Guinea 112, 169, 181n4
New People's Army 172–3
'no touch' approach to nature 73
North West Provinces 211, 218, 219–20
Northern Sierra Madre Nature Park 174
Nuaulu 156

Oliver, William 168
Orientalism 3
Orissa 211

Palembang 191, 192
pangolins 97
Pathans, analogy with tigers 223
Persoon, Gerard 9, 165–84, 192
Petelangan 170
Peterson, Jean 173
Philippines 2, 9, 167, 172–5
poaching, 140–1, punishment of, 44–5, 48
population; population growth 2; rural
    depopulation 246–7
Princess Mononoke, The 68, 250n6
Punjab 222
qi (chi'i) 18, 97

raccoon-dogs 97
Rajputana 212

Ramayana, The 115, 208
Rangarajan, Mahesh 125n10, 162n14,
    207–32
rhinos 189, 191, 198, 199
Ruttanadakul, Nukul 8, 112–28
Rye, Simon 171

Sado 247
sago palms 175–6
Sakai 170, 197, 198–9, 200, 202, 203n20
sambhar 215
Sanderson, G. P. 212–13
Schebesta, Paul 190, 191, 194–6, 203n7,
    n9–15, n17, n21, n23 and n25
Schefold, Reimar 154, 161, 162n11 and
    n15, 177
Schneider, Gustav 199–200, 203n17,
    204n33–34
'Sealing the Hills' edict 39–40, 41, 42
Seeland, Klaus, 8, 131–46
Semang 188–91, 200, 202
Senoi 193–4, 200, 201
shamans 190
Shimane 67
Shinto 68–9, 77, 78, 234–6
Sima Xiangru 21
Singapore 92
Sinha, R. K. 4
Song Jun 27
spears 196, 198, 199
Sponsel, Leslie 8, 112–28
Sri Lanka 4
Sterckx, Roel 5, 15–35
Sukumar, Raman 3–4
Sumatra 7, 8–9, 10, 118, 125n10, 147–61,
    165, 169, 170–2, 186, 187, 191, 193,
    202, 203n1
Sundarbans 216, 224
Sutō, Isao 65
Suzuki Bokushi 238, 239

Taiji 75–6, 78–9, **82**, 83
Talang Mamak 170
Tambiah, Stanley 31, 117
Taoism 96
Thailand 2, 4, 89, 98, 99, 103, 112–23
Tibet 2, 5, 6, 7, 36–51, 89, 144
tigers 2, 4, 6, 10, 23, 25, 92, 137, 139, 248;
    as ancestors 157–8, 159, 188, 203;
    bounties on 210, 215, 217, 224; *dukun*
    and 159; eaten by humans 196, 197,
    198; ecology of 217, fear of 158, 160; as

game animals 213; as game destroyers 214; as livestock predators 158, 216, 224; as Lord of the Forest 188; as mankillers 158, 188, 209, 214–15, 217, 224; as pest controllers 218, 223; poaching of 160; as protectors 202; religious associations of 218; revenge against 159; and ritual 159; traps for 158–9, 187, 194, 200, 201, 211; as vermin 213; weretigers 157–8, 196–7
traditional Chinese medicine 91–8, 105
TRAFFIC 97–8

United Liberation Front of Assam 144
United Provinces 219, 223
Uttar Pradesh 221

Vercors, Jean Bruller 112
Verkerk Pistorius, A. W. P. 192–3
Vietnam 2, 88, 89, 99, **100**, 103

Wakayama 244, 248
Wang Chong 25
water buffaloes 149–50, 158, 177
Waterschoot van der Gracht, W. A. J. M. van 192
Whale and Dolphin Conservation Society 80–1
whales 2, 7, 73–82, 84–5; as cute 80; as food 74–6; graves of 77; memorial service for 77, **78**, 79; as resource 74–6; revenge of 78–9; spirit of 77; and tourism 84
whale-watching 79–82
whaling 4–5, 6, 7, 73–9; accidents 78–9; criticism of 73; public opinion in Japan 82
wild boar/wild pigs 2, 6, 9, 23, 137, 138, 139, 150, 165–81, 199, 248; courage of 62; curse of 71n3; and dogs 60, 63; as fighter 60; and human courage 62–3; hunting of 150, 151–2, **152**; in India 182n9; Islamic views of 170–1, 178–9, 181n2; as mankiller 61–2; as markers of identity 169; and masculinity 63; as meat 63; as medicine 63; as pests 170–2, 177; 'pig bombs' 173–4; and ritual 176–7; traps for **154**, 173; varieties of 167–8; and vitality 63; as war enemy 60, 65; were-pigs 172
wild dogs 137, 138
wildlife as food 90–1; habitat 2; as medicine 89, 91–2; tourism 247
wildlife pestilence; compensation for 141–5; and 'fate' 138; survey of 135–9
wildlife trade 4–5, 7–8, 38, 49, 59, 89, 95–104
wolves 2, 11, 25, 186, 218–22; as benign animals 234; bounties on 219–21, 222, 239; and child-snatching 219, 222, 238; ecology of 220, 225, 251n13; extermination of 219–21; extinction 234; fear of 238, 244; hatred of 202; hunting of 239; Indian subspecies 218–22; Japanese subspecies 234; as livestock predators 219, 220, 222; as mankillers 219, 238; as pest controllers 235, 242–3; as protectors 235–6; reintroduction of 241–4; and rituals 236–7, **238**; Romulus and Remus legend 237; Tibetan subspecies 221; wolf shrines 234–6
World Wide Fund for Nature (WWF) 80, 143

Xenophon 15
Xunzi 18, 19
yaks 36
Yamaguchi 77
Yamanashi 68
Yanagita, Kunio 237
Yellowstone National Park 233
Yoneyama, Toshinao 247
Yunnan 92, 99

Zhan He 30

The Nordic Institute of Asian Studies (NIAS) is funded by
the governments of Denmark, Finland, Iceland, Norway
and Sweden via the Nordic Council of Ministers, and works
to encourage and support Asian studies in the Nordic
countries. In so doing, NIAS has been publishing books
since 1969, with more than one hundred titles produced in
the last ten years.

Nordic Council of Ministers